TPM for the Lean Factory

INNOVATIVE METHODS AND WORKSHEETS
FOR EQUIPMENT MANAGEMENT

TPM for the Lean Factory

INNOVATIVE METHODS AND WORKSHEETS
FOR EQUIPMENT MANAGEMENT

Ken'ichi Sekine
Keisuke Arai

PRODUCTIVITY PRESS | PORTLAND, OREGON

Originally published as *Kakushin TPM Manuaru.* © 1992 and 1998 by Ken'ichi Sekine and Keisuke Arai. Published by Techno Publishing, Tokyo, Japan.

English edition © 1998 by Productivity, Inc.
Translated by Karen Sandness.

Additional copies of this book are available from the publisher. Discounts are available for multiple copies through the Productivity Press Sales Department (800-394-6868). Address all other inquiries to:

Productivity Press
P.O. Box 13390
Portland, OR 97213-0390
United States of America
Telephone: 503-235-0600
Telefax: 503-235-0909
E-mail: service@ppress.com

Cover design by Mark Weinstein
Page, figure, and table composition by William H. Brunson, Typography Services
Additional figures by Lee Smith, Smith & Fredrick Graphics
Illustrations by Gordon Ekdahl, Fineline Graphics
Printed and bound by BookCrafters in the United States of America

Library of Congress Cataloging-in-Publication Data

Sekine, Ken'ichi, 1926–
    [Kakushin TPM manyuaru. English]
    TPM for the lean factory : innovative methods and worksheets for
equipment management / Ken'ichi Sekine, Keisuke Arai.
        p.   cm.
    Translation of: Kakushin TPM manyuaru.
    Includes bibliographical references and index.
    ISBN 1-56327-191-5 (hc.)
    1. Total productive maintenance.   I. Arai, Keisuke.   II. Title.
TS192.S4513   1998
658.2'02—dc21   98-36058
                                                                           CIP

03 02 01 00 99 98 10 9 8 7 6 5 4 3 2 1

Contents

Publisher's Message

Lean manufacturing cannot happen in a factory that lacks dependable, effective equipment. When equipment has a history of breakdowns and defective operation, a plant must maintain excess work-in-process and finished inventory "just in case." When minor stoppages eat into productive time, employees have to watch automated equipment that should run by itself.

Ideally, total productive maintenance (TPM) provides a framework for addressing these problems. During factory visits over the years, however, lean manufacturing experts Ken'ichi Sekine and Keisuke Arai have seen many companies that fail to reap the full benefit from their TPM activities. Sekine and Arai are careful observers of factory dynamics, and they have recognized the key equipment-related problems and misunderstandings that cause plants to miss their lean manufacturing goals. *TPM for the Lean Factory* shares their results-oriented "TPM Innovation" methods with a wider Western audience.

This book will be most useful to readers who have some experience with implementing the basic pillars of a TPM program, such as autonomous maintenance to inspect equipment and maintain operating conditions, or focused improvement to reduce equipment-related losses. The authors' unique TPM Innovation approach refocuses the purpose of such activities, using a Toyota Production System–style emphasis on elimination of waste. Chapter 1 introduces seven wastes that stem, in their view, from incomplete implementation of TPM. To eradicate these wastes, Sekine and Arai emphasize three main topics in the first six chapters: the new 5Ss, instant maintenance, and improved setup operations.

5S refers to a set of activities designed to improve the safety and functionality of the workplace itself; it is a critical first step in any improvement program. In Chapters 2 and 3, the authors describe a new approach to 5S to replace superficial campaigns that stop at posters, slogans, and "cosmetic" cleanup. The new 5Ss draw on workplace discipline as a foundation, but emphasize the first two Ss: Organization and Orderliness. These two principles promote standard locations for needed items and standard labels that help everyone find and store things properly. When implemented properly, the authors believe, the other three Ss— Cleanliness, Standardized Cleanup, and Discipline—virtually take care of themselves. An example from Mynac, a garment assembly factory, highlights some of the features of their approach.

The instant maintenance approach outlined in Chapters 4 and 5 is a system for making quick repairs of small failures, such as broken limit switches. It also includes suggestions for preventive modifications and standardization of repair methods so everyone can respond quickly. Although the idea of standardizing your breakdown maintenance routines may seem counter to the preventive and life-extending focus of TPM, the authors are really suggesting that we be prepared for the unexpected, like emergency medical technicians for machines. Unexpected failures can happen, and it's helpful to plan in advance how to fix them quickly.

Improving setup operations is a forte of Sekine and Arai, who give an extended presentation of their methods in *Kaizen for Quick Changeover* (Productivity Press, 1992). In Chapter 6 of this book, they focus on organizing the preparation area for setup in a U-shaped line for efficient gathering and storage of the appropriate materials and tools. This approach saves time and helps avoid errors that could cause breakdowns or defects.

Chapter 7 deals with planned downtime: the losses that come from inefficiently planned lines and underutilized equipment. This productivity loss falls "outside" the major equipment losses traditionally measured by overall equipment effectiveness (OEE), but it is especially significant from the larger lean manufacturing perspective of avoiding waste. The fix is not equipment improvement per se, but rethinking product families for efficient use of existing equipment.

Product defects are a form of waste no factory can afford to ignore. Defects often result from major or minor equipment failures, but the causes are often elusive and multifaceted. Chapter 8 outlines a method for identifying important defect and failure factors using testing methods such as orthogonal arrays.

Chapter 9 offers a wealth of resources on making daily equipment inspections more effective. The authors share steps for teaching employees how to conduct inspections and for managing spare parts. General steps are given for eliminating minor stoppages, and a set of checklists helps operators use their senses (sight, hearing, touch, and smell) to spot future trouble at its early stages.

Chapter 10 presents a set of "test yourself" questions and answers, similar to questions that appear in earlier chapters. You will notice that these questions don't always review the book's material; they are more in the nature of "extra credit" questions drawing on the reader's own background knowledge. The questions throw light on the breadth and depth of knowledge that is expected of manufacturing managers in Japan.

The topics in Chapters 1 through 9 are summarized in an implementation overview that sets out in condensed, flowchart form the basic steps for implementing each topic. The overview topics are keyed to a set of 50 worksheet examples and blanks that support implementation. We hope you will use these as models to build forms that meet the specific needs of your workplace.

Ken'ichi Sekine and Keisuke Arai are among the leading lean manufacturing consultants in Japan. In addition to this book and *Kaizen for Quick Changeover*, Productivity Press is privileged to publish English editions of two other books on their methods: Sekine, *One-Piece Flow*, and Sekine and Arai, *Design Team Revolution*. The authors' original thinking about important topics in factory management is sure to stimulate improvements at your own plant.

Production of the English edition of this book was a team effort, for which we thank everyone involved. Productivity appreciates the opportunity to work again with Mr. Sekine and Mr. Arai, who adapted and clarified the manuscript for us. Karen Sandness translated the text. At Productivity, the project team included Karen Jones, development editing; Miho Matsubara, translation support; Susan Swanson, production coordination and text design; and Mary Junewick, art development and proofreading. Copyediting was done by Sheryl Rose. Graphics and illustrations were created by Bill Brunson of Typography Services; Lee Smith of Smith & Fredrick Graphics; and Gordon Ekdahl of Fineline Graphics. Pages were designed and composed by Bill Brunson. The cover was designed by Mark Weinstein.

Steven Ott

President, Productivity, Inc.

Preface

The goal of our TPM Innovation approach is to maintain international competitiveness. Japanese know-how has taught the world how to make goods inexpensively by reforming the manufacturing process and using workers more efficiently.

Reforming an existing system means rejecting the status quo and exposing its weaknesses. Unless we expose these weaknesses, we will be unable to improve the way we do things. In that spirit of improvement, this guide to our TPM Innovation approach proposes three additions to the current understanding of the TPM system:

- The new 5Ss
- Instant maintenance
- Improved setup operations

Figure P-1 will give you a sense of how these elements work together in TPM Innovation.

The New 5Ss

Our development of the new 5Ss was prompted by observations of a particularly misguided interpretation of the 5S system in a South Korean company. There the 5Ss had been distorted into a system where managers ordered workers to do things like paint the shop floors to make the plant look clean, buy new shelves and storage boxes, and spend all day tidying up the plant. By converting the 5S method into a goal in itself, management had completely missed the point.

The overall goal of the 5S system is to increase the value added by each worker. To increase the value added, we must create well-adjusted and orderly lines based on the principles of the first two Ss: Organization and Orderliness. Above all, these two Ss bring standard positions and standard labeling to the factory.

Focusing on the first two Ss gives a new perspective on the meaning of the 5Ss. President Ichinose of Mynac, a clothing manufacturer, has implemented these new 5Ss at his plant. We introduce the foundations of the new 5Ss in Chapters 2 and 3.

Instant Maintenance

During plant visits in Europe and North America, we've seen operators send for the maintenance technicians whenever a breakdown occurs. The technicians come and perform the repairs while the operators take a break. This scenario divides the workforce into two types of people: those who cause breakdowns and those who fix them. It's almost cartoonlike.

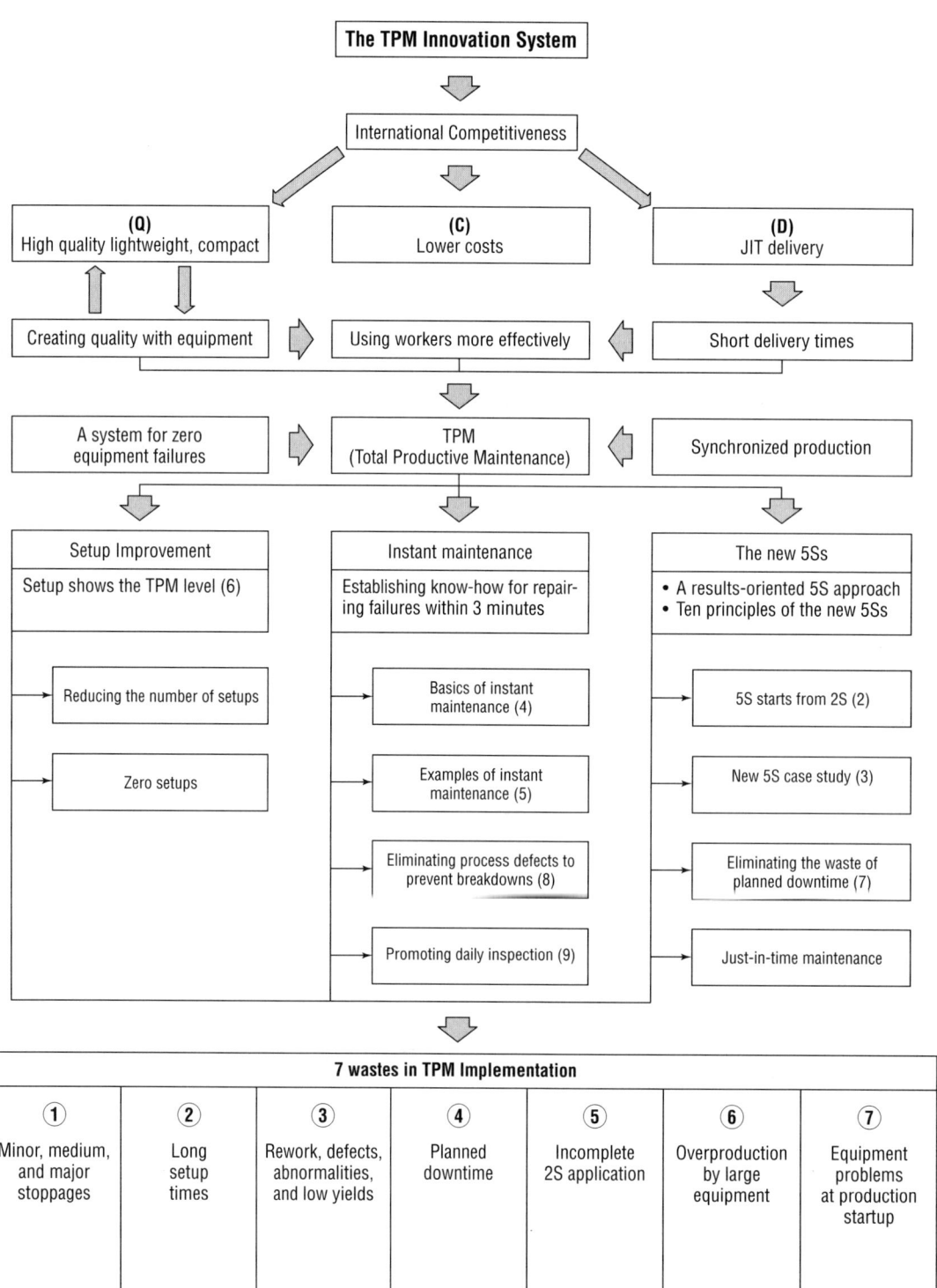

Figure P-1. The TPM Innovation System

The remedy for this waste is instant maintenance. Instant maintenance is a technology for restoring equipment to its former state within three minutes of a breakdown. We believe it's important to teach operators the skills they need for instant maintenance, and in Chapters 4 and 5 we present steps for implementing this important addition to their skills.

If we are diligent in implementing instant maintenance, it will naturally lead to improvements in daily inspection and maintenance and, in turn, to improvements in planned maintenance. It should save your company money, because you won't be so dependent on the equipment manufacturers for maintenance and repairs.

Improved Setup Operations

Improving setup means reducing changeover operations to a single step or even eliminating them altogether. Plants that implement setup correctly can eliminate the problem of minor stoppages as well as trial and adjustment after the changeover. Chapter 6 gives steps for this type of improvement.

This new TPM Innovation approach is built on these three pillars, as well as on other methods that root out the seven types of waste connected with TPM. You must pull those "weeds" on a daily basis to get your plant to reach its full lean potential.

Ken'ichi Sekine

Keisuke Arai

TPM for the Lean Factory

INNOVATIVE METHODS AND WORKSHEETS

FOR EQUIPMENT MANAGEMENT

The Seven Types of
Waste in TPM

Why TPM Now?

In recent years, TPM at many plants has turned into "display TPM." It consists of plastering the walls of the plant with management graphs and telling the workers to look at them. The original style of TPM, however, was much more rough and ready.

In the mid-1980s, we led a study group from France around the Toyota plant in Japan. As we were walking past a transfer machine line, one of the motors began to smoke. Immediately, a whistle blew, a person who appeared to be the leader shut off the power, and two operators came running.

The leader swiftly issued some orders to the two operators and then began to remove the burned-out motor. By the time the operators brought the replacement motor, the bolts were already loose, and the team of three quickly got the new motor up and running. The members of the study group burst into applause. They were so astounded to see this direct approach that they stood transfixed, not even hearing their guide's directions to move to the next area.

At their company, group members said, it took at least four hours and perhaps as much as a full day to replace the motor of a transfer machine. That was why they were so impressed at seeing the Toyota workers accomplish the task in a mere nine minutes.

This plant didn't have a single management graph on its walls, but thanks to its well-trained workers, the rough and ready style of maintenance was alive and well. We were so impressed that we began calling this approach "instant maintenance."

TPM has three basic foundations:

1. Everyday maintenance: Workers take care of their machines and production lines themselves.
2. Periodic maintenance: Certain maintenance tasks are scheduled to happen annually or semiannually, like a complete physical exam for machines.
3. Instant maintenance: Small problems with equipment are fixed within three minutes.

This third method is the most important one at the job site.

The Seven Types of Waste

A certain machine products factory, a medium-sized company, had a monthly output worth ¥100 million (approximately $800,000).* At the beginning of 1990, top management made a momentous decision to construct a new plant and institute a high-priced form of factory automation (FA) within three years.

After a few months, the new plant was complete and was partly able to run on an automated basis. Machining precision improved, and the factory appeared to have undergone a revolution.

However, a closer look revealed that the company was using the new facilities in old ways. The younger operators took care of the automated operations, while the older operators merely ran single-function machines, just as they always had. They had no interest in acquiring the new technology skills, and they seemed unable to keep up with the changes. Despite their years of experience, they weren't much help in supervising the younger operators, because their expertise was limited to one single-function machine. This meant that the lower-paid younger operators were several times more productive than the highly paid older operators. Furthermore, people stood in front of the automated machines and just watched them work, so more machines were idle than ever before.

Why did this happen? It was because most of the on-site managers knew only the old ways of using the new machines. In particular, since the FA equipment had to be continually reprogrammed, changeover took a long time, and mistakes in programming led to large numbers of defects. For that reason, only part of the work was delegated to the automated equipment and the rest was done on single-function machines or by hand.

In addition, the automated equipment sometimes broke down, leading to both major and minor stoppages. There were no personnel within the company who could maintain the FA equipment, so they depended on the manufacturer's service representatives. This meant not only that their production technology did not develop properly, but also that their know-how flowed to outsiders, the efficiency of their equipment declined, production slowed down, and parts inventories increased. All these factors gradually led to lower quality and increased production costs.

The introduction of FA had failed to increase output. The company needed to raise the proportion of automated equipment, and the key to increasing efficiency was maintenance and retooling.

Most of the waste in this plant stemmed from incomplete implementation of TPM, as shown in Table 1-1.

The first fundamental source of these kinds of wasted effort is not installing equipment appropriate to the amount of production. This means, for example, producing small lots on specialized machines and producing large lots in machining centers.

---

* Amounts given in yen in the original text have been converted to approximate U.S. dollar amounts at a rate of ¥125/US$1 throughout this book.

Table 1-1. The Seven Types of Waste in TPM

1. Minor stoppages, medium stoppages, major stoppages
2. Lengthy setup times
3. Manual rework, defects, faulty products, and low yields
4. Planned downtime
5. Incomplete 2S application
6. Overproduction by large equipment
7. Equipment problems at production startup

The second source is rushing to automate fully without first introducing the simpler forms of automation. After all, even with automation, eliminating human workers is difficult. For example, even when you streamline the work so it requires only 0.3 worker, you still have to have one worker, but the other 0.7 of that person's time is like excess inventory. This excess is the reason automation doesn't reduce costs as much as it is intended to. The ultimate goal should be lowered costs, not just modernization of equipment.

The third source of waste is failure to automate changeover, even when the rest of the process uses automation effectively. This leads to deviations from standards and variations in product quality. When specialized lines undergo simple automation, it's an easy process, but since changeover is performed by people, retooling the more complicated equipment takes a long time. Quality is a reflection of the number of processes. If changeover takes too long, it inevitably leads to mechanical breakdowns and defects.

The fourth source of waste is the lack of synchronization between processes. When process A is performed on high-speed equipment and process B is performed on low-speed equipment, unfinished inventory piles up between processes. For all practical purposes, unfinished inventory is useless inventory.

The fifth source of waste is a lack of mechanisms for ensuring that employees follow safety rules. Safety is determined by the number of safety devices on the equipment. In the recent trend to pursue speed in manufacturing, operators are too often left unprotected, even though most machines have some inherent hazards. The result is an unexpectedly large number of injuries attributable to the lack of safeguards.

Defining and Eliminating Waste

Let's define and examine the various types of waste and then consider strategies for eliminating them.

Waste Due to Stoppages

Too often, manufacturing machines just stop working for one reason or another. These stoppages are classified as minor, medium, or major, depending on how long they last.

Minor Stoppages

A minor stoppage is the kind of waste in which the machine keeps stopping and starting, breaking the rhythm of the process. Minor stoppages last less than four minutes and tend to consist of the following types of mishaps:

1. Work gets caught on something, jams up, sticks together, or gets lodged in a tube during conveyance. There may be a parts shortage or an oversupply or undersupply of work. Work may fall off the conveyor or not be inserted at the proper time.

2. During assembly, work may be crushed or destroyed, two pieces may be picked up at once, and there may be problems with chucking. Pins may not line up, and ejection may not happen properly.

3. Sensors may give a faulty reading due to problems with the computer or display.

These problems have different causes and remedies; our focus here is on what they have in common.

Causes	*Suggestions for Elimination*
1. Letting someone else worry about minor stoppages (leaving it up to the operators)	1. Point out the importance of eliminating minor stoppages as an overall policy. The following three tactics are the most fundamental:
2. Not having a complete grasp of what is actually going on with the processes, machines, and plants that experience minor stoppages	• Autonomous maintenance • Use of customized jigs • Simple automation
	2. Make on-site observations of minor stoppages.
	3. Create a chart to track and summarize minor stoppages.
	4. Figure out the mechanisms behind the occurrences.

Medium and Major Stoppages

These kinds of waste occur when equipment or machines break down, forcing the line to stop for a longer time. Medium stoppages last 4 to 30 minutes, whereas major stoppages last more than half an hour. Major stoppages are particularly likely to happen during production startup.

Causes	*Suggestions for Elimination*
1. Both sporadic and chronic break-downs occurring in succession 2. Not understanding the weak points of the equipment 3. Not making a list of severely worn or deteriorated parts	1. Watch equipment usage conditions and basic conditions closely (cleaning, lubricating, and tightening). 2. Determine the equipment's MTBF (mean time between failures): • Record parts replaced and replacement dates in the maintenance log. • Find and improve weak points in equipment design. • Improve technology for operations and setups. 3. Create an instant maintenance system. 4. Create a concurrent engineering system for developing permanent solutions.

Waste Due to Long Setup Times

This type of loss occurs either when setup or changeover goes wrong, or when adjustments are needed because the changeover is not standardized. Making too many adjustments causes deviation from the standards, which in turn causes vibration and leads to breakdowns.

Cause	Suggestions for Elimination
1. Leaving setup up to the operators' individual methods	1.1. Implement visual management in the form of production status boards and setup charts.
	1.2. Make pre-setup part of the line.
2. Not knowing machining standards and recognizing deviations from standards	2.1. Use fixtures, molds, and blades to prevent deviation from the machining standards.
	2.2. Clearly indicate the standards on the machining chart.
3. Taking too much time for adjustments	3. Introduce zero changeover (the principle of immovable standards).
4. Not using completely standardized parts	4. Standardize parts by plant.

Waste Due to Rework, Defects, Faulty Products, and Low Yields

In these wastes, problems with the equipment, machines, or apparatus lead to rework, defects, and abnormalities, and consequently to lower yields.

Causes	Suggestions for Elimination
1. Chronic defects due to chronic abnormalities	1. Quantify the number of chronic defects, their frequency, and the losses they cause. Introduce a zero-defect production system in conjunction with elimination of tiny flaws.
2. Haphazard setups (including adjustments)	2. Make improvements according to zero changeover procedures and foundations. Apply the principle of unchanging machining standards.
3. Unnecessarily strenuous or dangerous work	3. Apply concurrent engineering methods to revise the work.
4. Losses from more frequent changeovers	4. Based on past records, the line leader creates production systems week by week and rearranges the production sequence, taking the processes with the easiest changeovers first.

Waste Due to Planned Downtime

This waste occurs when a plant has a lot of specialized machines and specialized lines, but the goods produced on them no longer sell in large numbers. Even though modern, automated equipment has been introduced, no one knows how to use it, nor does the plant have the technology for maintaining the programming. Such situations cause a kind of waste that reduces the operating rate of the machine. Idle machines don't earn any money for the company.

Cause	Suggestions for Elimination
1. On-site managers who don't know how to use new equipment and older technicians who don't know how to use computerized equipment.	1. Improve line setup procedures using: • takt time production • multifunctional training • standardization of parts
2. Imprecise equipment specifications or no parts-specific or machine-specific settings for standard time or speed	2. Eliminate speed loss and improve any bottlenecks in the system.
3. One operator per machine or FMS.	3.1. Set up FMS in U-shaped lines. 3.2. Implement process razing.
4. More than 50 percent planned downtime.	4. Study the current situation, discover causes (start with the most problematic machines and lines), then devise countermeasures one by one, gradually improving the entire situation.

Waste Due to Incomplete 2S Application

This waste occurs in factories that have debris around the machines, poor environmental conditions, and fixtures, tools, or parts that are never put back where they belong. Younger workers turn up their noses at this kind of workplace because they see it as dirty, dangerous, and degrading. The 2Ss are the place to begin to make real changes in the workplace.

Cause	Suggestions for Elimination
1. The 2Ss have not been completely implemented.	1. Implement the 2Ss (Organization and Orderliness):
2. The 3Ss have not been completely implemented.	• Infrequently used items: Discard or store remotely.
3. Training and discipline have not been completely implemented.	• Frequently used items: Keep close by or wear them.
	• Remember the goal of Orderliness: Be able to retrieve anything within 30 seconds.
	• Implement standard placement, standard labeling, standard amount.
	2. Implement the 3Ss (Cleanliness, Standardized Cleanup, and Discipline):
	• Make cleaning a form of spot check.
	• Look at lubrication, tightening, temperature, and damage from debris.
	• Maintain discipline: Do or delegate everything that has been decided.
	• Pay attention to and improve safety and sanitation.
	• Check nonobvious and hard-to-reach places.

Waste Due to Overproduction by Large Equipment

This waste involves production of excess unfinished and finished inventories, as well as non-value-adding conveyance before and after processing.

Cause	Suggestions for Elimination
1. Being resigned to the problem because you think using large equipment is inevitable	1. Implement zero changeover to make more effective use of the large equipment:
2. Increasingly, plans have to be suspended, causing other problems	• Produce in small lots or one item at a time, eliminating obstacles one by one.
	• Bring minor stoppages to zero.
	• Implement zero changeover.
	• Implement takt time production.
	2.1. Use small-scale specialized machines.
	2.2. Create common-use lines.

Waste Due to Equipment Problems at Production Startup

At many companies, TPM seems to mean creating graphs to show off to visitors rather than using graphs for actual management. Some companies post a lot of improvement deployment charts, but actual performance doesn't always improve as a result. One key element that's missing is an exchange of ideas between the people who develop and design the machines and the people who use the machines. Only when this communication is established will equipment startup problems get resolved.

Test Yourself

After reading about the seven types of waste in TPM (see Table 1-1), which of the seven do you think is the most wasteful, and why?

Our Answers

It's number 3, waste due to rework and defects. The overriding cause of this kind of waste is the need for more a more orderly changeover process. The following five factors contribute to disorderly changeover.

1. People believe that Orderliness happens naturally in high-precision processes.
2. Standard surfaces in designs, changeovers, fixtures, and machines are imprecise, so measurements from the standard surfaces yield varying values.
3. Debris, vapor, and other environmental contaminants get into the equipment.
4. The operators don't follow basic changeover principles and procedures; each person has his or her own methods.

5. Since the equipment and replacement parts lack precision and durability, workers have to improvise their way around problems.

The more adjustments people have to make, the sooner the equipment will deteriorate and the more quality will fall. This is because the machines, equipment, molds, fixtures and other parts are no longer the way the original designers envisioned them. Table 1-2 shows nine rules for improving the setup process.

Table 1-2. Nine Rules for Improving the Setup Process

Eliminating Waste During Preparations:

Rule 1: Prepare everything you can ahead of time.

Rule 2: It's all right to move your hands, but don't move your feet.
Put everything you need within easy reach.

Eliminating Waste During Replacement:

Rule 3: Make it unnecessary to remove or loosen bolts.

Rule 4: If you see unnecessary bolts, eliminate them.

Eliminating Waste in Adjustment:

Rule 5: Don't change the standards for dies and fittings (pressing).
Don't change the standards for the process (machining).

Rule 6: Adjustment is waste, so don't move the base.

Rule 7: Use a block gauge for all adjustments that require reference to a scale.

Rule 8: Set standards for stoppers and guides.

Rule 9: Provide cradles and guides for attaching molds to vertical surfaces (pressing).
Provide intermediate fixtures that can be used for a number of processes (machining).

The 5Ss Begin with the 2Ss

2

Proclaiming the New 5Ss

The most important part of the 5Ss is thorough training and discipline about the first 2Ss: Organization and Orderliness (see Figure 2-1).

Suppose that after a 5S campaign, it looks as if every item in the workplace has been assigned a proper location and everything is being done according to established standards. However, within a month, we start seeing piles of unfinished goods, and the plant is no better off than it was before. This is because training and discipline have been neglected.

In this era of individual freedom, training and discipline have just barely survived in the workplace. People seem resigned to the idea of being trained and disciplined

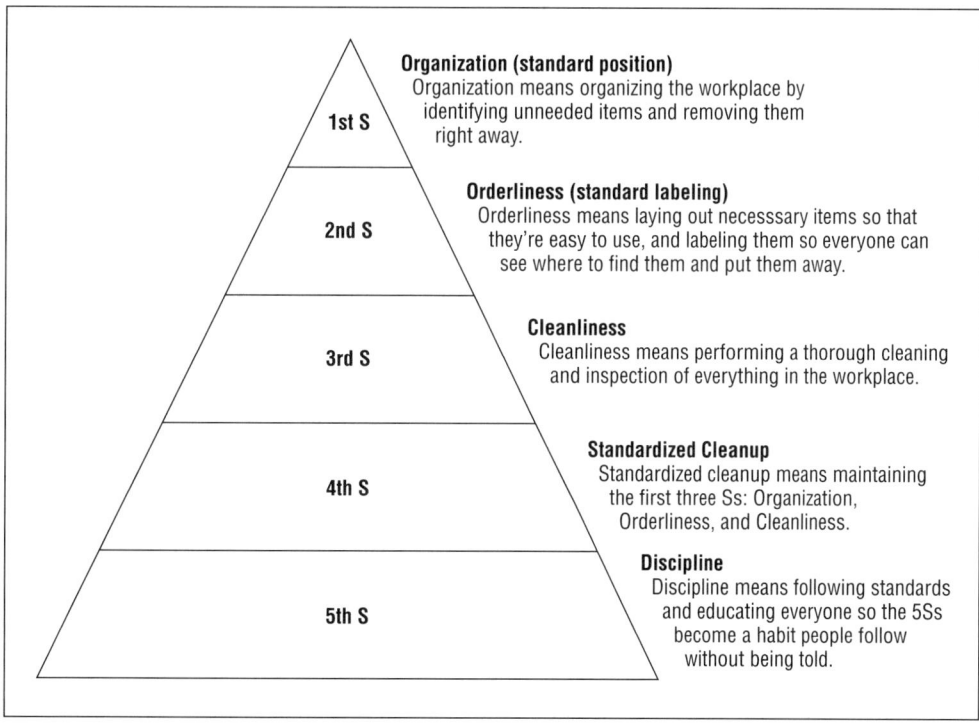

1st S — **Organization (standard position)**
Organization means organizing the workplace by identifying unneeded items and removing them right away.

2nd S — **Orderliness (standard labeling)**
Orderliness means laying out necesssary items so that they're easy to use, and labeling them so everyone can see where to find them and put them away.

3rd S — **Cleanliness**
Cleanliness means performing a thorough cleaning and inspection of everything in the workplace.

4th S — **Standardized Cleanup**
Standardized cleanup means maintaining the first three Ss: Organization, Orderliness, and Cleanliness.

5th S — **Discipline**
Discipline means following standards and educating everyone so the 5Ss become a habit people follow without being told.

Figure 2-1. The New 5Ss Are Sustained Through Discipline

as a condition for earning their wages and salaries. We would like to see workplace discipline become an internal approach that doesn't depend on outside pressure. Thus, our definition of discipline is training the employees so thoroughly in the first 2Ss that they follow these principles even when no one is watching.

Three points must be followed for success in this approach. First, make sure the top managers have their own act together. If they hope to achieve the 2S goals of Organization and Orderliness, managers must first set a good example. Otherwise, they can talk all they like, but their employees won't listen to them.

Second, create a system that fosters repetition and review. Since training and discipline are supposed to make good habits automatic, they require a system that fosters repetition of activities and review of how well they are performed. To do this, you may need to throw out the current situation and procedures, then rethink everything from the ground up.

Third, devise a motivating strategy that keeps everyone aware of the results of their efforts. The saying goes that you get what you deserve, but rewards and criticism are often controlled by the whims of powerful people. Rather than just repeating that phrase, it's better to create economic structures that incorporate the principle of getting good results for doing good things, so people really do get what they deserve.

We think these three points are the essence of discipline and training. With that, we offer an in-depth reflection on the 5S system as it has been practiced in the past, and propose ways of implementing the New 5Ss.

5S Means More Than Posters on the Wall

5S has become a very popular workplace program. Campaigns are going on in plants everywhere. However, nagging doubts are arising: Why is it that the more 5S signs and slogans a company has on display, the less likely it is to have truly implemented the 5Ss, or even the 2Ss?

We also see managers eagerly touting 5S campaigns, implementing workplace checklists, and putting evaluation charts up on the walls (Figure 2-2). Yet the 5Ss end up only partly implemented—despite management's vigorous advocacy, an abundance of how-to books, and the widespread belief that these principles are the foundation for improvement.

Many 5S campaigns fail because they are implemented as "look at us" campaigns, designed to impress visitors more than to make real improvements in the workplace. This approach has three flaws:

1. Workers do not gain anything from the 5S activities.

   Even if 5S goals have been achieved in a certain plant, the people working there know very well that they will not benefit from them. They're smart enough to realize that if they put everything away five minutes before the end of their shift, they'll just have to take it all out again the next day. They also know from experience that if they leave cleanup duties to others, they'll have to look for everything the next day, and this will eat into their work time.

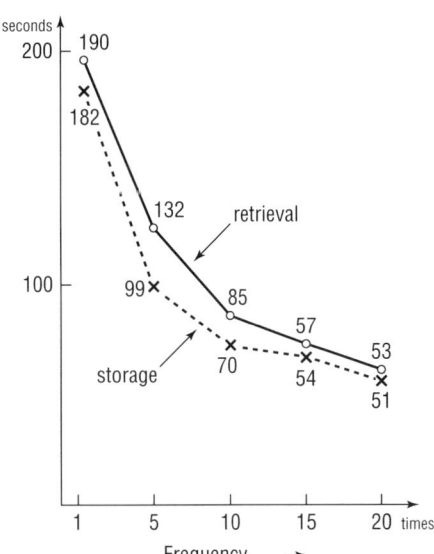

Figure 2-2. Graph of Retrieval and Storage Times

2. 5S campaigns become goals in themselves.

There's no doubt that putting unneeded items away, cleaning every corner, and getting rid of junk all make for a more pleasant working environment. But workers come under a lot of stress when they know their supervisors will issue reprimands for letting just one little piece of debris slip by.

In a situation like this, the 5S campaign itself has become the goal. We call this warped application the "Jive Ss": Store things, Stick to rules, Superficially clean, Switch to new fixtures, Serve reluctantly.

3. 5S activities happen only during a campaign.

In this situation, the plant may be neat and tidy during the actual campaign, but it returns to its previous disordered state the minute the campaign is over.

An authentic 5S system stays true to the end purpose, which is not only improving the plant but improving the manufacturing process.

Discipline Is the Heart of 5S

As we observed before, the practice of discipline and etiquette has diminished in recent times. For example, people now speak with family members using very informal language, sometimes even rudely. At work, however, it's important to be polite and pleasant to coworkers, no matter what mood one is in. People try to stay on good terms with others at work because it affects their economic well-being.

What should we do to ensure that discipline and good relations survive in the workplace? In answering that, we'd like to elaborate on the three points we raised at the beginning of the chapter.

1. Make sure top managers have their own act together.

   It is important for the people at the top to set a good example. Too many executives preach about reducing costs and yet use expensive company stationery for jotting random notes. They preach about improving human relations, but they bad-mouth fellow managers and talk curtly to their staff members. How can they expect to create a disciplined work force when their own desks are disaster areas and they are seen loafing with their feet up during the day? Employees notice hypocrisy and use it as an excuse to let discipline and morale slip even further.

2. Create a structure that fosters repetition and review.

   The concept of throwing out your current system and making only the necessary products at the necessary time in the necessary amount was devised to prevent problems with missing components. The way to do that is to create a production line that requires few personnel and eliminates slowdowns, defects, and the accumulation of excess parts. This is known as razing (breaking down) the process. Once a U-shaped line has been set up, the managers themselves try making it work according to standards. Then they have their employees practice the process over and over.

   If part of the process doesn't work, the on-site managers should review it until they come up with something that does work. The whole point is to create an environment in which the techniques of repetition and review result in a line that functions with reduced personnel and no waste.

3. Devise a motivating strategy.

   A famous general once said, "Some people will not budge unless you show them how to do it, make them do it, and then praise them for doing it." In addition, you have to offer your employees guidance when they're doing something wrong; don't let mistakes slip by. If they ignore your advice and persist in their mistakes, you need to correct their actions. The important thing is to create an environment in which this can happen.

   To motivate people, managers need to have generous and tolerant personalities. Cultural anthropologist Chie Nakane has said that it's best if the people at the top are not geniuses. When the people on top do too much of the thinking and work by themselves, their employees feel alienated and think they have little to contribute.

   The company won't function smoothly unless everyone recognizes that managers depend on their employees just as the employees depend on their managers. As part of this recognition, managers should share their ideas, which will help gain their employees' understanding. They also need to offer long-term guidance as policies are implemented.

Implementing the 2Ss for the Pre-Setup Stage

The kind of work that goes on in the typical discrete products plant usually follows the flow of stages shown in Figure 2-3. This is true for every aspect of manufacturing: machining, assembly, installation, or design.

The pre-setup stage (sometimes called external setup) is most important because it determines whether changeover and other operations will occur without any waste. Ironically, improvement of the pre-setup stage is often postponed, because waste is often hard to spot then. An inefficient pre-setup process, however, can lead to delayed startups and accumulation of in-process inventory during opera-

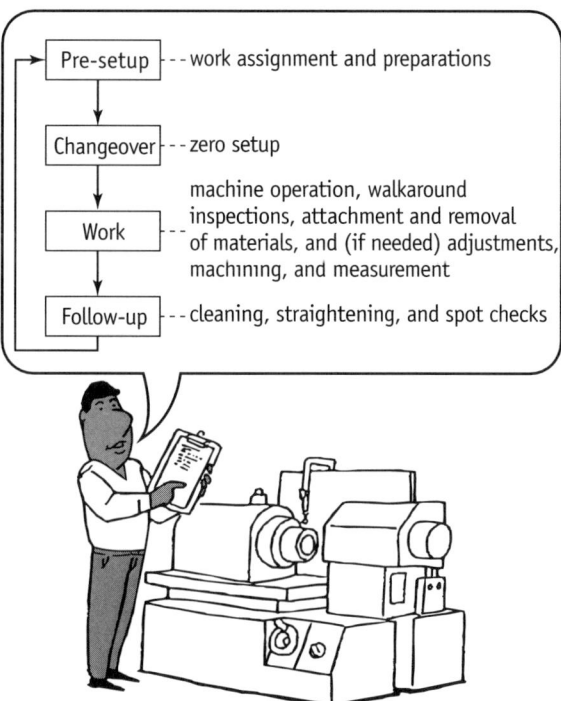

```
Pre-setup ---work assignment and preparations
    │
    ↓
Changeover ---zero setup
    │
    ↓               machine operation, walkaround
                    inspections, attachment and removal
  Work      ---     of materials, and (if needed) adjustments,
                    machining, and measurement
    │
    ↓
Follow-up ---cleaning, straightening, and spot checks
```

Figure 2-3. Stages of Manufacturing Work

tions. Thus, improving the pre-setup stage with clear visual management should be a priority.

We witnessed the following pre-setup-related exchange at plant N. Operator M was running around and asking, "Where did the replacement parts go? I can't find them."

His supervisor happened to pass by. "What are you fooling around for when we're so busy?"

"We've been having a lot of minor stoppages," M replied, "so I thought we should change the limit switch, and there was one on the shelf here just last week."

After searching for the switch for 30 minutes, M finally decided to disassemble and repair the old one. He returned to his machine. "Whoa! Who put these here?" he exclaimed, spotting some tote boxes piled up next to a pillar. They concluded that someone must have brought replacement parts to the area and just piled the empty boxes by the pillar, since there was no designated place to put them.

That scene was one of the first things we observed when we first set foot on the factory floor during our visit to this motor parts plant. When we asked whether that kind of thing happened often, the supervisor replied, "It happened three days ago during retooling. Minor stoppages of three minutes turned into a major stoppage. A jig had gotten lost. We were just about to give up and make a new one when we got a phone call from a painter who subcontracts on our line. It turned out that the jig had fallen into one of his tote boxes. I had to rush over there to pick it up."

The primary cause of these problems was that the 2Ss—Organization and Orderliness—had not been implemented properly. Anyone could have figured that out by asking "why?" a few times, but it was our first visit and we were drawn to the 5S signs hanging nearby. We concluded to ourselves that they hadn't really implemented Organization and Orderliness, and we told them so.

Was this pronouncement correct? Well, let's look at what happened later. The machine that received the new limit switch functioned smoothly, at least during the time we were there. Once things had settled down, we looked at the limit switch that had been removed and saw that it was completely caked with oil and debris. When we removed its cover, we saw that even the inside was dirty. We concluded and announced that this was because the 3Ss—Cleanliness, Standardized Cleanup, and Discipline—had not been observed.

To confirm our conclusions, we decided to extend our intensive on-site observations at the plant. Figure 2-4 shows the layout of a parts machining area at the plant. The machines were not arranged in a logical horizontal array, but similar pieces of equipment were placed together. Furthermore, the inner part consisted of mobile workbenches located in semi-fixed positions in front of the machines. Nearly all the machines required only a single operator.

To make sure everyone on site was of the same mind, we began by eliminating the waste that was currently going on. These types of waste are listed in Table 2-1. It's clear that when we look at ways to eliminate these wastes, we naturally end up considering the 2Ss and 3Ss.

During this exercise, we had a sudden realization. The 2Ss (Organization and Orderliness) and 3Ss (Cleanliness, Standardized Cleanup, and Discipline) show up separately in plans for eliminating waste, and we began to see a contradiction here. The 5S program is seen as the foundation for improvement, and many factories hang up 5S posters—but we think of the 5Ss as a single entity, even though the five elements can't be implemented all at once. In fact, we run into trouble when we think of these separate concepts as a unit and try to improve all of them at once.

The definitions of the 2Ss boil down to making sure that there are no unnecessary items in the area and that all the necessary items are within reach for immediate use. Applying this idea to the machines and other items in the plant, we can say

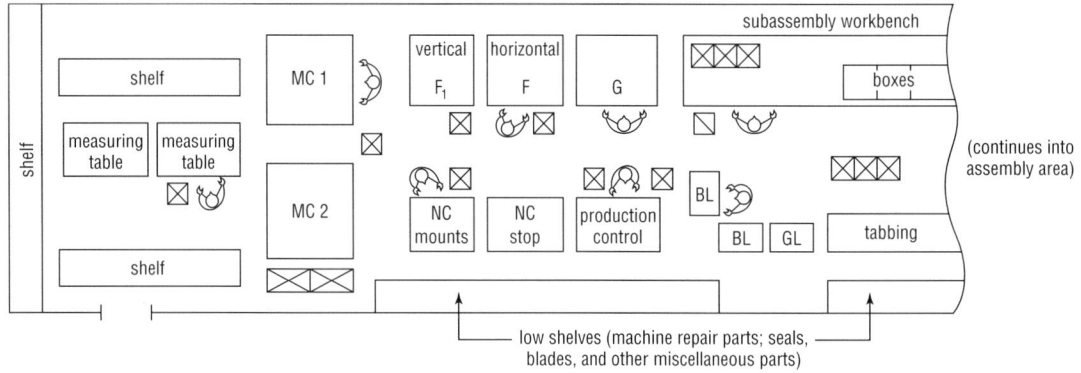

Figure 2-4. Machining Layout at Plant N

Table 2-1. Wastes Observed at Plant N

Waste Discovered	Priority	Causes	Waste Elimination Approaches
Waste from minor stoppages	2	1. Dirty machines and parts 2. Stopping machines before going to get parts 3. Cutting debris caught in the machine	1. Thoroughly implement the 3Ss (Cleanliness, Standardized Cleanup, and Discipline). 2.1. Make preparations during pre-setup. 2.2. Arrange the machines in the order of the machining process. 3. Fix the blade so that the cutting chips are smaller. Also reconsider the height of the work, the jigs, oil circulation, and other factors.
Waste from medium and minor stoppages	1	1. A lot of running around looking for things (pre-setup process is inadequate) 2. Sporadic breakdowns (due to accelerated deterioration caused by debris)	1. Thoroughly implement the 2Ss (Organization and Orderliness) including standard position and standard labeling. 2. With natural deterioration, MTBF allows prediction of replacement time; since the cause is debris, however, thorough implementation of the 3Ss (Cleanliness, Standardized Cleanup, and Discipline) is needed. Workers also need to learn instant maintenance methods.
Waste due to operators' standing idle	3	1. A one person, one machine approach 2. Changeovers of 20 to 40 minutes	1. Convert to multi-machine operation. 2. Apply zero setup principles and techniques to reduce time
Waste due to defects and readjustments	4	(remainder of chart omitted due to length)	

that the 2Ss mean having everything needed for the process arranged in the order in which it will be used.

Stated another way, implementing the 2Ss means placing work and tools according to a plan based on the flow of the process; the question of what to place where and in what amounts should also be determined according to this flow. We often call these principles "standard position" and "standard labeling."

In light of what we had observed, we decided to apply the 2S principles to the pre-setup process to focus the improvement effort at this plant.

Steps for Eliminating Waste in Pre-Setup

Step 1. Getting a Grasp of the Current Situation

We begin by trying to grasp what is currently going on. The proper procedure is to learn about the frequency and duration of the current setup procedures through operation analysis techniques such as work sampling; since time was limited in our situation, however, we chose to use a simpler survey approach.

At this plant, employee surveys revealed that the setup occurs about ten times a day and that the pre-setup stage takes 30 to 40 minutes each time, with the entire setup process taking one hour. This means that an average of 350 minutes were wasted each day—nearly the equivalent of an entire shift for one operator.

Step 2. Making On-Site Observations

Next we go into the plant to implement our waste reduction plans. First we make a plant layout chart like the one in Figure 2-4 and observe the process of gathering the items needed for pre-setup, either moving with the operators or following them. Table 2-2 records the observations made at plant N.

Step 3. Eliminating Waste in Searching

The key to pre-setup improvement is eliminating the waste involved in looking for things. Assuming that this was due to failure to implement standard positioning

Table 2-2. Pre-Setup Waste Elimination Approaches

Aspect	Waste Discovered	Priority	Causes	Waste Elimination Approaches
Written operating instructions	Waste in thinking	3	1. Different from numerical values in production plans 2. Amount of inventory cannot be read correctly (cannot be seen) 3. Not clear how current they are	1. Long-term production plans are the problem. You don't have to think if you create a system in which you make only week-by-week plans or work only on what comes in. 2. An extreme measure is to eliminate the warehouse and move to a storage bin system. Until then, reduce inventories by half and conduct an inventory process breakdown. 3. Install storage spaces; make production dispatch boards and post them.
Jigs, implements	Waste in looking for things	1	1. Things not put back after use; no specific places to put things 2. Instruction signs are out of date 3. Items resemble each other so closely that it's hard to tell them apart	1. Thoroughly implement Organization and Orderliness. 2. Someone should take responsibility for making rules about how often signs should be reevaluated and assign people to make sure that they are correct. 3. Items should be differentiated by features such as color, as well as having standard positions and standard labels.
Parts and materials	Waste due to walking around	2	1. Items are stored in several places, not just one 2. Old and new items are mixed together, leading to an inadvertent last-in, first-out system	1. Have a single place for each item and eliminate wasted search time by placing storage bins in a U shape so they are easy to reach. 2. Load items into a slanted "flow rack" for a first-in, first-out retrieval system.

and standard labeling, we added more detail to our layout diagram, and we realized that there was no logical flow to the way people looked for things. We therefore took the needed items and rearranged them in the order required for setup.

Step 4. Designing the Layout of the Pre-Setup Line

Figure 2-5 is a drawing of the pre-setup line we designed at plant N. Responsibility was assigned to a single operator, who goes around on the day before setup and confirms that the necessary raw materials, parts, other materials, and jigs are available. The operator writes missing items on a list and orders them to arrive for the beginning of work the next day. By the time work begins, the setup items have been laid out on carts in quantities slightly larger than needed.

Step 5. Putting the Principles to Work on the Spot

Since the pre-setup person at plant N is one of the most experienced operators, things usually go smoothly. However, if unforeseen circumstances arise, the operator has the leeway to change the locations of certain items, as long as the basic U-shaped arrangement is kept in place.

Step 6. Documenting the Results

After a while, we can compare the results achieved under the new pre-setup system with the results from before. At plant N, they even cleaned and repaired the jigs on the pre-setup line, a task usually relegated to the cleanup phase.

As an additional benefit, the productivity of the machining lines rose 30 percent. This is because the operators no longer spent time searching for things and thinking about what to do next. Instead, they were able to devote more time to operations that added value to their product.

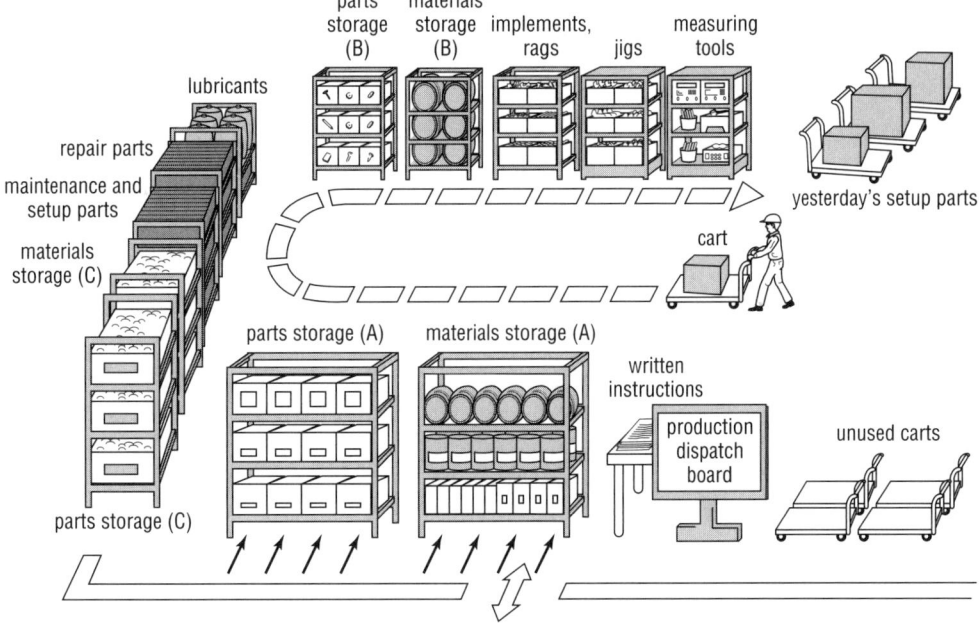

Figure 2-5. The Pre-Setup Line at Plant N

By following these steps, employees completely implemented standard positioning and standard labeling, and it was obvious that this plant had simultaneously rid itself of debris and procedural foul-ups.

You may be noticing that the steps we have described are similar to the technique of process razing. In fact, really applying the 2Ss is equivalent to process razing. Furthermore, once you implement the 2Ss for pre-setup and go through a changeover without any slipups, you won't have problems with minor stoppages.

Maintaining Improved Conditions

What can you do to ensure that things aren't carelessly misplaced and unfinished goods don't pile up? We see overproduction as the cause of these problems. Chapter 3 describes 5S activities at a garment assembly plant that ultimately used process redesign to maintain neat conditions by avoiding overproduction. Two techniques from that plant can help maintain Organization and Orderliness:

- Lay out each process as a line and set up storage bins (standard position, standard labeling, standard amount).

- Post schematic diagrams on the storage bins showing what to use when, to prevent overproduction, reduce the number of unfinished goods, and promote organization and orderliness.

In other words, create a smoothly flowing system, setting up different lines for different internal customers throughout the entire plant so these types of problems do not recur.

It's a good idea to take before and after pictures of work areas that have been especially successful in implementing the 2Ss and display them in the plant. This increases employees' interest in the campaign and helps motivate them to participate.

Photographs 2-1 through 2-4 are examples from 2S inspections at company Y. Photograph 2-1 shows how raw materials were piled up around the machines in greater quantities than necessary. Photograph 2-2 shows how the amount of materials was reduced through pre-setup activities to only the amount needed for one day's work.

Photograph 2-3 shows how the wiring was arranged alongside the machines: Before improvement, the operators were constantly stepping on it as they worked. Photograph 2-4 shows the improvement: A plastic conduit was purchased, and all the wires were run through it, ending the clutter and tripping hazard on the floor.

Benefits of Proper 2S Implementation

Properly implementing 2S for the pre-setup stage has two advantages:

1. It's not necessary to make a special effort to clean the jigs and molds, because the 3Ss are incorporated in the basic work of the pre-setup stage. Following the 2Ss for this stage is enough to ensure cleaning.

2. If cleaning is standardized, you'll end up with a clean and tidy work site where things don't have a chance to get dirty. Furthermore, if you're careful about teaching employees to work smoothly, you'll get increased safety.

Photo 2-1. Excessive Raw Materials

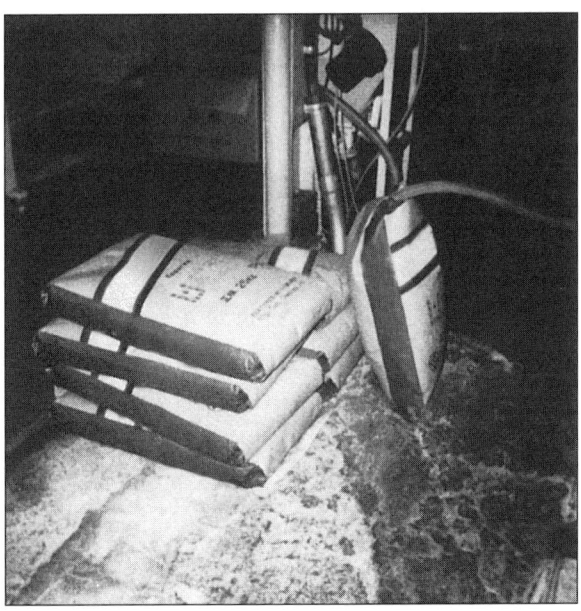

Photo 2-2. Quantity After Pre-Setup Improvements

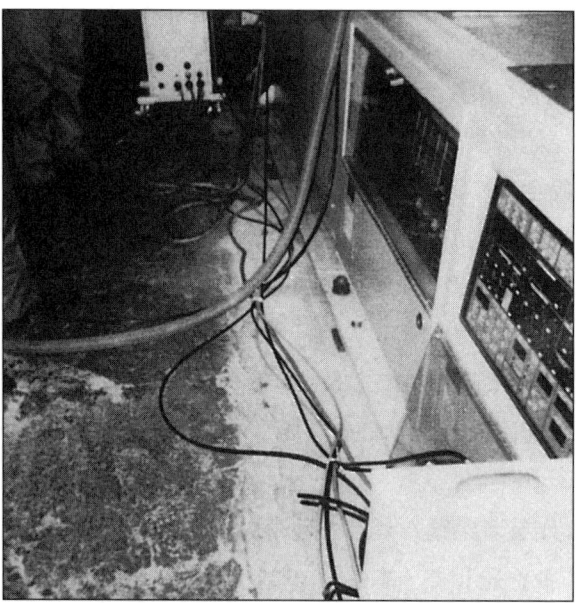

Photo 2-3. Hoses and Wires Causing a Safety Hazard

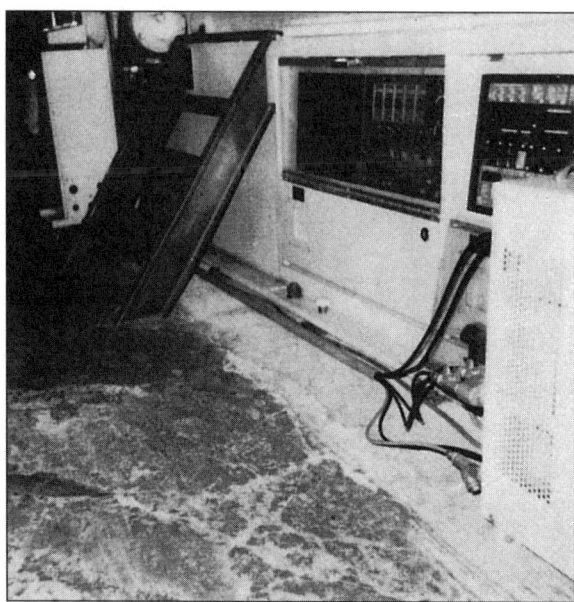

Photo 2-4. Wires Threaded through a Conduit

A New 5S Example:
Mynac Company Case Study

3

The 5Ss (Organization, Orderliness, Cleanliness, Standardized Cleanup, and Discipline) promise that factories can be improved without spending much money, so managers find them extremely attractive. They imagine that if they appoint someone to head implementation and set up a 5S command center, they'll end up with a sparkling, efficient plant in no time at all. Furthermore, they tend to believe reports that other companies have succeeded with the 5Ss, even if the reports are untrue, and they urge all their employees to get cracking on the project.

Once the project has been kicked off, the newly appointed 5S manager goes at the task hesitantly, not really sure what to do. He or she starts by appointing an Organization Manager and an Orderliness Manager, creating posters, and soliciting ideas for slogans. Orders go out to paint the aisles to cover up the pitted surfaces. Broken shelves are replaced with new storage boxes, so worn-out, junky parts and tools can't be seen from the outside. At the end, the managers conduct a 5S inspection. They take photos of any sites that fail the inspection and hang them up in the cafeteria to shame the workers responsible.

When a 5S campaign is treated as an end in itself, it means that management has forgotten the overall goal and become caught up in showing off their techniques. Even if they happen to succeed in implementing some sort of 5S program, no value is added to the product, and profits remain stagnant. That is why we have decided to write about a company called Mynac, which succeeded in implementing the 5Ss by reconceiving them as "the creation of well-ordered production lines."

The 5Ss Have Become the Jive Ss

Executive B, managing director of a South Korean company, came right out and admitted, "I'm not so sure that implementing the 5Ss will really increase our productivity."

According to B, his company had received advice on the 5Ss from a consultant. He admitted that the plant was now cleaner and more pleasant than before, but the methodology reminded him of the types of propaganda campaigns common under the regime of a former South Korean president. The methods included

• Removing and picking up debris

• Throwing out everything that wasn't needed

Figure 3-1. 5Ss vs. Jive Ss

- Making pictures of the implements and arranging them in a row
- Replacing dirty shelves and pallets with new ones
- Painting safety aisles on the floor to make it look new
- Storing all the parts, materials, documents, and machining specifications
- Making employees clean the plant constantly

Although the campaign appeared to succeed, productivity remained flat, and discontented rumblings were heard from the employees.

Executive B wondered why the 5Ss hadn't succeeded. We believe that they failed at his company because the goals of the campaign were unclear. The company's manufacturing methods had not been improved. The 5S campaign had been conducted simply for the sake of conducting a campaign. When the technique became an end in itself, the 5Ss turned into the Jive Ss (see Figure 3-1).

Successful 5S Application Requires Serious and Methodical Top Managers

The 5Ss at Mynac

Mynac is a manufacturer of upscale women's apparel located in Nagano prefecture, in the city of Iida, an area of Japan famous for its apple orchards. The company specializes in small-lot, multiple product lines and boasts that its workers can sew anything that it is possible to sew. The company is particularly well known for its quick turnaround of special orders.

Founded in 1970, the company has a total capital of ¥60 million (approximately $480,000). With 130 employees, whose average age of 27 gives the company a youthful air, the company is unconcerned about the labor shortages that plague some other Japanese companies.

The name of the company is an acronym for the English phrase "My Nagano Apparel Company." The "my" part of the name is very important, since the employees are all stockholders. The president, Mr. Ichinose, is more of a facilitator than a manager, and that is only one of the ways in which he differs from the usual company president. For one thing, he is extremely punctual, habitually arriving five minutes before an appointment. He is also scrupulous about keeping promises, and he prefers to settle his accounts with cash rather than running up a tab.

Mr. Ichinose is also very focused on his objectives. He likes discussions that concern whatever he is focusing on at the moment, but he quashes any irrelevant remarks. He has a strong sense of right and wrong, to the point of being considered stodgy. No shady businessman with dubious political connections, he is known for breaking off with people who act unethically, even if they're regular customers. In addition, he has good manners, and, as befits someone in the fashion business, he dresses smartly.

Combine all of these traits and you have one hardworking, methodical manager, and this is the quality that is most important to his subcontractors. The reason is that manufacturers live or die depending on whether they can lower the cost per item or reduce the time required to produce each item. If the managers are not serious about this, their subcontractors, in turn, cannot survive.

Mr. Ichinose is from the Nagoya area, with friends at Toyota and its affiliates, and he is familiar with Toyota's production methods. Figure 3-2 shows the relationship between Toyota's seven types of waste and the 5Ss.

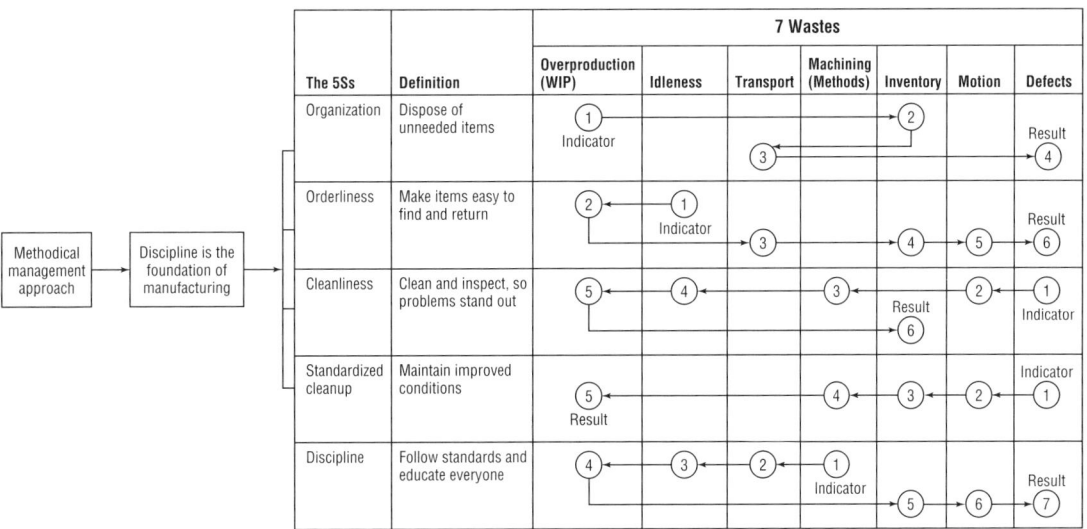

Figure 3-2. The 5Ss and Toyota's Seven Types of Waste

The first S, Organization, refers to getting rid of anything that's unnecessary. This applies especially to unfinished goods and idle inventory. A company should first get rid of waste in the form of overproduction during processing and overpurchase of materials.

The second S, Orderliness, refers to having everything close at hand where it is easy to use. Indicators of deficiencies in this area are time wasted looking for misplaced items, idle workers standing around because parts are missing, and time searching for repair or setup parts. Only a company not plagued by such unproductive periods can avoid working on Orderliness.

The third S, Cleanliness, means absence of debris and dirt through cleaning and inspection. For a clothing manufacturer, one of the indicators is defects. Spring clothes, in particular, are made with bright and light-colored cloth, and if they are shipped with stains, the company will lose customers.

The fourth S, Standardized Cleanup, refers to maintaining neat and clean conditions. Again, defects are an indicator that this area needs work. In this case, keeping the factory clean and attractive is also critical in attracting young people to work there.

The fifth S, Discipline, means following definite rules. One of the Japanese words for this concept refers to the basting thread used in garment sewing. Put in place by a master tailor, the basting thread holds the cloth in place and creates standard seam lines for the other sewing machine operators to follow.

At Mynac, Mr. Ichinose sets this example. As president, he devotes a lot of effort to setting an example of standards, unlike many executives who are attracted to the 5Ss because they seem to be an inexpensive improvement method. Such executives typically establish a 5S command center and leave everything up to the employees, which inevitably leads to failure. They then bring in consultants for guidance, but their initial failure to set and enforce the example makes the campaign fail again in a short time.

Discipline is the basis of effective manufacturing, and it is essential for the managers to set the example here. The rest of this chapter explains how Mr. Ichinose went about it.

A New Way of Thinking about the 5Ss

Figure 3-3 shows Mr. Ichinose's plans for implementing 5S. They're ultimately based on the relation between the 5Ss and the Toyota production system, and the first order of business is eliminating waste.

The ideal form of organization is a well-ordered, smoothly flowing production line. This means that the line produces one item from start to finish with no unfinished goods lying around. Parts are stored neatly, and no time or motion is wasted in looking for, retrieving, or putting away parts and supplies.

The floor is always clean, with no debris or stains, which makes for a pleasant workplace (see Photograph 3-1). In other words, the 5Ss have been distilled into the single goal of creating a well-ordered, smoothly flowing line. If, on the other hand, the techniques become ends in themselves, we get the familiar Jive Ss and the inevitable failure that follows.

Overall goal: well-organized, smoothly flowing production lines

5S	Goal	Steps for achieving the goal		
		Stage 1	Stage 2	Stage 3
Organization	Creating well-organized, smoothly flowing production lines	Most of the unneeded items are excess inventory and unfinished goods	Creation of compact U-shaped lines	U-shaped 2S lines that are simplified and automated
Orderliness	Everything in its place	Creating storage bins for parts	Improvement of the pre-setup process (reorder charts)	Just-in-time supply production with kits
Cleanliness	Creating a clean, pleasant workplace	Improving the floor surface	The managers are responsible for picking up scraps	Improving the ceiling; a debris-free workplace
Standardized Cleanup	Maintaining a clean manufacturing process	Improving the cafeteria and rest areas	Remodeling the toilet facilities	Improving the lighting (2000 lux)
Discipline	A well-mannered workplace	Courtesy campaign	Everything done with spit and polish	Campaign to instill a sense of style

Figure 3-3. 5S Implementation at Mynac

Mr. Ichinose chose to begin his reorganization of the production lines by applying the process-razing approach to break down the various processes. (To learn more about Mynac's process razing, see Chapters 8 and 13 of Ken'ichi Sekine, *One-Piece Flow*, Productivity Press, 1992.)

Take another look at Figure 3-3. Although some factories have a lot of randomly placed items to deal with in Stage 1 of Organization, at Mynac most of the unnecessary items that hinder the creation of a smoothly functioning line are excess inventory or unfinished goods. We need to focus on what to do with these items.

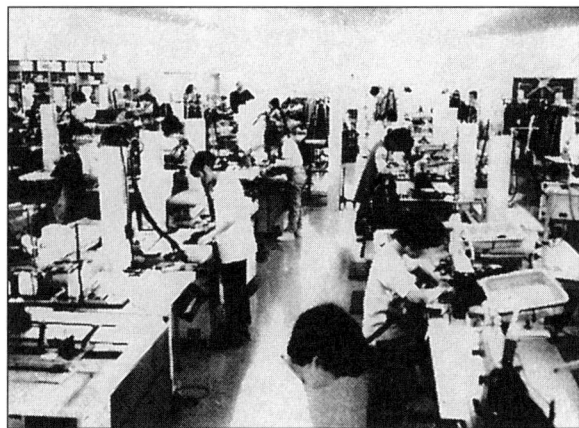

Photo 3-1. Well-Organized, Smoothly Flowing Production Lines

The most effective method for dealing with excess inventory is to create a U-shaped production line based on the results of process razing. This technique is also called process design; at Toyota, it is referred to as management you can understand by looking.

In Stage 2, a compact U-shaped line is created. In Stage 3, the line is simplified and automated as much as possible. Once the second S, Orderliness, is implemented, using tidy parts storage and other innovations that eliminate waste, the result is a 2S-style U-shaped line. All that's needed to create a well-ordered, smoothly flowing line is to add the last three Ss: Cleanliness, Standardized Cleanup, and Discipline.

Mynac's 5S Know-How: The Ten Principles

According to an old saying, if you try to teach yourself to play a game, such as chess or even golf, it will take you ten thousand repetitions to achieve even the lowest-ranking professional status. That means that if you played once a day, you'd need 28 years to reach that level. Yet if you get some coaching in the basics, you can probably reach a professional level after a thousand games, or only 3 years. Since learning the fundamentals from a pro reduces your effort by a factor of ten, an iron rule for acquiring any new skill is to master the basics at the beginning.

The 5Ss are like that, too. If your initial program is wrong-headed, the Jive Ss will take root in the company, and it is hard to break these bad habits later. That's why we've taken Mr. Ichinose's way of implementing the 5Ss and restated it in terms of ten principles. Table 3-1 lists the principles, and we will discuss each of them in detail.

Table 3-1. Ten Principles for Implementing 5S

1. Keep in mind that your true goal is a well-ordered, smoothly flowing production line. Never let the 5S process itself become the goal, or you'll end up with the Jive Ss.
2. To keep your processes from getting bogged down, first get rid of all excess work-in-process. If you set up and maintain U-shaped production lines, organization and orderliness will follow naturally.
3. Distinguish necessary items from unnecessary items and get rid of the unnecessary ones immediately. Excess inventory is a major unneeded item to target.
4. Eliminate the waste that arises from looking for things. The secret of orderliness is to position items according to their frequency of use and to make sure they can be returned easily to their proper places.
5. Everyone needs to be responsible for picking up debris, including top management. This will keep the plant neat and clean.
6. Cleaning is an occasion for spot checks. Pieces of equipment should be labeled in the order of their tendency to break down, and they should be cleaned and inspected every day.
7. Toilet facilities should be better than the ones in the employees' homes. This creates a clean and hygienic atmosphere throughout the plant.
8. No electrical wires should dangle from the ceiling. You'll have a more streamlined production line if they enter or exit the machines from the side.
9. Clerical and administrative departments should also be set up in U-shaped lines to provide a readily observable 5S example.
10. If management sets the example with the first four Ss, the fifth S, discipline, will follow naturally.

Principle 1: Keep in mind that your true goal is a well-ordered, smoothly flowing production line. Never let the 5S process itself become the goal, or you'll end up with the Jive Ss.

This is the first and most important rule. To this end, setting up U-shaped lines is absolutely necessary (see Photograph 3-2). Even if your 5S consultant only advises you to set up a 5S command center and start a campaign, you must think about the end result you want with your lines and inventory. Otherwise, you'll end up with only superficial improvements that add no value to your product, and your whole company will be afflicted with the notorious Jive Ss.

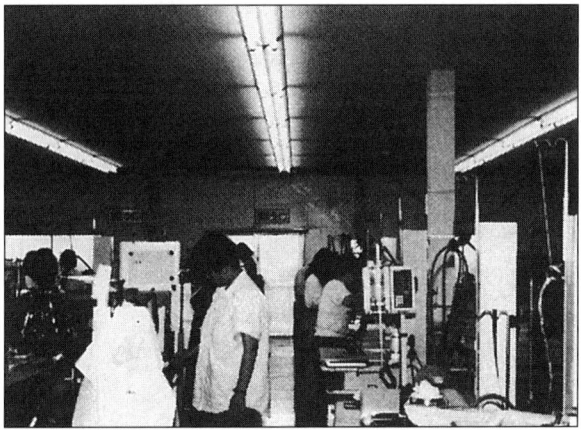

Photo 3-2. Well-Organized, Smoothly Flowing Line Production

Principle 2: To keep your processes from getting bogged down, first get rid of all unfinished goods. If you set up U-shaped production lines, organization and orderliness will follow naturally.

If you set up U-shaped lines in which each item moves continuously through the manufacturing process, you won't have unfinished goods lying around and operations won't get bogged down. Set up parts storage bins to eliminate the waste that comes from looking for and replacing items; use open steel shelves on wheels that let you see if anything's missing (see Photograph 3-3). Tactics like these can eliminate the need to set aside special time for organization and orderliness. For all practical purposes, creating a U-shaped line is the same as creating a well-ordered, smoothly flowing line.

Principle 3: Distinguish necessary items from unnecessary items and get rid of the unnecessary ones immediately. Excess inventory is a major unneeded item to target.

This distinction is a major principle of the 5Ss, and it is also a fundamental concept for grading products. When looking for items to get rid of, start with excess inventory and unfinished goods.

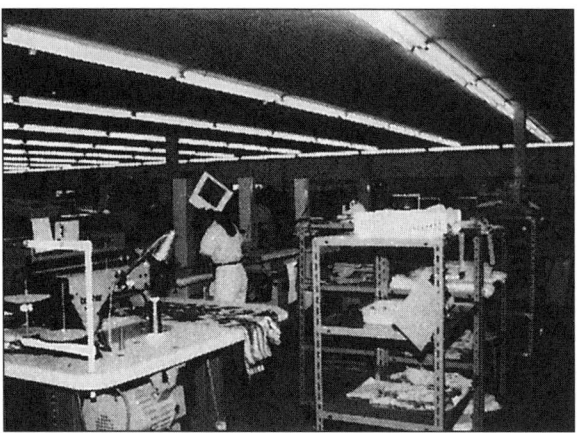

Photo 3-3. Open-Shelved Carts Keep Operations Moving Smoothly

1. Attach red tags to any unneeded items.

2. Decide how to dispose of the items or put them to better use.

3. Immediately get rid of anything outdated.

4. Look at any unused equipment and determine how to dispose of it or use it elsewhere.

5. Carry out process razing on any line that has planned downtime more than 50 percent of the time.

6. Make sure that the managers are aligned with the program. You may find that the dead wood on your managerial staff forms the largest category of "unneeded items." Such managers

   • aren't motivated to improve.

   • can't envision ways of improving any waste that they discover.

   • don't understand the concept of waste and don't recognize it when they see it.

   • get caught up in the techniques and can't carry them through to the actual goals.

   • scold their subordinates in front of other employees.

   • give up easily.

   The top executives should eliminate such dead wood.

   • First of all, they shouldn't promote people who are likely to become dead wood.

   • They shouldn't create corporate structures that make it easy for incompetent managers to get by.

   • They need to trim around the edges occasionally, just as a gardener prunes a tree.

Mr. Ichinose likes to boast that Mynac has no dead wood on its payroll. Incidentally, as Photograph 3-4 shows, Mynac doesn't have much in the way of inventory, either.

Photo 3-4. Mobile Inventory Carts

Principle 4: Eliminate the waste that arises from looking for things. The secret of orderliness is to position items according to their frequency of use and to make sure they can be returned easily to their proper places.

Looking at how a plant carries out setup and changeover gives you a vivid understanding of how well it has implemented the 5Ss. A company's level of orderliness can be measured by how much or how little time employees need to spend looking for things.

An observer can check to see whether all the materials, parts, implements, processing specifications, molds, and other necessary equipment are

• kept in standard locations with standard labels in standard amounts

• laid out in a U-shaped line

• placed so that they are both easy to take out and easy to put back in their proper place (see the diagonal stripes on the shelved items in Photograph 3-5)

• neatly put away after use

Photo 3-5. Diagonally Sloping Document Shelves

If you take a look at the section for evaluating Orderliness in Mynac's 5S Evaluation Sheet (Table 3-3, on pages 44–47), you will see the following levels of waste caused by having to search for things at setup time:

Level 1: Employees wander aimlessly searching for parts and materials.

Level 2: Employees move in a zigzag pattern gathering up what they need.

Level 3: Some effort is wasted searching for things.

Level 4: Setup carts are used to assemble everything needed.

Level 5: No time is lost searching for things (U-shaped formation).

The idea is to arrange the supplies needed for setup so the employees can walk around a U-shaped line picking up everything they need. At Mynac, they arrange the thread, dress patterns, processing charts, specifications, and fabric in such a U-shaped array; everything is in a standard location with a standard label in standard amounts.

Furthermore, the documents that used to clutter up the plant are stored on open shelves, as are the footwear the employees use on the job to keep their feet clean. The footwear storage shelves are set up so each pair is easy to take out and to put back. The same is true for the thread, implements, and document files.

> *Principle 5:* **Everyone needs to be responsible for picking up debris, including top management. This will keep the plant neat and clean.**

There are three reasons for picking up scraps. The first is overall tidiness. The second is getting rid of trash. The third is preventing excess friction on the casters of the supply carts. When the operators are busy doing their jobs, only the managers pick up debris. If you have a disciplined work force, this happens as a matter of course.

> *Principle 6:* **Cleaning is an occasion for spot checks. Pieces of equipment should be labeled in the order of their tendency to break down, and they should be cleaned and inspected every day.**

Cleaning can be a kind of spot check. For example, an unusual amount of dirt on a piece of equipment is a sign of trouble. Every morning, and after all planned downtime, the equipment, sewing machines, and other installations are cleaned according to their inspection labels. The goals are to prevent the buildup of spattered oil and to ward off mechanical breakdowns stemming from debris getting caught in the equipment. These cleaning sessions don't seem like such a burden once the employees are used to them.

> *Principle 7:* **Toilet facilities should be better than the ones in the employees' homes. This creates a clean and hygienic atmosphere throughout the plant.**

The employee restrooms at Mynac are almost fun to use. Walking in, one is greeted with lively Latin dance music. Air dryers take the place of paper towels, avoiding mess and waste.

Photo 3-6. Part of a Rest Room at Mynac

The restrooms are also sparkling clean, which was undoubtedly one of Mr. Ichinose's intentions (see Photograph 3-6). It stands to reason that you can start implementing the 5Ss by improving the restroom facilities, particularly at small- and medium-sized companies.

Principle 8: **No electrical wires should dangle from the ceiling. You'll have a more streamlined production line if they enter or exit the machines from the side.**

At one time, plant designers were advised to hang all wiring and pipes from the ceiling. The result was U-shaped production lines with jungles of wires and pipes overhead.

When companies conducted process razing or developed new lines, the worksites were so enveloped in spiderwebs of electrical wires that the workers couldn't see what was happening in other parts of the plant.

One day, Mr. Ichinose tried to see what would happen if the wiring of the sewing machines were linked horizontally. When this principle was applied to all the lines, the result was a bright, clean-looking, well-illuminated plant, nearly 2000 lux (see Photograph 3-7). The key to maintaining purity may be a brighter work-place, even though it may cost a little extra.

Principle 9: **Clerical and administrative departments should also be set up in U-shaped lines to provide a readily observable 5S example.**

A private office is a status symbol. Executives and clerical workers often demand their own offices, complaining that they can't concentrate in a noisy environment or hear their phone calls clearly. Once each department has its own office, how-

Photo 3-7. Wires Run Horizontally

ever, psychological barriers come up and 5S efforts fall to pieces. People let papers accumulate on their desks, clearing off little spaces to perform clerical duties, draw up designs, or study charts when they need to.

The worst part is that the clerical staff and executives are off in a separate world, smoking, drinking beverages, or chatting while the employees on the shop floor are hard at work. This goes against the spirit of 5S.

At Mynac there's no distinction between the industrial and clerical staffs. All the employees are in one big room and can see what everyone else is doing (see Photographs 3-8 and 3-9.). That's why Mynac can get along without special 5S campaigns and doesn't need to have nitpicking general inspections.

> *Principle 10:* **If management sets the example with the first four Ss, the fifth S, discipline, will follow naturally.**

Success in sustaining 5S results requires employee education and good examples set by top management. One critical point on which employees need training and

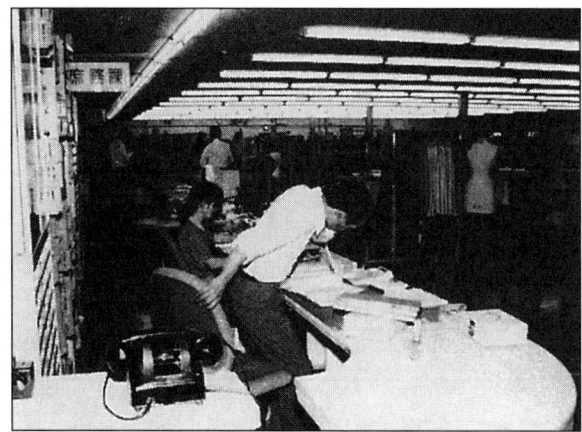

Photo 3-8. A U-shaped Production Line for Sewing

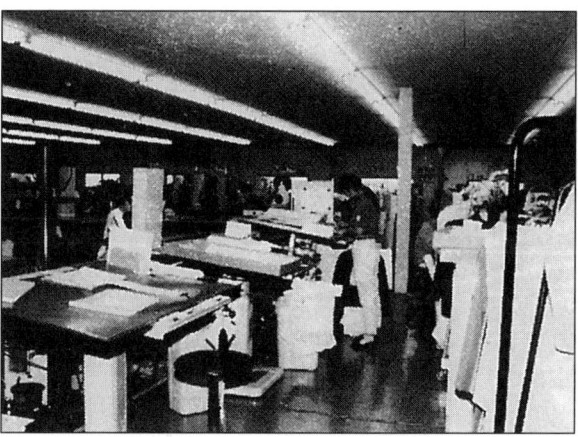

Photo 3-9. Administrative Functions Have Also Been Restructured into Production Lines

support is learning to follow standard procedures. People tend to be unsteady for a while until they learn the job; they need management support and feedback on the way.

Another point for employee training for just-in-time production is to work according to a regular rhythm, rather than simply working as fast as one can. A constant speed is more important for meeting a standard cycle time and avoiding production of excessive inventory.

Managers must also inform workers of safety hazards they might encounter on the job, as well as precautions they should take. Certainly, managers should also recognize employees for learning and for doing their jobs well.

The top executives of the company should take personal responsibility for setting up the training program for new hires. Table 3-2, which shows how Mynac trains its new sewing machine operators, is included for reference.

Discipline is imparted from experienced workers to inexperienced workers and from supervisors to other employees. Just as a sewing machine operator follows the basting thread already laid down, the new hires carefully follow the example set by experienced workers and supervisors. This discipline is the key to high-quality manufactured goods, and it should be both modeled and enforced by top management.

Here are some of the things we observed at Mynac:

1. Politeness is very important at the company. The supervisors greet the employees first. The minute Mr. Ichinose encounters an employee, he greets him or her, which creates a pleasant atmosphere in the plant. The employee usually returns the greeting with a smile.

2. Mr. Ichinose follows the military custom of arriving at an appointment five minutes early, whether inside or outside the company.

3. He is also scrupulous about keeping promises, whether to the union or to individual employees or in his personal life. His custom of paying all debts immediately is another indication of his nature.

4. In general, employees do not go out of their way to greet visitors, the idea being that they should concentrate on their work. If their eyes happen to meet those of a visitor, they acknowledge the visitor with a brief nod.

5. Younger employees are expected to treat senior employees with respect.

6. Messages, information, and confirmations are transmitted immediately, whether by telephone, by fax, or in person, even if the recipient is in a meeting.

7. Periodic walk-around inspections allow Mynac to get by without any 5S campaigns. The managers check the fabric end processing shelves, which tend to get messy, and also look to see whether fabric ends or patterns have fallen under the work tables. They make sure that the color-matching posters and company notices are properly fastened to the wall and they order repairs on any pitted areas in the floor surface.

Steps for Interesting and Enjoyable 5S Activities

Figure 3-4 outlines activities at Mynac that promoted a friendly competition to raise the 5S level in work areas throughout the plant.

Step 1: Announcing the Program from the Top

At Mynac, Mr. Ichinose called the managers together and announced that there would be a 2S campaign for "sweating out the grime" at the plant. The inspection team members were mostly sewing instructors. They came up with a humorous campaign in which they joked that they realized that the employees probably bathed every day, but couldn't everyone use an occasional trip to the sauna to sweat out hidden grime? About two hours per month were allotted for the promotion activities.

Step 2: Determining the Inspection Sequence

The managers met and determined the order in which the various worksites would undergo inspection. They made up a schedule that had them inspecting the entire plant over a six-month period.

Step 3: Sweating Out the Grime

Saunas in some countries have attendants who rub down the customers with towels as they sit in the steam. Flecks of grime will roll off even the cleanest-looking person under such circumstances. Like sauna attendants, the inspection teams looked for the hidden "grime" in each department.

They wrote up the problems as one-line summaries on a Waste Elimination Form, noting, for example, that "unneeded molds are lying around near Machine 160-1." Then each manager ranked these problems in terms of their importance.

Step 4: Determining the Extent of the Problems

The chair of the committee collected the Waste Elimination Forms and determined the seriousness of each problem by assigning point values to each committee

Table 3-2. Daily Schedule for New Hires at Mynac

Goals of the Training Session

1. Mastering continuous sewing.
2. Making a skirt on their own.
3. Learning to evaluate their own work.

March 28 (Thursday): Opening Ceremonies 10:00

Daily Times / Dates	8:25-9:00: Meetings 9:00-10:00: Meetings 10:10-11:30: Meetings 11:30-12:10: Reading text	12:55-14:55: Sewing machine practice 15:05-16:00: Video instruction 16:00-17:00: Sewing machine practice 17:00-17:20: Meeting
March 29 (Fri.)	President Ichinose: "Welcome to the Real World;" "The International Scene" Managing Director: "Labor Saving Practices in Sewing"	Sewing machine practice: (special menu) threading, using the pedal
March 30 (Sat.)	Department Manager: "Explanation of Wages and Benefits" Assistant Director N: "Rules and Procedures"	Section Manager: "The Hardware Side of Sewing Machines" Sewing machine practice: inserting the bobbin, adjusting the stitches
April 1 (Mon.)	Department Manager: "Working in Groups" Assistant Director N: "Rules and Procedures"	Sewing machine practice: high-speed sewing, continuous sewing
April 2 (Tues.)	Department Manager: "The Current State of the Apparel Industry" Assistant Director Nakajima: "Rules and Procedures"	Sewing machine practice: sewing in a straight line
April 3 (Weds.)	Department Manager: "CAD and CAM at Mynac" Assistant Director N: "Rules and Procedures"	Sewing machine practice: high-speed extended sewing
April 4 (Thurs.)	Assistant Director C: "Patterns and Designs" Assistant Director N: "Explanation of Benefits"	Sewing machine practice: straight line extended sewing
April 5 (Fri.)	Sewing machine practice: straight line sewing, backstitching	Sewing machine practice: sewing curved lines (large and small) (dinner follows)
April 8 (Mon.)	Sewing machine practice: straight line sewing, backstitching	Sewing machine practice: sewing curved lines (large and small) (dinner follows)
April 9 (Tues.)	Sewing machine practice: high-speed continuous sewing (including backstitching)	Section Manager: "The Hardware Aspects of the Machine" Sewing machine practice: overlock threading, review of basic techniques
April 10 (Weds.)	Sewing machine practice: overlock threading, review of basic stitches	13:30-16:00: Motivational lectures at the Prefectural Cultural Center
April 11 (Thurs.)	Sewing machine practice: overlock threading, review of basic techniques	Sewing machine practice: overlock adjustment, sewing with an overlock

(continued)

Table 3-2. Daily Schedule for New Hires at Mynac, continued

Daily Times / Dates	8:25-9:00: Meetings 9:00-10:00: Meetings 10:10-11:30: Meetings 11:30-12:10: Reading text	12:55-14:55: Sewing machine practice 15:05-16:00: Video instruction 16:00-17:00: Sewing machine practice 17:00-17:20: Meeting
April 12 (Fri.)	Sewing machine practice: sewing darts	Sewing machine practice: sewing darts, ironing
April 13 (Sat.)	Sewing machine practice: attaching fasteners	Sewing machine practice: attaching fasteners, ironing
April 15 (Mon.)	Explanation of instruction manuals Sewing machine practice: attaching fasteners	Explanation of process charts Sewing machine practice: attaching fasteners
April 16 (Tues.)	Sewing skirt lining	Sewing skirt lining
April 17 (Weds.)	Sewing skirt	Sewing skirt
April 18 (Thurs.)	Blind stitching Final practice	Final practice
April 19 (Fri.)	Follow-up meeting The skirt is yours to keep!	Managing Director: Closing ceremonies and dinner

member's rankings. An item ranked "1" received five points, an item ranked "2" got four points, and so on. In the case of a tie, the chair made the decision.

Step 5: Labeling the Worst Problem

Once the committee ranked the various problems, it returned to the site in question, confirmed its earlier evaluation, and labeled the worst problem. During this visit, the committee took a baseline snapshot of the trouble spot and posted it in a conspicuous place, such as the company cafeteria. The committee also promised to return at some specified time to take another snapshot. Committee members were careful, however, not to take this technique to the point of harassment. They used humor, but they were sensitive to employees' intelligence and feelings.

Step 6: Working to Remove the Label

The people responsible for the area that was labeled "worst" worked on improving organization and orderliness at their site. They called their coworkers together and explained why they were given the label and what it meant. Then everyone got rid of unnecessary items, resolved to pick up debris every day, and cleaned and straightened up the site.

Step 7: Confirming That All the "Grime" Has Been Removed

The committee later revisited the site. After walking around and inspecting the area, the committee members determined whether organization and orderliness had indeed been implemented. If they had, the committee made a public show of

Figure 3-4. An Organization and Orderliness Campaign

removing the "worst" label in front of the workers from the area. By using the Waste Reduction Forms and the labels, the plant got rid of its accumulated "grime" in no time.

Step 8: Setting Up Standard Storage Locations with Standard Labels

The committee promoted the use of standard storage locations and standard labels so that the "grime" wouldn't return when employees absent-mindedly laid things down in random places around the work site. They promoted the principle of "a place for everything and everything in its place," recognizing that unfinished goods and excess inventory are major factors in "grime."

How can a plant avoid being stuck with unfinished goods and excess inventory? Both these problems stem from overproduction and overpurchasing. Everything needed for a process should be stored—in standard places with standard labels in standard amounts—near the production line so employees can keep track of parts and supplies better.

Step 9: Creating 2S Production Lines

The company then turned its attention to creating well-organized, orderly production lines, or "2S lines" (see Photograph 3-10). The shift began with process razing, which involved drawing up plans for moving from preventing overproduction to reducing unfinished goods to promoting the 2Ss in the storage units, based on the "full work" concept. The idea is to create a smooth flow throughout the entire plant so that no more "grime" can accumulate. Creating this flow is also called process creation.

The managers turned the 2S campaign into an enjoyable competition. After six months, they awarded a special plaque to the department that had done the best job. In addition, they handed out individual prizes, such as the "right on" award, the "against all odds" award, and the "sparkling clean" award.

Photo 3-10. The Designer's Production Line

Methods for Maintaining the 5Ss as a System

It is extremely difficult to maintain the 5Ss, as the state of our own desks indicates. It may be possible for a conscientious individual to follow the 5Ss successfully, but as a company grows and builds more plants, matters get out of hand. Sometimes it can take as long as ten years for a company to make the 5Ss part of customary procedures, a habit employees observe as a matter of course.

This is why we created the 5S Evaluation Checklist shown in Table 3 3. When we evaluated Mynac according to these standards and tallied the results on Table 3-4, it scored 90.7 points out of 100. How would your company score?

The navy refers to its version of the 5Ss as "tradition." We challenge you to make the 5Ss part of your company's routine, turn them into an honorable tradition, and take pride in them.

Table 3-3. 5S Evaluation Checklist

Worksite	
Section	
Line	

Date	
Members	

Checkpoints	Evaluation Level				
	1	2	3	4	5

Organization

	1	2	3	4	5
Warehouse (inventory management)	The storage area is too cluttered to walk around freely.	Items are placed irrationally.	Items have designated storage places, but they're often ignored.	Items are stored in proper locations, but no standard criteria indicate when to reorder.	Items are managed for a just-in-time supply and tracked on inventory boards.
Aisles	Work-in-process and other items stand in a roped-off area in the aisles.	Items are set on the sides of the aisles so employees can pass, but carts and dollies cannot pass.	Items protrude into the aisles.	Items protrude into the aisles but have warning labels.	There is no work-in-process, so the aisles are completely clear.
Work areas	Items lie scattered around for months, in no particular order.	Items lie around for months, but they don't get in the way.	Unneeded items have been red-tagged and a disposal date has been set.	Only items to be used within the week are kept around.	Only items needed the same day are kept around.
Machine parts storage	Parts are jumbled together with paper scraps and rags.	Broken and unusable parts are being stored.	Frequently used parts are stored separately from those that will not be used soon.	All parts are stored in standard places with standard labels according to an easily understood system.	Nothing is found out of place.
Workbenches and tables	Tables are covered with unneeded materials.	Tables hold materials that are only used once every two weeks.	Tables hold extra pencils and other unneeded stationery items.	Items remain on the tables for as long as a week.	Only the minimum items needed are kept on the tables.
Equipment (mainly sewing machines)	Equipment is placed in no particular order, some of it rusted and unusable.	Usable and unusable equipment are kept together.	Unusable and unneeded equipment has been thrown out.	Equipment is managed according to its frequency of use and degree of importance.	The equipment is set up so anyone can find what they need to use at any time.
Line organization	The line is in disarray, and planned downtime occurs more than 40 percent of the time.	Planned downtime is as high as 40 percent, and the flow of the line is unclear.	Planned downtime is around 30 percent, and there is waste involved in transporting material.	Planned downtime is around 20 percent, and unfinished goods are kept on the line.	The line is well-organized and flows smoothly, with no more than 10 percent planned downtime.

Orderliness

	1	2	3	4	5
Reorder level (items arrive when needed, in the required quantities)	Parts are reordered when 90 percent have been used up, and parts shortages still occur.	Parts are reordered when 90 percent have been used up, standby occurs at assembly processes.	Parts are reordered when 95 percent have been used up. Problems sometimes occur during model changes.	Parts are reordered when 99 percent have been used up. New product orders generally arrive on time.	The work site practices just-in-time manufacturing, reordering only as inventory is used up.
Waste from searching during setups	Employees wander aimlessly searching for parts and materials.	Employees move in a zigzag pattern gathering up what they need.	Some effort is wasted in searching for things.	Setup carts are used.	No time is lost in searching for things.

(continued)

Table 3-3. 5S Evaluation Checklist, *continued*

Checkpoints	Evaluation Level				
	1	2	3	4	5

Orderliness, *continued*

	1	2	3	4	5
Jig and tool storage	Jigs and tools are scattered all over the place.	Jigs and tools are stored in boxes.	Different kinds of jigs and tools are stored separately.	Jigs and tools are stored on shadow boards.	Jigs and tools are arranged by frequency and order of use.
Parts and materials	Defective and good parts are stored together.	Defective parts are kept on a separate shelf.	Only good parts are kept in storage; nicks, humidity damage, and other problems are avoided.	Storage shelves are clearly labeled and well organized.	Parts are delivered just-in-time, using kanban and tracking boards.
Drawings and charts	Current drawings are jumbled together with torn, outdated charts.	Drawings are organized and filed by category.	Drawings that are hard to read have been replaced with new ones.	Drawings are stored so they're easy to retrieve.	The system allows anyone to return drawings to their proper places.
Documents and other written materials	Documents are scattered randomly on tables and shelves; old documents lie forgotten in storage.	Documents have been straightened enough so that someone who looks long enough will find them eventually.	Documents of the same type are stored in the same place.	Documents and written materials are classified and color coded.	Visual storage is fully implemented. Anyone can easily retrieve documents and return them to their proper place.

Cleanliness

	1	2	3	4	5
Wires and pipes on the ceiling	Dusty wires and pipes dangle haphazardly from the ceiling.	Wires and pipes are laid out for each line, but they are hard to clean around.	Wiring for each line is bundled, making cleaning easier.	Few pipes or wires are evident, and there's no debris from overhead.	No pipes or wires hang from the ceiling.
Aisles	Aisles are littered with cigarette butts, thread, and metal shavings.	There are no large pieces of trash, but small paper scraps, debris, and dust are present.	The aisles are cleaned only in the morning.	Surface defects discovered during cleaning are quickly repaired.	Efforts are made to keep the aisles from getting dirty in the first place.
Machines and equipment	Machines and equipment are dirty and are used in that state.	Visible parts of the equipment appear to be cleaned occasionally.	Equipment is cleaned during setup and changeover.	Operators clean the equipment once a day.	Machines and equipment have inspection labels and are cleaned every morning.
Cleanliness of work areas	Pieces of thread, scraps of cloth, and cutting dust are scattered around.	There's no large debris, but smaller debris and dust are present.	The area has been cleaned.	The area is cleaned every day at the end of the shift.	Debris and dust are caught automatically to keep the area clean.
Work tables and desks	Surfaces are piled so high with documents, tools, and parts that they can't be cleaned.	Dust and debris have accumulated under the work tables.	Work tables and desks are cleaned once a day.	Even the legs of the tables and desks are clean, and all nicks and scratches have been repaired.	Everything is kept clean at all times.
Windows, window frames, and walls	Windowpanes are missing or cracked, with haphazard repairs.	The panes are dirty, and dust and debris have accumulated on the frames.	The panes are dirty, but the frames are occasionally cleaned.	Both the panes and frames are kept clean.	Window shades are used; walls are kept clean and uncluttered, and no extraneous items are attached to the walls, giving the area a pleasant atmosphere.

(continued)

Table 3-3. 5S Evaluation Checklist, *continued*

Checkpoints	Evaluation Level				
	1	2	3	4	5

Cleanliness, *continued*

Tools, jigs, and molds	Some items are rusted.	No items are rusted, but some are covered with oil or dirt.	Only the parts that users actually touch are clean.	Grinding dust and other debris have been cleaned off, making tools and implements pleasant to handle.	Devices prevent accumulation of debris in the first place, and any debris is quickly cleaned off.

Standardized Cleanup

Restrooms	Facilities are dirty; supplies run short; unpleasant to use.	Fixtures are rinsed, but they are still dirty and supplies run out sometimes.	Restrooms are cleaned once a day, but they're still a bit dirty.	Restrooms are clean and hygienic, and supply shortages do not occur.	Restrooms are pleasant and well lit, with music piped in.
Cafeteria and employee lounges	These areas are so dirty one doesn't want to sit there.	The areas have been cleaned somewhat, but they're still dirty.	One wouldn't mind sitting there in work clothes, but not when dressed up.	Chairs have been cleaned so clothes don't get dirty.	The areas are extremely clean, sanitary, and attractively decorated; guests can be taken there.
Implements and jigs (low-cost improvements)	Items have been patched together with tape and the like.	Signs and signal light stands are made of flimsy materials such as cardboard.	Makeshift implements look weak and fragile.	Some equipment is crude-looking and handmade.	Well-made mechanisms use simple automation.
Overall layout	The room is dark, making detailed work difficult.	Light and illumination have been provided, reducing the possibility of on-the-job injuries.	The room is bright and safe, with lighting, illumination, and shades.	Ventilation is sufficient, giving the work area a refreshingly airy feel.	The plant is obviously a healthy working environment.

Discipline

Workplace attitude	Employees avoid eye contact and don't say anything, even when they bump into others.	People say "Excuse me" when they bump into someone, but otherwise ignore others.	Employees make eye contact with and greet only about 10 percent of the other people.	Employees acknowledge about half of the others they encounter.	Everyone is polite and pleasant and at least smiles and nods to others.
Smoking	Employees smoke openly, even in front of first-time visitors. More than 50 percent smoke on the job.	Employees smoke even while being addressed by a supervisor. Less than 50 percent smoke on the job.	Employees light up immediately after meals, without asking permission of visitors. Less than 40 percent smoke on the job.	Employees don't smoke if their visitor doesn't smoke. Less than 30 percent smoke on the job.	Employees don't smoke. Less than 20 percent smoke on the job.
Manner of speech	People use unnecessary jargon. They generally ignore others' ideas and are not good listeners.	People have a know-it-all attitude toward others and do not actively pay attention when others are speaking.	People occasionally speak politely and are receptive to others' ideas about half of the time, paying attention when others are speaking.	People always speak politely, with respect for their leaders. They are receptive to others' ideas about 60 percent of the time and nod in assent while actively listening.	People generally speak politely and respectfully to everyone. They are receptive to others' ideas about 70 percent of the time and are positively supportive of others who are speaking.

Table 3-3. 5S Evaluation Checklist, continued

Checkpoints	Evaluation Level				
	1	2	3	4	5
Discipline, continued					
Clothing	Many employees wear soiled clothes	Many people have buttons missing or wrinkled clothes.	Most people are cleanly but carelessly dressed and don't wear their name tags.	The employees look stylish and well put together.	Most employees wear even their work uniforms with pride and flair.
Punctuality	Employees are often more than 30 minutes late. No one is concerned about time.	Employees are often up to 20 minutes late. Some people are often lax about time.	Employees phone ahead when they'll be 10 minutes late.	Employees often hurry to keep appointments or to arrive within 5 minutes of the scheduled time.	Employees arrive 5 minutes ahead of the scheduled time and are never late.
Instilling the 5S spirit	Employees are unaware of 5S conditions and ignore colleagues' mistakes or infractions.	They notice when conditions aren't maintained, but mention it casually if at all.	They don't talk to the person responsible, but try to fix the situation themselves or tell the other person's boss about it.	They talk to the other person on the spot, speaking softly. The other person does not accept the advice positively.	They talk to the other person on the spot, and he or she accepts the advice positively.
Subtotals	points	points	points	points	points
Total	Goal: Try to raise the total by 20 points within the next evaluation period.				
Comments					

The Fundamentals of Instant Maintenance

4

Several different kinds of related problems can arise with the equipment, machines, and installations on the factory floor:

1. Unplanned minor stoppages (3 minutes or less) that disrupt the rhythm of the production line.

2. Unplanned medium stoppages (4 to 30 minutes) and major stoppages (more than 30 minutes) that halt production. (People usually think of these as fundamentally different from minor stoppages, but we're convinced that all stoppages have the same ultimate causes.)

3. Waste from stoppages for pre-setup, parts replacement, or misadjustments during setup and changeover.

4. Mechanical stoppages when blades are replaced.

5. Waste from rework, defects, or poor yield, caused by variations in the quality of parts and materials obtained from outside suppliers.

6. Waste from machines and equipment standing idle due to shortages of parts.

7. Waste from lines disrupted by employee absences.

We could also talk about planned stoppages, but these have more to do with what happens between the time the customer places the order and the product is manufactured. They will be discussed in Chapter 7.

The seven types of waste listed here are all significant, but from the point of view of TPM, the most important kind of waste is defects due to equipment problems.

In this chapter we look at some of the procedures for eliminating minor, medium, and major stoppages.

Step 1. Study Current Conditions

Some problems are immediately obvious, but others are not so obvious. You have to begin by taking a good look at the workplace. Prepare some of the following survey instruments from industrial engineering:

1. a PQ analysis chart (see example in Table 5-1, p. 66)

2. a layout flow diagram (see example in Figure 5-1, p. 67)

3. a monthly line efficiency report (see Worksheet 19, p. 264)

4. an equipment operation analysis chart (see Table 4-1)

5. a minor stoppage cause sheet (see Table 4-2)

Table 4-1. Equipment Operation Analysis Chart

No.	Measure	Machine #2	Machine #3	Machine #4	Machine #5	Machine #6	Total
1	Working days	21	24	24	24	24	117
2	Possible operating time (minutes)	9,480	33,670	33,615	33,625	33,615	144,005
3	Standard strokes per minute	160.0	230.0	210.0	230.0	210.0	216.1
4	Actual operating time (minutes)	8,432	29,486	28,577	29,114	27,759	123,368
5	Unit output at 100% capacity (2) × (3)	1,516,800	7,744,100	7,059,150	7,733,750	7,059,150	31,112,950
6	Actual good units produced (P number)	1,376,792	6,425,936	5,553,048	5,897,052	5,185,716	24,438,544
7	Machine counter	1,425,144	6,433,756	5,763,670	6,125,548	5,393,327	25,141,445
8	Intermediate counter	1,391,353	6,031,785	5,589,185	5,916,928	5,208,858	24,138,109
9	Packaging machine counter	0	0	0	0	0	0
10	Main unit weight loss (kg)	0.0	0.0	0.0	0.0	0.0	0.0
11	Packaging machine weight loss (kg)	0.0	0.0	0.0	0.0	0.0	0.0
12	Availability (operating rate) (4) ÷ (2)	88.9%	87.6%	85.0%	86.6%	82.6%	85.7%
13	Efficiency rate (6) ÷ (7)	96.6%	99.9%	96.3%	96.3%	96.2%	97.2%
14	Quality rate (6) ÷ (5)	90.8%	83.0%	78.7%	76.3%	73.5%	78.5%
15	Main unit performance loss rate	2.4%	6.2%	3.0%	3.4%	3.4%	4.0%
16	Packaging machine performance loss rate	1.0%	-6.1%	0.6%	0.3%	0.4%	-1.2%
17	Total loss rate	3.4%	0.1%	3.7%	3.7%	3.8%	2.8%

Table 4-2. Minor Stoppage Cause Sheet

Machine number	Machine #1		Machine #2		Machine #3		Machine #4		Machine #5		Total		Average per machine per day	
Working days per month	24		24		24		24		24		120			
	Freq.	Time	Freq.	Time	Freq.	Time	Freq.	Time	Freq.	Time	Freq.	Time	Freq.	Time
Mechanical problems (by machine area)														
Dust collector	1	10	1	10	5	45	5	45	0	0	12	110	0.1	0.9
Processor 1 and 2	52	712	30	255	39	228	43	251	39	261	203	1,707	1.7	14.2
P part	5	78	3	27	1	20	2	2	3	38	14	165	0.1	1.4
Liner	9	25	17	71	12	44	11	19	8	16	57	175	0.5	1.5
P-1 part	11	265	5	98	2	7	2	7	8	21	28	398	0.2	3.3
DE	0	0	2	5	28	136	2	18	28	170	60	329	0.5	2.7
Bending part	19	247	28	143	36	149	16	46	30	128	129	713	1.1	5.9
P-8 part	0	0	0	0	3	5	2	5	6	27	11	37	0.1	0.3
NL part	4	11	37	131	13	20	13	38	13	36	80	236	0.7	2.0
Cutter	4	36	6	104	4	62	2	9	3	20	19	231	0.2	1.9
FS part	5	83	25	298	2	53	4	192	2	18	38	644	0.3	5.4
Preheater	4	62	39	480	3	56	1	5	2	25	49	628	0.4	5.2
F part	25	601	32	247	31	275	14	60	19	160	121	1,343	1.0	11.2
FC part	2	125	17	262	5	45	8	48	4	7	36	487	0.3	4.1
F coating	0	0	1	6	0	0	1	3	1	10	3	19	0.0	0.2
FS part	1	5	6	70	24	198	3	6	10	198	44	477	0.4	4.0
S out	8	52	11	33	25	129	35	199	17	54	96	467	0.8	3.9
B feed	40	157	27	112	45	161	36	99	9	84	157	613	1.3	5.1
SRB feed	16	328	20	53	3	11	8	15	11	35	58	442	0.5	3.7
F feed	31	84	55	108	48	55	34	105	23	55	191	407	1.6	3.4
SE	8	185	0	0	2	42	0	0	0	0	10	227	0.1	1.9
SC	3	9	1	4	1	20	0	0	2	4	7	37	0.1	0.3
W part	3	15	12	63	2	17	0	0	0	0	17	95	0.1	0.8
Cutter	3	19	3	13	2	12	2	15	5	43	15	102	0.1	0.9
Bending part	62	766	27	297	24	114	9	83	30	272	152	1,532	1.3	12.8
Main unit	5	12	5	282	1	3	3	118	15	426	29	841	0.2	7.0
C part	6	292	21	470	3	32	9	91	12	197	51	1,082	0.4	9.0
Electricity	4	7	7	106	29	176	9	76	3	73	52	438	0.4	3.7
Air	1	70	0	0	0	0	1	30	3	117	5	217	0.0	1.8
Defects in materials														
Cut off	118	398	109	354	49	125	87	230	90	298	453	1,405	3.8	11.7
Defective	0	0	2	10	5	13	23	57	26	76	56	156	0.5	1.3
LT defect	6	17	15	103	2	20	17	43	8	30	48	213	0.4	1.8
N defect	15	53	23	216	19	59	50	143	8	18	115	489	1.0	4.1
S defect	0	0	6	109	4	22	2	12	36	259	48	402	0.4	3.4
B sheet defect	0	0	2	125	1	2	0	0	7	47	10	174	0.1	1.5
P defect	0	0	0	0	0	0	0	0	1	8	1	8	0.0	0.1
F defect	2	4	0	0	4	7	1	2	0	0	7	13	0.1	0.1
FP polyethylene defect	1	40	1	5	0	0	0	0	0	0	2	45	0.0	0.4
Human error														
Forgetfulness	12	73	31	142	10	23	69	192	24	88	146	518	1.2	4.3
Error	30	65	82	312	83	284	114	233	92	211	401	1,105	3.3	9.2
Other	33	460	12	311	14	121	7	30	10	163	76	1,085	0.6	9.0
Downtime														
Morning, noon, end of shift	1	30	1	30	1	30	1	30	1	30	5	150	0.0	1.3
Sampling	0	0	0	0	1	10	0	0	1	10	2	20	0.0	0.2
Power outage	0	0	0	0	0	0	0	0	0	0	0	0	0.0	0.0
Total	549	5,366	721	5,435	584	2,791	645	2,527	608	3,693	3,107	19,812	26	165
Average per day	22.9	223.6	30.0	226.5	24.3	116.3	26.9	105.3	25.3	153.9	25.9	165.1		

Table 4-3. Check Sheet for Equipment Problems by Machine

Line: SP2B-AP3

Time period: October 5–7

Equipment	Phenomenon						Cause				Frequency	Notes (things noticed, causes, countermeasures)
	Won't start up	Stops during operation	Abnormal indications	Defects	Won't shut down		Operating method	Inspection error	Abnormality	Unclear		
Shaft pressure press												
Combined gear assembly												
Drier												
Outer diameter grinder				1						✓	1	
Burr remover												
Washing machine												
Million coater												
B.B. Press fitter	### ### ### ### ###							✓			25	Work sensor loose
Accumulated pressure tester												
Screw tightener		### ### ### ### ### ### ### ### ###						✓			61	
Magnetization, pressure resistance	### ### ### ### ### ### ### ### ### ###										50	
Thrust tester												
Characteristic tester												
Pulley press fitter												

Group leader	Manager	Supervisor
MD	RT	SZ

Line and by Machine

A-1	A-2	A-3	A-4	A-5	Total
P$_1$	P$_1$	P$_2$	P$_1$	P$_2$	
1,271	1,152	886	803	683	4,795
			1	1	2
9	4	10	10	4	37
1					1
	1				1
0.8%	0.3%	1.1%	1.2%	0.6%	0.8%

Step 2: Summarize the Problems with Each Line and Machine

1. As a general rule, you can create a matrix-type chart in which the columns show the type of problem and the rows show the name of the line. However, you'll get clearer results if you create a separate chart for each line and each piece of equipment, as shown in Tables 4-3 and 4-4.

2. You'll discover differences in performance among the pieces of equipment in your plant. You can also create a list indicating these differences, as in Table 4-5.

3. It may be helpful to create a cumulative summary of defects for each line for further reference (see Table 4-6). You may be surprised to see how often mechanical problems and defective products go together.

Step 3: Analyze Minor Stoppage Mechanisms and Their Causes

Confirm the Situation

You can confirm the actual situations underlying many minor stoppages with your own eyes. If you are faced with low-frequency minor stoppages, you can install a high-speed video camera to capture them. Look at:

1. The true nature of the stoppage. Analyze any electrical and mechanical abnormalities and make your judgment based on those findings.

2. The methods of operation.

3. How equipment is inspected.

4. The cleanliness of the equipment.

5. Comparisons with other lines.

Table 4-5. Differences Between Pieces of Equipment

Work _____   Equipment: Screw tightener

Nature of phenomenon		Line (manufacturer/installation date)		
		B–2 (A/April)	B–3 (B/October)	Difference (O)
Drive	X-Y drive	Air	Air + Stepping	O
Air	Main unit air pressure	6.0 kg/cm²	6.0 kg/cm²	
	Driver drop pressure	2.5 kg/cm²	2.5 kg/cm²	
	Work lamp pressure	3.5 kg/cm²	3.2 kg/cm²	O
	Screw tightening torque			
Screw tigntening	Power	4TK10GK-A 4GK9K	4TK10GN-A 4GN9K	O
	Screw adsorption	Minicomputer	Minicomputer	
Parts feeder	Main unit	NITTO	NITTO	
	Screw pressure supply	2.0 kg/cm²	2.0 kg/cm²	
	Pressure hose	Dirt, nicks	Dirt	O
	Vibration	5.0	5.5	O
Parts	Screws			
Other	Vacuum filter	Dirt	Dirt	Metal shavings
	Rubber retainer	Debris adhering	Debris adhering	
Work in question				
Past break-downs				

Differences discovered:
1. Nicks on the inside of the pressure hose that hinder the screws
2. Differences in vibration inside the parts feeder
3. Differences in the drive formula

Table 4-6. Summary of Defects by Line

Line	1	2	3	4	5	6	7	Subtotal	8	9	10
Number of inspections	1,200	866	531	797	500	180	624	4,698	356	961	524
Number of defects	12	7	1	11	2	1	3	37	8	13	9
Rate (%)	1.00	0.80	0.18	1.38	0.40	0.55	0.48	0.78	2.24	1.35	1.71
Ball sound				1				1	1		1
Getting stuck	1	1	1	2		1		6			
Nicks in the gears				1			1	2	1	2	
Housing		1						1	1	3	
Differences in current											
Coating peeling off											
Generator defect											
Absent-minded errors	2	1		1				4	1	1	2
Errors in operation	2	3		1	1			7	1		
Parts defects	2							2	1	2	2
Shaft vibration									2	1	2
Voltage	1			3	1			5			
Broken wires	1	1		1				3		1	1
Defect in lead wire											
Gears meshing	2			1				3			
Thrust	1						2	3		1	1
Unclear										2	

Line	28	29	30	31	Subtotal	32	33	Subtotal	Total
Number of inspections	605	485	515	475	2,528	568	614	1,182	21,921
Number of defects	4	4	4	2	20	8	8	16	247
Rate (%)	0.66	0.82	0.77	0.42	0.79	1.40	1.30	1.35	1.12
Ball sound							1	1	10
Getting stuck	2				4	1	1	2	25
Nicks in the gears							1	1	21
Housing									13
Differences in current									24
Coating peeling off									1
Generator defect									
Absent-minded errors					2		2	2	27
Errors in operation				1	1				12
Parts defects			1		1		2	2	17
Shaft vibration						-		1	20
Voltage						-		1	7
Broken wires									6
Defect in lead wire	1			1	2				6
Gears meshing									3
Thrust						1	1	2	14
Unclear	1	4	3		8	4	1	5	39

Analyze the Mechanism and Make a Hypothesis about the Cause

In mechanism analysis, you guess the causes based on analysis of the phenomenon. Then you construct a theory of why the stoppage occurred (see Figure 4-1).

Step 4: Form a Clear Picture of the Phenomenon, Mechanisms, and Causes

Apply Why-Why Analysis

Use Why-Why Analysis to figure out the root causes of the problem. Ask the question as many times as it takes to get at the root cause. Figure 4-2 shows how Why-Why analysis might be applied to the problem described in Figure 4-1. Sometimes it is helpful to ask the Four Ws and One H (who, what, when, where, and how) to identify basic facts about the problem.

Phenomenon: Screws getting stuck

Causes: Changes in hose pressure

Root cause: Friction with the internal guide

Plan for improvement: Switching to an exterior guide will allow easier replacement when friction occurs

---

Step 5: Set Up an Instant Maintenance System

Educate People about Instant Maintenance Items

Determine the items that need to be handled with instant maintenance, train the employees in the proper procedures, and publicize the procedures within the plant.

---

1. Replace the pressure hose.
2. Bend the hose into the proper shape.
3. Replace the guide.
4. Adjust the amount of vibration in the parts feeder.

Designate Major Maintenance Items

Take care of designated major maintenance items systematically to ensure thorough maintenance. Lists of improvement plans such as Table 4-7 will help organize your activities. Labels should indicate which items operators should check and which items the chief engineer will check. Operators should check their assigned items thoroughly before each shift, following illustrated instructions displayed in the workplace.

Items for operators to check:

- filters
- pressure plate
- fallen screws

Items for chief engineer to check:

- friction on the O-rings within the bit
- hoses
- torque meter adjustment

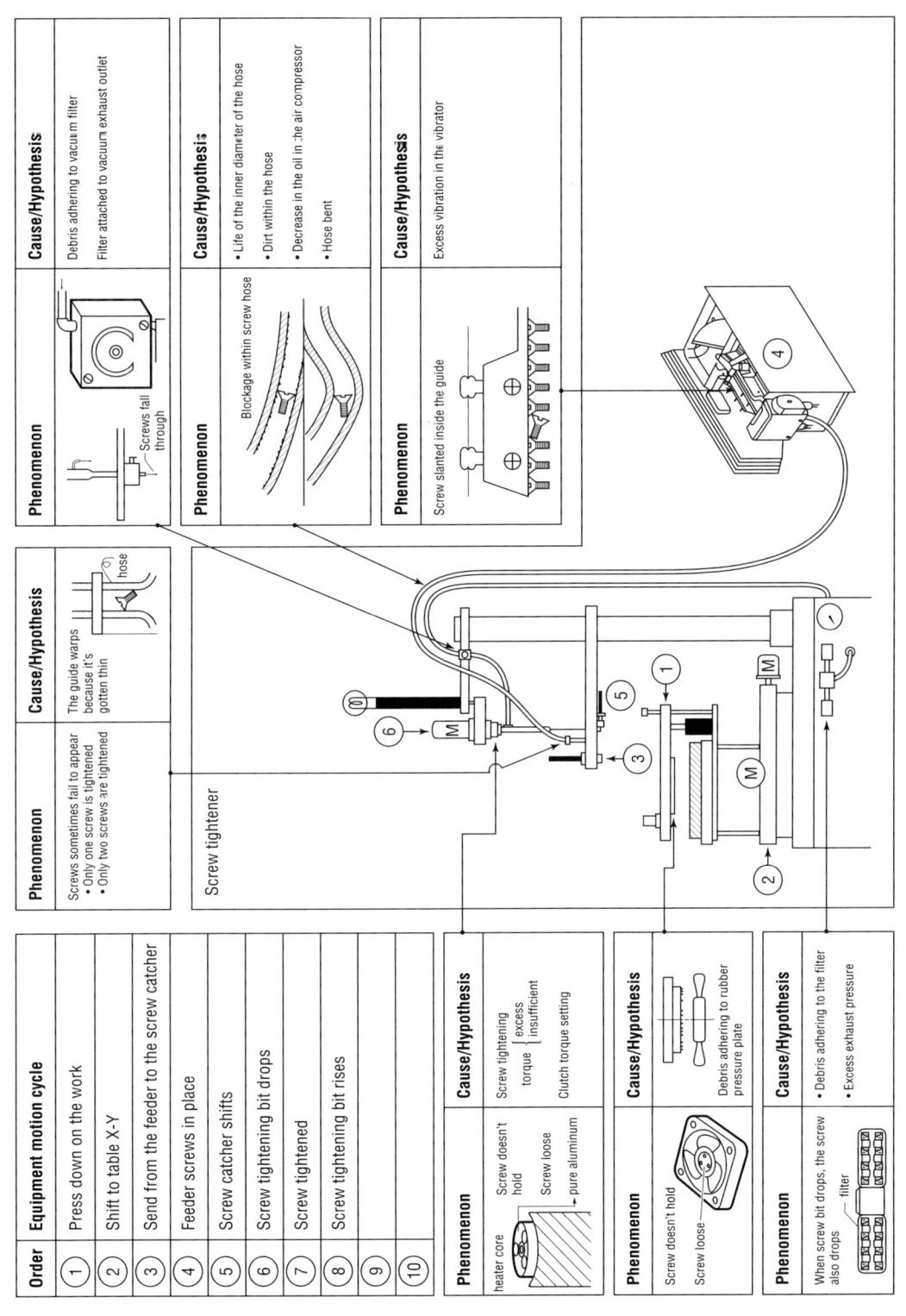

Order	Equipment motion cycle
①	Press down on the work
②	Shift to table X-Y
③	Send from the feeder to the screw catcher
④	Feeder screws in place
⑤	Screw catcher shifts
⑥	Screw tightening bit drops
⑦	Screw tightened
⑧	Screw tightening bit rises
⑨	
⑩	

Phenomenon	Cause/Hypothesis
Screws sometimes fail to appear • Only one screw is tightened • Only two screws are tightened	The guide warps because it's gotten thin

Screw tightener

Phenomenon	Cause/Hypothesis
Screws fall through	Debris adhering to vacuum filter Filter attached to vacuum exhaust outlet

Phenomenon	Cause/Hypothesis
Blockage within screw hose	• Life of the inner diameter of the hose • Dirt within the hose • Decrease in the oil in the air compressor • Hose bent

Phenomenon	Cause/Hypothesis
Screw slanted inside the guide	Excess vibration in the vibrator

Phenomenon	Cause/Hypothesis
heater core — Screw doesn't hold Screw loose — pure aluminum	Screw tightening torque { excess / insufficient Clutch torque setting

Phenomenon	Cause/Hypothesis
Screw doesn't hold Screw loose	Debris adhering to rubber pressure plate

Phenomenon	Cause/Hypothesis
When screw bit drops, the screw also drops — filter	• Debris adhering to the filter • Excess exhaust pressure

Figure 4-1. Mechanism Analysis

57

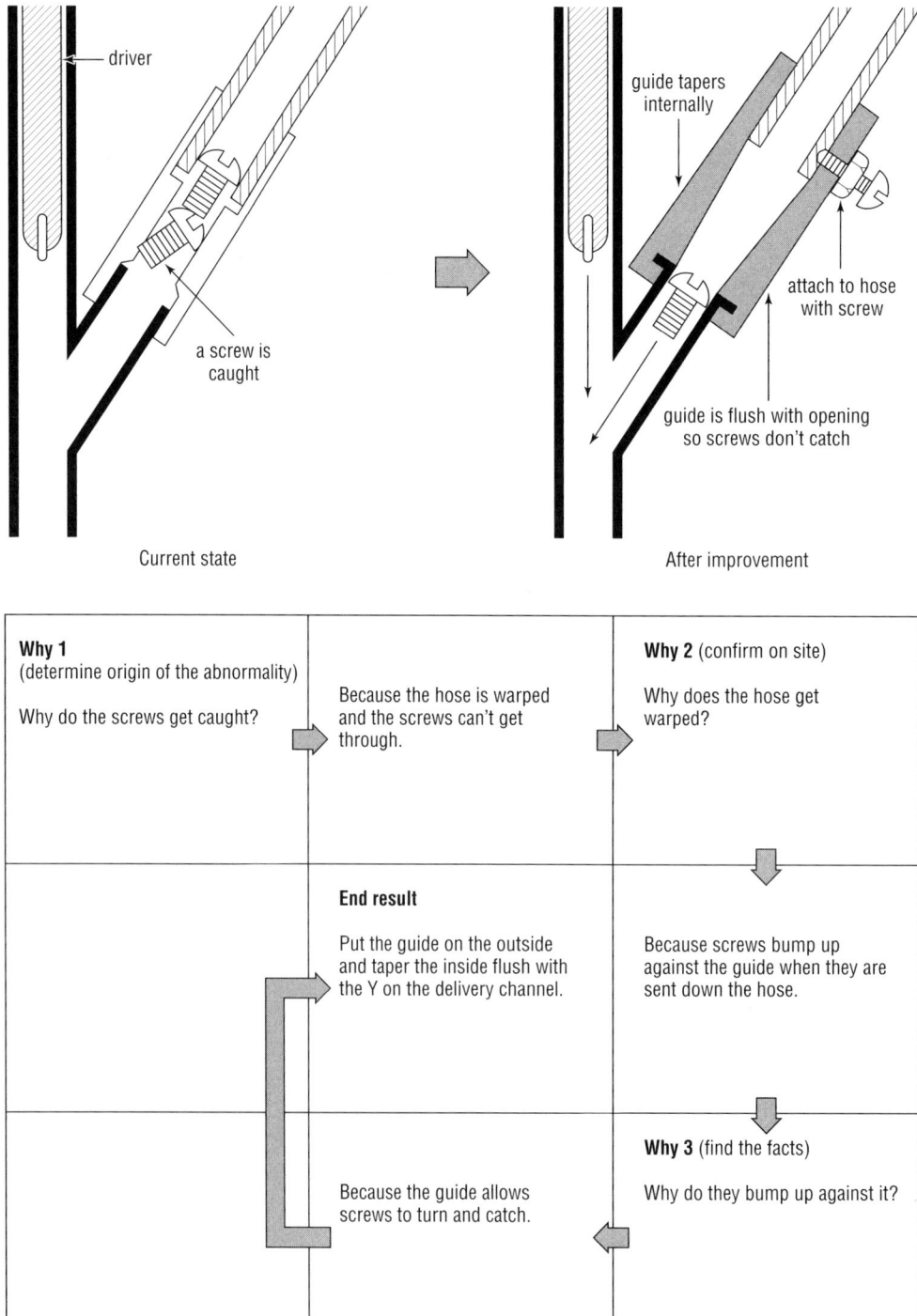

Figure 4-2. Why-Why Analysis for Caught Screws

Table 4-7. Equipment Problem-Solving Plan

Phenomenon	Plan / Cause	Minor improvements	Who	When	Effect	Medium and major improvements	Who	When	Effect
1. One of the screws doesn't tighten.	1. Different type of screw mixed in 2. Change in hose coupler	Confer with supplier and arrange to exchange current item.	H			Change shape of coupler since nothing can be done with mechanism.	O		
2. Screws fall off.	1. Blocked filter opening 2. Friction on O-ring of bit 3. Clogged air filter on secondary compressor	Regular cleaning Instant maintenance (3 minutes) Regular cleaning	T		O	Get rid of metal shavings. I'd like to discontinue this.	O		
3. Stops within the tube	1. Damage to the hose 2. Hose curvature	Periodically replace hose. Increase curvature of hose.	T T		O O				
4. Screws caught inside feeder don't flow smoothly.	Vibration, guide position	Adjust the amount of vibration.	T						
5. Screws don't hold.	Debris on the pressure plate	Periodically clean.	T						
6. Screws loose	Torque measurement	Adjust to correct value with torque meter.	T		O				

Step 6: Provide Support for Autonomous Maintenance Activities

1. Create a daily maintenance sheet for autonomous maintenance, as in Figure 4-3.
2. Attach inspection labels to each part of the machine that should be checked (see Figure 4-4). During regular autonomous maintenance activities, the operator follows the checkpoints on the red front side of the label, then flips it over to the blue side when finished. Table 4-8 shows a checklist of points to inspect before each shift.
3. Classify the inspection items. To avoid overburdening the operators, limit their duties to small-scale spot checks, as illustrated in the manual in Figure 4-3.

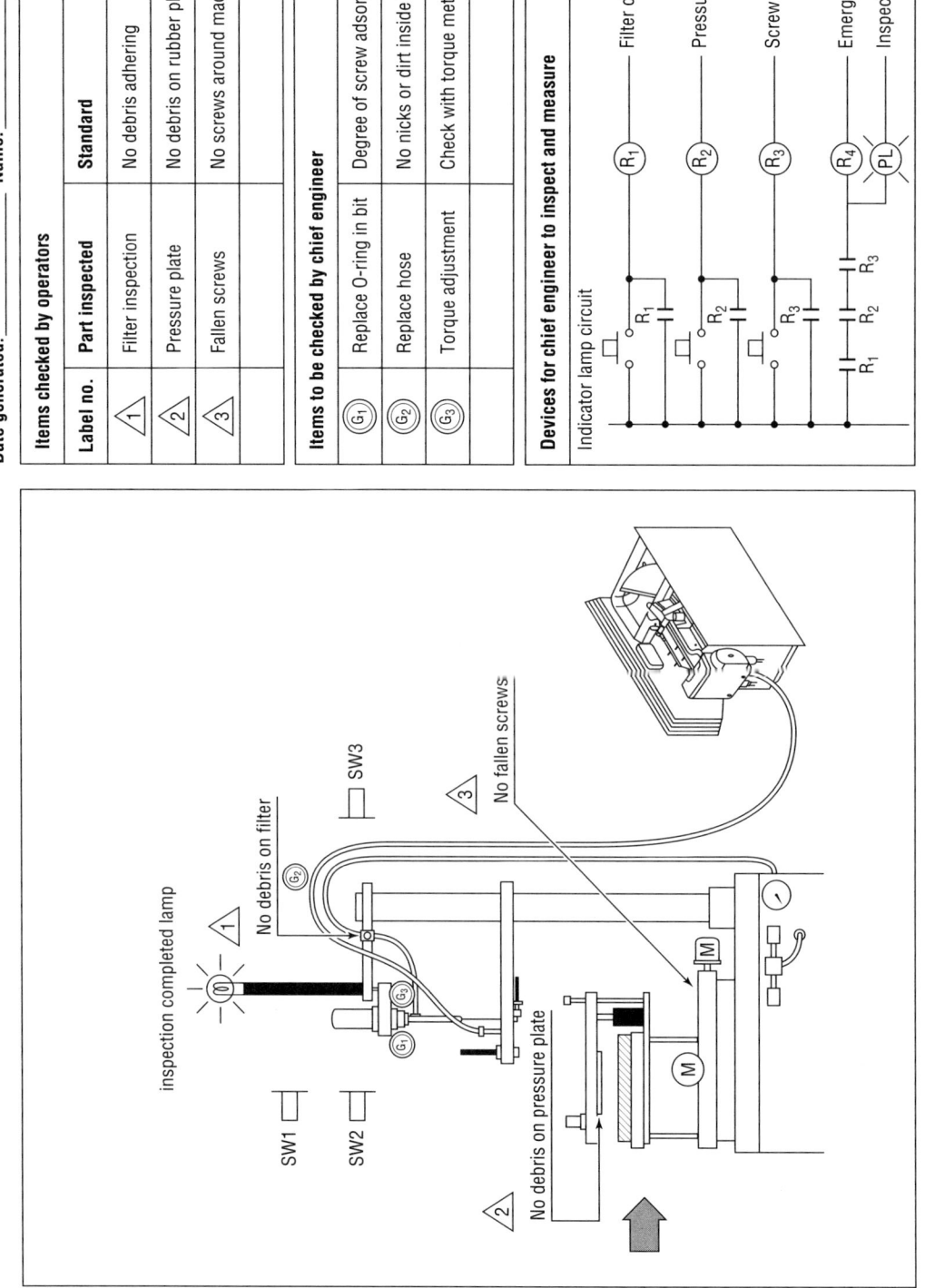

Date generated: _____ Name: _____

Items checked by operators

Label no.	Part inspected	Standard
△1	Filter inspection	No debris adhering
△2	Pressure plate	No debris on rubber plate
△3	Fallen screws	No screws around machine

Items to be checked by chief engineer

G₁	Replace O-ring in bit	Degree of screw adsorption
G₂	Replace hose	No nicks or dirt inside hose
G₃	Torque adjustment	Check with torque meter

Devices for chief engineer to inspect and measure

Indicator lamp circuit

R_1 — Filter check

R_2 — Pressure plate check

R_3 — Screw sensor

R_4 — Emergency stoppage detection

PL — Inspection completed lamp

inspection completed lamp

No debris on filter

No fallen screws

No debris on pressure plate

SW1 SW2 SW3

Figure 4-3. Daily Maintenance Sheet

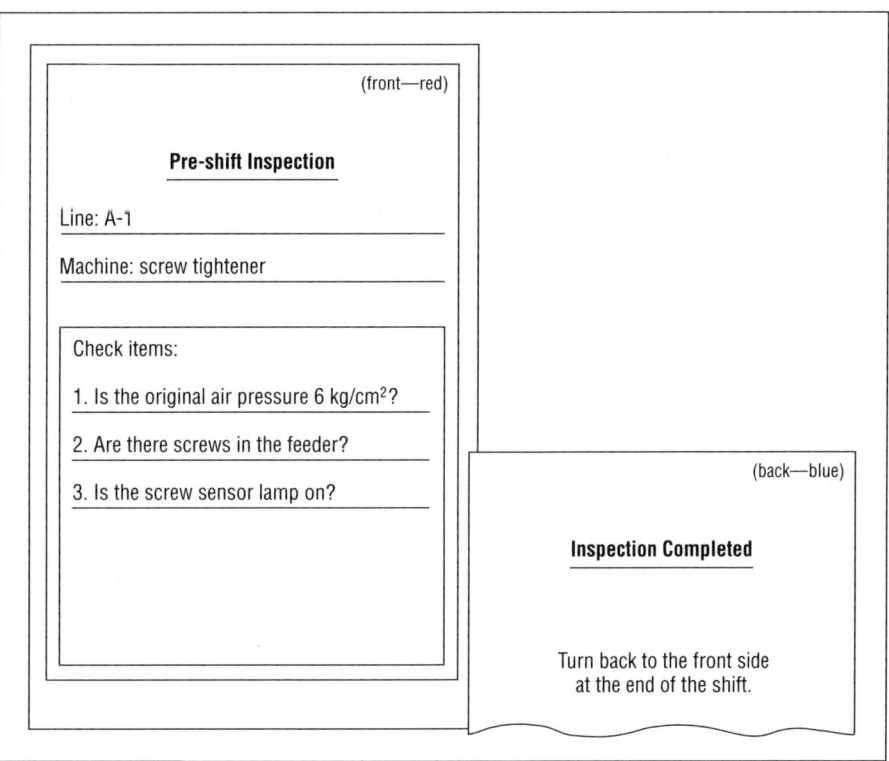

Figure 4-4. Inspection Label

Step 7: Create Instant Maintenance Manuals and Demonstrate Techniques

1. When equipment problems occur, jot down what happened, possible causes, and possible solutions, as in Table 4-9.

2. Once repairs have been completed, record the incident on a maintenance note as in Figure 4-5.

3. Compile the records of similar incidents and incorporate the best repair methods into an instant maintenance manual. Table 4-10 shows one approach for a sheet for this manual.

4. Make instant maintenance available to the entire plant. Everyone can be both a teacher and a pupil. (For more about this approach, see Sekine and Arai, *Kaizen for Quick Changeover*, Productivity Press, 1992.)

With all these procedures in place, it should take no more than three minutes to examine a problem, consider possible causes, analyze them, replace parts, and make adjustments.

Instant maintenance is your first and most critical line of defense against the inevitable problems with machines and equipment.

Table 4-8. Pre-Shift Check Sheet

```
Safety department → Safety office → Supervisor
                    (maintenance)
```

Pre-shift Check Sheet

Section manager	Center manager	Team leader	Supervisor

Month _____ Machine number _____

Daily inspection items

Inspection item

Summary	Inspection date

Inspection item / Summary	1	2	3	4	5	6	7	8	9	10	11	12	13	14	15	16	17	18	19	20	21	22	23	24	25	26	27	28	29	30	31
1. Filter inspection			✓	✓	✓	✓	✓	✓																							
2. Pressure plate			✓	✓	✓	✓	✓	✓																							
3. Fallen screws			✓	✓	✓	✓	✓	✓																							
4. Original air pressure, 6 kg/cm^2			✓	✓	✓	✓	✓	✓																							
5. Presence of screws in feeder			✓	✓	✓	✓	✓	✓																							
6. Screw sensor lamp lit			✓	✓	✓	✓	✓	✓																							

Defective areas	Special notes about defect locations
	Processing conditions
1	Screws blocked

Summary of notations

X = Needs repair
⊗ = Repair adjustment completed
✓ = No abnormality
(Report any abnormalities discovered to the team leader.)

How to use this check sheet

1. The supervisor will record the daily inspections.
2. Team leaders will approve the check sheets once a week.
3. The safety office will keep this record for one year.

Table 4-9. Record of Equipment Problems

Date	Line	Equipment	Phenomenon	Cause	Countermeasures	Downtime	Person in charge
10/3	A 4	Burr remover	Brush came off	Loose screw, insufficient inspection at beginning of shift	Switch from M3 to M5 screws	10 minutes	A
"	"	Screw tightener	Abnormal display of amount of screws	Different types of screws used together	Clean sorting	2 minutes	A
"	"	Screw tightener	Sudden stoppage	Grabbing	Replace the guide	8 minutes	A
"	"	Screw tightener	Avalanche of screws	Unclear	Instant maintenance	10 minutes	B
"	"	Drier	Started up even though no work was in place	Fiber sensor broken, blockage → poking	1. For now, turn on the startup switch 2. Buy a new fiber sensor	15 minutes	C
"	"	Grinder	Roller doesn't stop in original position	Sequencer defect	1. Ask manufacturer to replace 2. For now, change the programming	Repairs, 3 hours	D

Repair Record (Maintenance Note)

Equipment: Dryer

Line: A4

Situation: Work cannot be confirmed; mechanism operates continuously due to sensor stuck in "on" position.

Countermeasures: Replace sensor

Repair time: 15 minutes

Person in charge: C

Figure 4-5. Repair Record (Maintenance Note)

Table 4-10. Instant Maintenance Sheet

Equipment: Screw tightener

Problem	Phenomenon	Factors	Disposition	Instant maintenance procedures
Doesn't start	• Doesn't start when the "start" button is pressed	• The safety stop signal is stuck in the "on" position • Not in original position	• Readjustment • Lead wire processing • Check each sensor	Doesn't start when the switch is turned on → Check the starting point sensor → (Off) Adjust sensor; (On) Is there a screw on the end of the bit? → (No) Screw supply; (Yes) Check screw sensor → (Off) Adjust sensor → (On) OK
Sudden stoppage	• Stops even though the work lamp is lit	• Screws loose in the rising tip sensor • The workpiece is defective and doesn't get pressed in • Low pressing pressure	• Tighten and attach after position adjustment • Eliminate defect • Air pressure adjustment	Stops suddenly after screws drop → Check original air pressure → (OK) Adjust pressing pressure; (OK) Check bit tip → (Damaged) Replace bit; (OK) Check for blockage in air nozzle → (Blocked) Clean nozzle; (Clear) OK
Defects	• Screw head crushed (difference in screw position)	• Table lock cylinder loose	• Tighten lock screw; modify programming	Defect (screw fails to hold) → Check bit tip → (Damaged) Replace bit; (OK) Check driver tightening force scale → (Strong) Adjust driver tightening force; (OK) Driver speed → (Fast) Adjust → (OK) OK
		• Work feed cylinder lock screw is loose	• Tighten lock screw; modify programming	
		• Bit bent	• Replace bit	
		• Cap ring damaged	• Replace	
	• Screw broken	• Bit dropping speed too fast	• Adjust speed	

A Case Study in
Instant Maintenance

5

This chapter tells how section manager A from company K, a communications equipment manufacturer, worked very hard to implement instant maintenance by following the steps described in the previous chapter. His steps don't track the steps presented in Chapter 4 exactly, but pragmatic readers can still benefit from his experiences, so we are presenting his steps almost verbatim.

Step 1: Study Current Conditions

If most of your equipment problems have obvious causes, you can get by with simple observation of the work site. For analyzing less obvious problems, the orthodox method is to collect data in the following forms:

1. a PQ analysis chart (see Table 5-1)
2. a layout flow diagram (see Figure 5-1)
3. a process path analysis (see Figure 5-2)
4. a process capacity table (see Worksheet 42, p. 310)
5. an equipment operating rate chart (see Table 4-1, p. 50)
6. a chart showing causes of minor stoppages (see Table 4-2, p. 51)

These data give the observer a grasp of current conditions. If possible, the observer should develop the ability to analyze raw data without having to convert them into graphic form such as pie charts or Pareto charts.

Step 2: Summarize the Problems with Each Line and Machine

Company K already had a form for recording data on machine stoppages, developed as part of another program (see Table 5-2). Figure 5-3 is a month-by-month record of problems with the inserter machine, classified by type. The graph clearly shows that problems decreased after improvement.

1. Once you have classified the machines as ones with few problems and ones with many problems, go to the work site and figure out how they differ.
2. In this case, the overwhelming majority of minor stoppages are due to running out of parts. Thus, what appears at first to be a mechanical problem may turn out to be essentially a miscalculation.

Table 5-1. PQ Analysis

Machine Type	Number of Units	UNI	R1	R2	PIN	AL	BP
1	59,442	32	19	11	9	24	
2	42,492	54	65	37	9		57
3	42,492	75	76	44	10		
4	36,542	61	84	48	16		
5	34,292	118	126	73	10	29	
6	32,292	57	51	29	6	16	
7	22,900	45	61	35	14		
8	20,200	24	29	17	16		14
9	20,200	62	42	24	13	27	
10	20,200	107	124	72	19		
11	16,030	72	102	59	7	40	
12	14,979	131	100	58	7	24	
13	14,029	74	71	41	9	13	
14	14,029	78	45	26	6	53	
15	12,718	1	0	0	0		
16	11,800	79	93	54	14		14
17	11,800	27	45	26	4	12	
18	10,500	65	52	30	6	19	
19	10,500	113	104	60	6		
20	9,750	43	72	42	10		83
21	7,850	158	113	65	17	16	
22	7,132	6	14	8	1		
23	6,900	50	54	31	6		
24	5,950	87	85	49	19	17	
25	5,830	105	111	64	6		
26	5,800	21	35	20	10		
27	5,800	122	115	66	24		24
28	5,380	46	75	43	19		
29	5,250	170	170	98	31		
30	5,150	63	79	46	7		39
31	4,950	69	28	16	6	21	
32	3,300	89	71	41	6		34
33	2,625	10	34	19	7		
34	1,793	8	0	0	1	10	
35	1,600	18	52	30	6	5	
36	1,275	15	25	15	3	5	
37	1,240	2	4	2	1		
38	1,150	50	54	31	6		
39	1,150	45	74	43	10		83
40	1,050	85	95	55	6		34
41	900	35	31	18	13		
42	850	40	58	34	16		
43	650	79	69	40	14		87
44	650	91	88	51	7		24
45	500	43	61	35	14		
46	450	77	61	35	16		100
47	450	12	0	0	13		
48	300	15	18	11	6		
49	300	156	180	104	21		
50	300	120	103	60	24		
51	300	56	46	27	6		24
52	300	139	85	49	13		

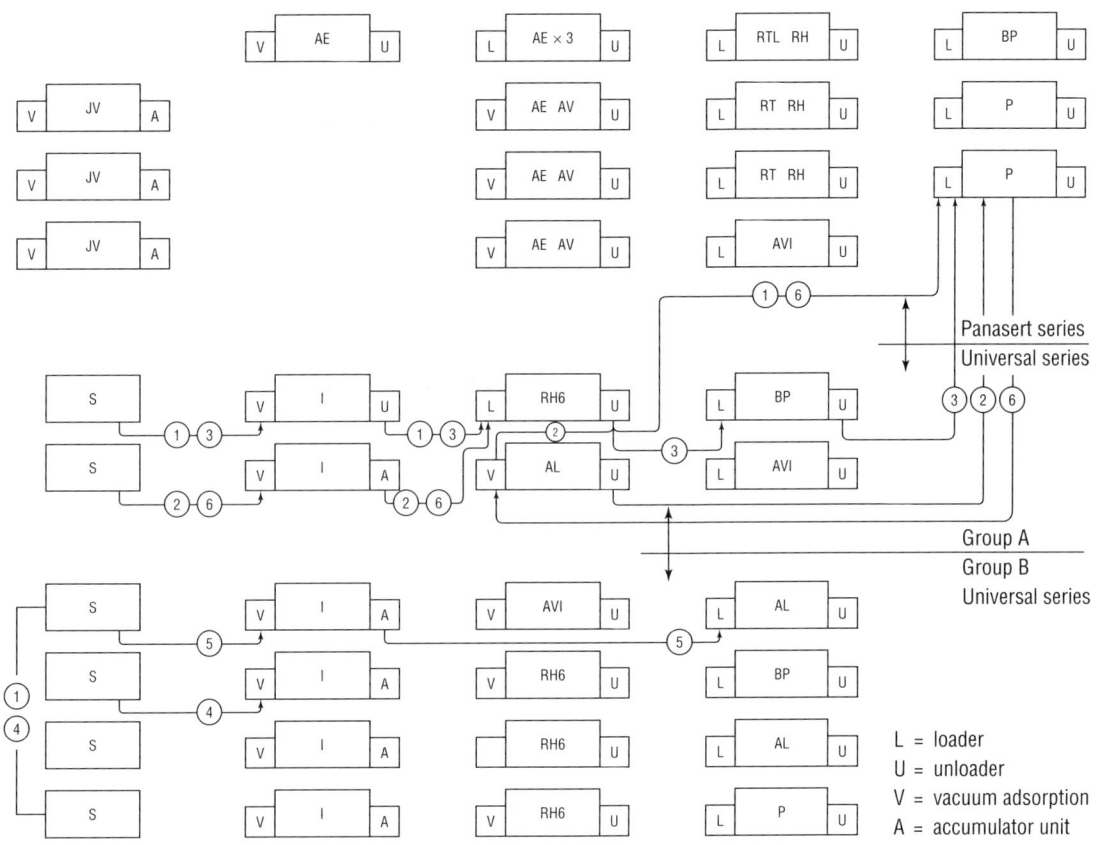

Figure 5-1. Layout Flow Diagram

Step 3: Analyze Stoppage Mechanisms and Hypothesize Causes

1. First, make observations in the workplace.

   • Look at the stoppages. Under what circumstances does the machine shut down?

   • Check for foreign matter. Are shavings, burrs, or other debris adhering to the equipment?

   • Look at how everything fits together. Note any deviations from the standards.

2. Analyze the mechanism by filling out PM Notes, as shown in Table 5-3.

Step 4: Form a Clear Picture of the Phenomenon, Mechanisms, and Causes

Apply Why-Why Analysis

Use Why-Why Analysis to figure out the root causes of the problem and how to eliminate it. For example, assume that a minor stoppage has occurred due to a diode insertion error. First draw a sketch of the situation on the upper left side of the Why-Why Analysis form, as in Figure 5-4. In this case, the tip of the molding lever and the parts lead are slightly worn.

Universal series

No.	1 S	2 I	5 R	10 AL	20 BP	50 P	Process series no.*	Family
1	◯	◯	◯			◯	58	C
2	◯	◯	◯	◯		◯	68	B
3	◯	◯	◯		◯	◯	78	C
4	◯	◯					3	A
5	◯	◯		◯		◯	63	D
6	◯	◯	◯	(◯)		◯	68	B

Panasert series

No.	1 JV	2 A10	5 A5	10 R	100 AL	200 BP	500 P	Process series no.*	Family
1			◯		◯			105	①
2		◯			◯			102	
3		◯	◯		◯			107	
4	◯	◯			◯			103	
5	◯		◯		◯			106	
6	◯	◯	◯		◯			108	
7			◯	◯			◯	515	②
8		◯		◯			◯	512	
9		◯	◯	◯			◯	517	
10	◯		◯	◯			◯	516	
11	◯	◯		◯			◯	513	
12	◯	◯	◯	◯			◯	518	
13			◯	◯	◯		◯	615	③
14		◯		◯	◯		◯	612	
15		◯	◯	◯	◯		◯	617	
16	◯		◯	◯	◯		◯	616	
17	◯	◯		◯	◯		◯	613	
18	◯	◯	◯	◯	◯		◯	618	
19	◯	◯	◯	◯		◯	◯	718	④

*This system assigns a unique number to each machine in a line that processes different types of parts. For each type of part, the unique numbers for the machines used to make it are added together; the result is a unique process series number that identifies the part and tells in shorthand what machines are used.—Ed.

Figure 5-2. Process Path Analysis Chart

Next, begin filling in the nine-square grid, beginning in the upper right corner. In Why 1, the point is to confirm the problem on site. Ask where the diode insertion error was discovered, using a method called the Four Ws and One H:

Who made the discovery? RT

What was discovered? A 5 mm diode (DR1)

Where? At insertion

How? The diode didn't go in

When? On December 27

Table 5-2. Minor Stoppage Analysis Chart

NPS Sheet Changeover Study Group	Minor Stoppage Analysis Chart
Plant: _____ Line: _____	Prepared by: _____ 1/13

Date	Shift	Insertion-point no.	Parts shortages	Error rate (ppm)	Operating rate (%)	A	B	C	D	E	F	G	H	I	J	K	L	M	Total errors/shortages	Number of machine shutdowns
12/24	1 / 2	183,945	53	359	73	6	2	2	1		1				1				66	4
12/25	1 / 2	166,910	61	617	80	11	6	7	5		6	2		5					103	6
12/26	1 / 2	185,220	68	588	85	12	5	9	8		4	1			2				109	3
12/27	② 1	160,140	61	449	90	6	3	1	1										72	1
12/28	① 2	87,510	23	548	83	10	4	5	3		2	1							48	1
1/6	1 / 2	184,650	51	417	71	13	4	6	3										77	6
1/7	1 / 2	128,300	38	545	88	15	3	3	5		5	1							70	2
1/8	1 / 2	96,332	22	529	89	13	3	5	2		2	2			2				51	4
1/9	1 / 2	143,662	43	382	82	4			2		3	3							55	4
Total		1,338,669	420	493 (mean)		90	30	38	30		23	10		5	5				651	31

Causes of stoppages

A	Head molding lever	G	Adsorption rubber
B	Rubber pusher	H	Anvil detection
C	Insertion guide	I	Cutter clincher
D	Play in the main unit	J	Other
E	Chain slackness	K	Tape position
F	Jaw sharpness, attachment	L	Parts defects (taping)
		M	Print plate hole off-center

Machine number

Comments

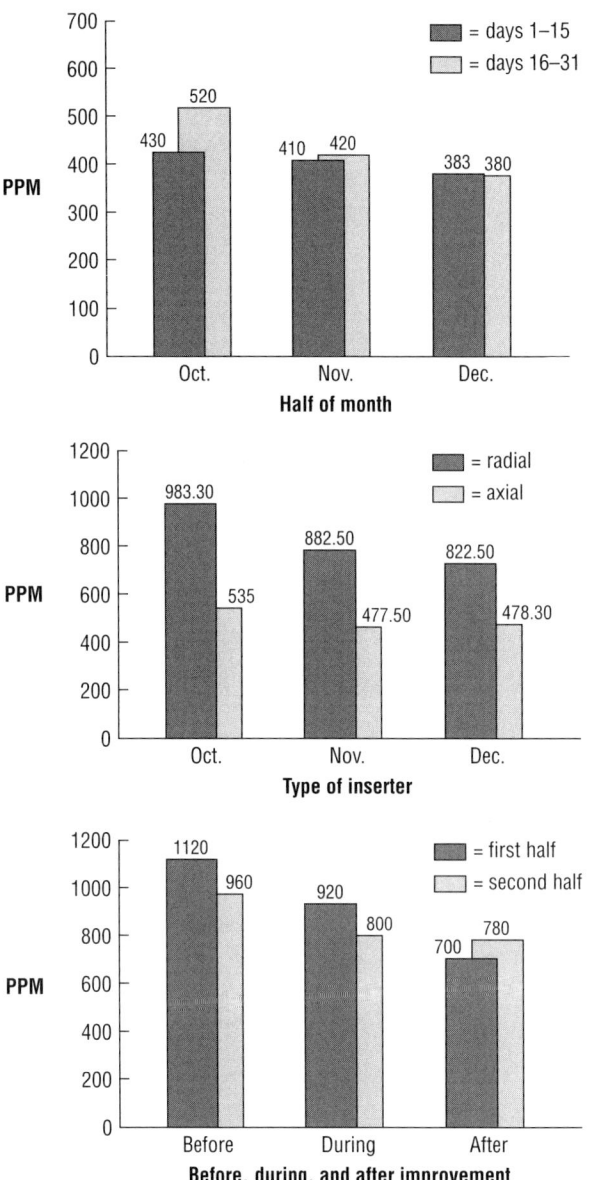

Figure 5-3. Inserter Machine Problems by Date, Machine Type, and Improvement Stage

Using this method is like investigating a crime. When the solution to a problem isn't immediately obvious, your only option is to go around asking questions.

In Why 2, the question is, "Why did the insertion error with the 5 mm diode occur?" Your answer should describe the phenomenon exactly as it appears. That is, it occurred because the part doesn't fit into the hole in the base. In other words, it's a position error.

In Why 3, proceed from the answer to Why 2 to dig deeper into the problem. "Why did the position error occur?" The simplest answer is that the base of the

Table 5-3. PM Notes

December	Department	PM Notes		Supervisor	Approval	Approval
Date	**PM Record**	**Why-Why Analysis Theme**	**Why-Why Analysis (Conclusion)**	**Items Referred to Other Departments**		
1	Individual differences in anvil detection methods	10 mm missing JW parts	Notation on the record sheet			
2						
3	Due to alternating operations, differences in visual power	Insufficient pins (YK 146 H 1704, 7 incidents)	Insufficient pins, checker manufacture	Ask Manufacturer A		
4						
⑤						
⑥						
⑦						
8	Program tape overshoots the block	Program tape defect	Replacement of NC generator puncher	Repair by Company M		
9						
10	Taping NG, excessive accumulation	Insertion error in EA 106016	Warning about parts handling	Referred to the materials manager		
11						
12	Inconsistency in the holes in the base due to LOT	YK 260 T 1003 error (2%)	Insertion at the average position of the inconsistent holes			
⑬						
⑭						
15						
16	Program read-in NG	10 mm missing JW parts old-style machine	Capstan roller failure (wear)	Order parts from Company M		
17						
18	Taping off center	OF10505011 insertion error	Parts handling, accumulate ten or more shelves	To the materials handler		
19						
⑳	5 mm insertion guide failure	5 mm diode error	Failure occurred two months early	Complain to manufacturer		
㉑						
22	Implement monthly cleaning of the parts cassette	RH2 capacitor insertion error	Parts cassette feed defect			
23						
24						
25						
26	RT machine position error, 2-3 pulse	Insertion error (radial)	Bearing failure in DC servomotor	Get motor repaired		
27						
㉘						
㉙						
㉚						
㉛						

Comments: • Taping failure is especially noticeable.

• Titanium coating process quality test on the 5 mm axial insertion guide.

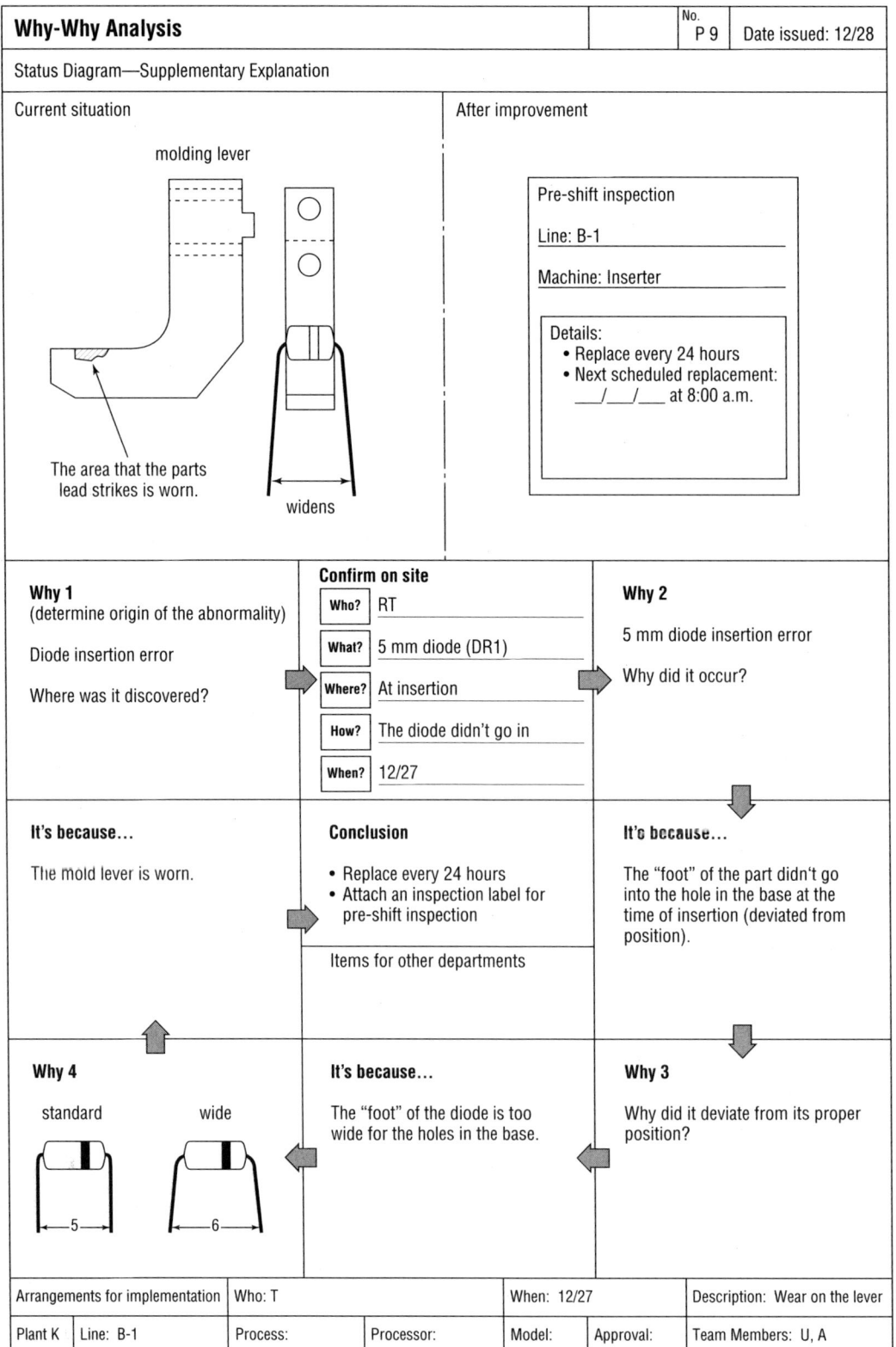

Figure 5-4. Why-Why Analysis for Diode Insertion Error

diode is too big for the hole. Investigation reveals that it's actually 6 mm instead of the standard 5 mm.

In Why 4, proceed from the previous answer to ask why a supposedly 5 mm diode is actually 6 mm. (Some may ask why it's necessary to keep asking questions when we already know the answer, but in this case it's a chance to practice the theoretical concepts.)

Here, the answer to Why 4 is that the molding lever is worn and so the base of the diode wasn't molded correctly. When something is worn, all you can do is replace it. For now, the best countermeasure is to replace the part every 24 hours and to attach an inspection label so that it will be checked at the start of every shift.

Figure 5-5 is a Why-Why Analysis for a capacitor insertion error. Figure 5-6 shows a Why-Why Analysis for an insertion guide.

Step 5: Set Up an Instant Maintenance System

Determine Instant Maintenance Parts

First, determine the parts that will be handled with instant maintenance by studying the cause and effect (fishbone) diagrams. This step helps you organize your own thoughts. Section manager A first divided the errors that occurred in his plant into radial errors and axial errors, then produced a separate diagram for each type (see Figures 5-7 and 5-8). On the charts, he marked the parts that were especially subject to severe wear:

Radial insertion errors	*Axial insertion errors*

Head
    Wear on rubber tip
	Head
	    Molding lever
	    Insertion guide
	    Rubber pusher feed
	    Unit
	    Cutter 1.2

Anvil
    Lead guide pin
    Cut and clincher	

	Anvil
	    Anvil body
	    Cut and clincher

Parts supply
    Wear in urethane rubber
    Parts cassette
    Tape cutter
    Lead cutter
	Parts supply
	    Parts cassette
	    Z shaft

Develop a Two-Bin System for Parts

Next, list all the parts designated for instant maintenance on a board, according to the two-bin formula. Carry out an anticipated PQ analysis and divide the parts into three categories: those used in large quantities (A), those used in medium quantities (B), and those not used very much (C).

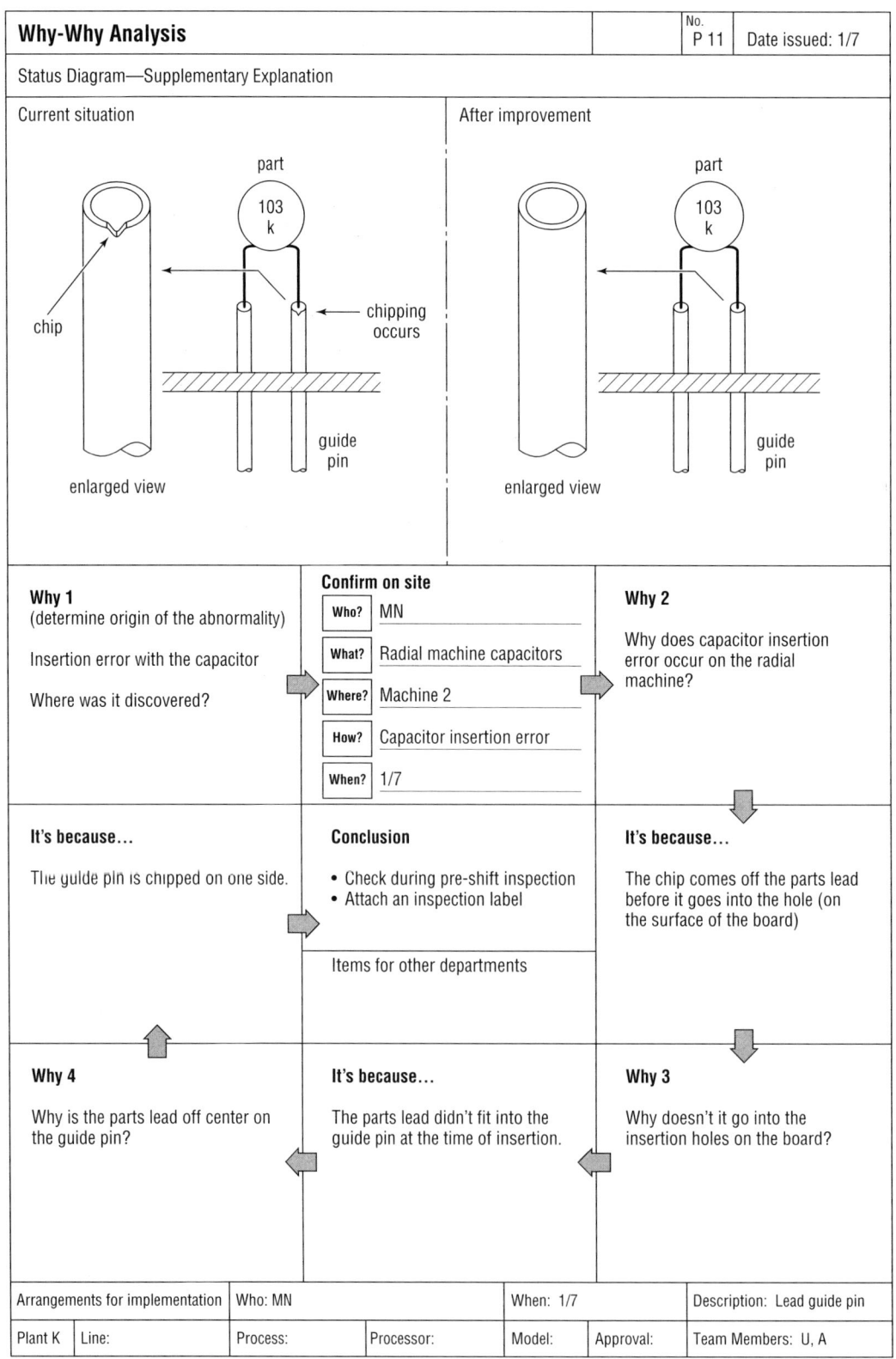

Figure 5-5. Why-Why Analysis for Capacitor Insertion Error

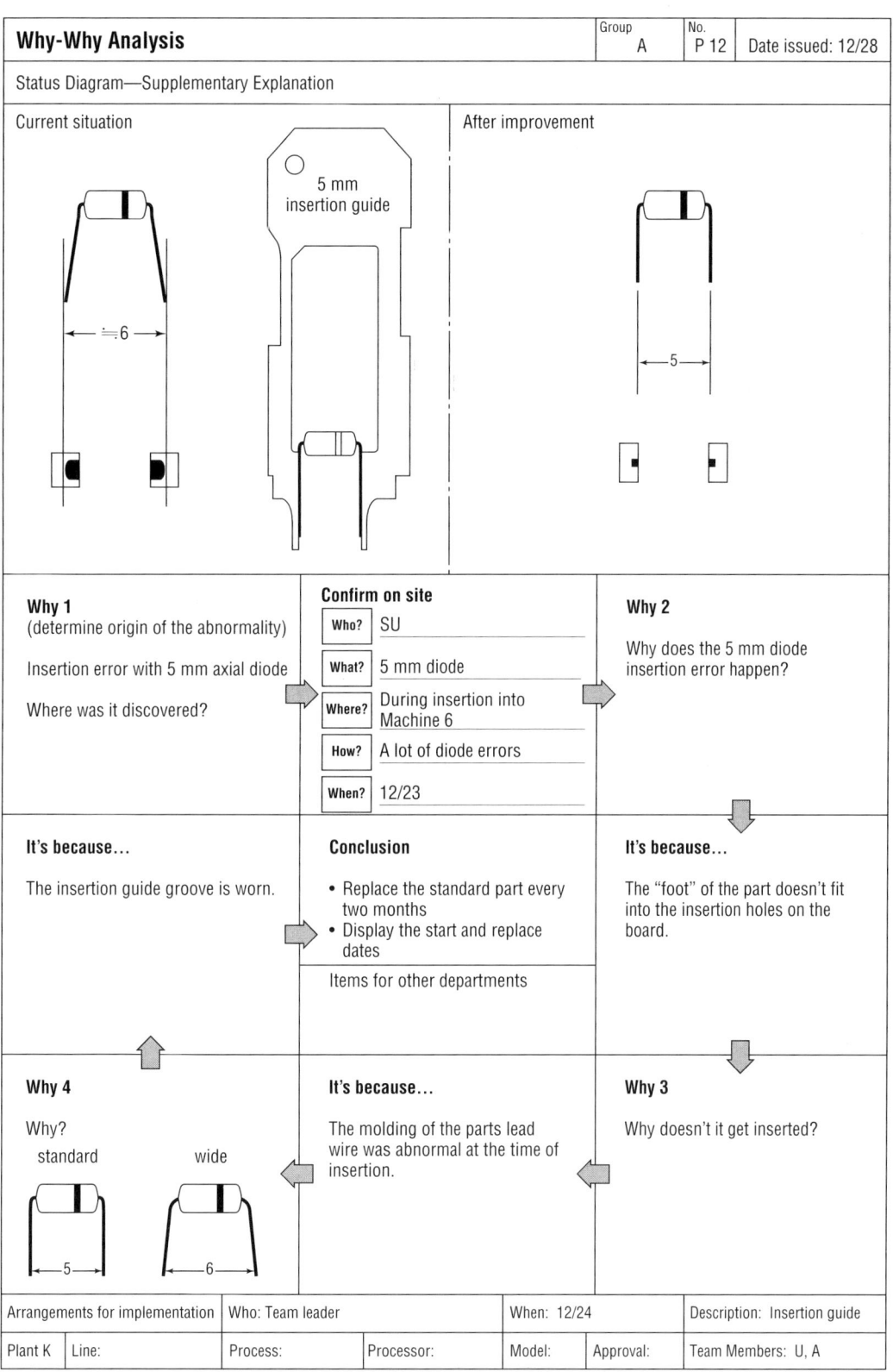

Figure 5-6. Why-Why Analysis for an Insertion Guide

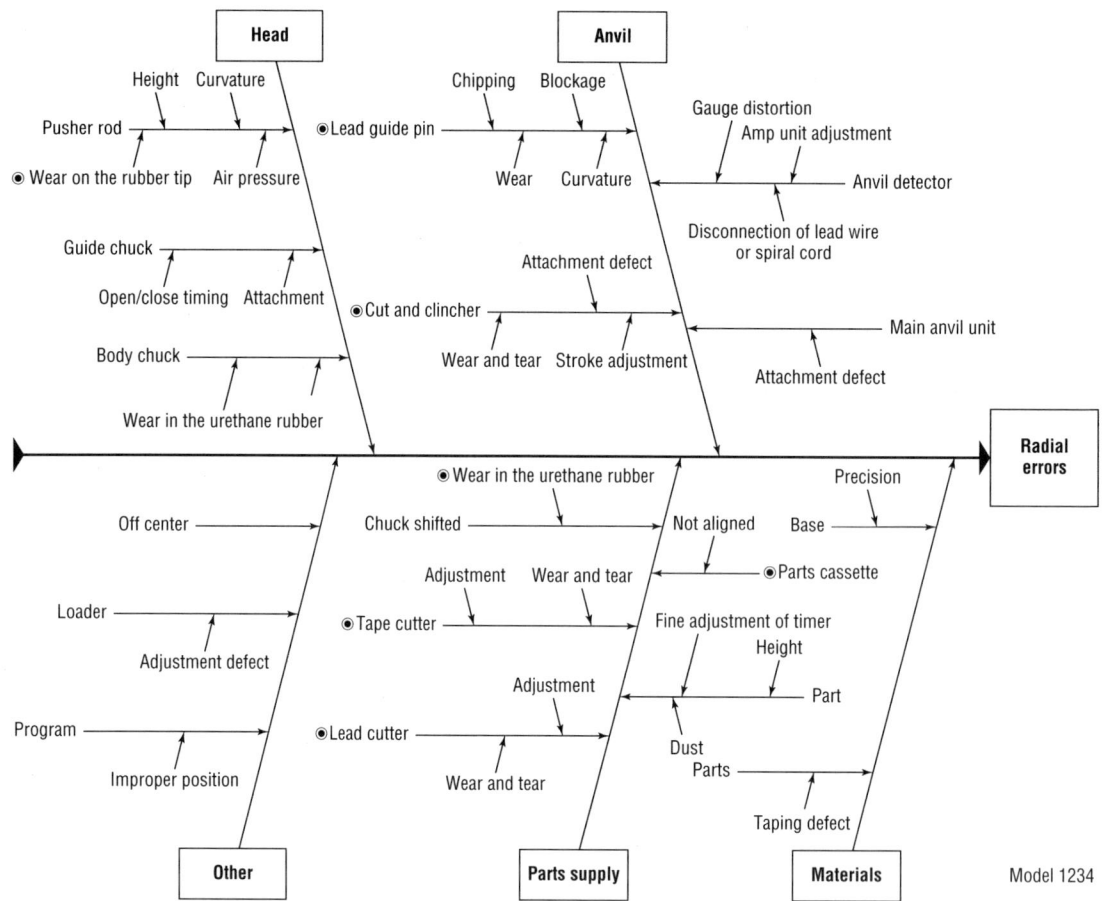

Figure 5-7. Cause and Effect Diagram for Instant Maintenance of Radial-Type Errors

The two-bin system is a kanban method for ordering standard quantities of parts at irregular intervals. Prepare sets of parts bins and use them as pictured in Figure 5-9. If you follow the system conscientiously, you will not run out of parts.

Create Instant Maintenance Sheets

Figures 5-10 through 5-14 offer examples involving the lead guide pin, tape lead cutter, urethane rubber parts, the lever, and the insertion guide.

Practice the Maintenance Routines

Before publicizing the instant maintenance routines in your plant, practice them on your own. Your goal is to complete each routine in three minutes. To achieve this goal, you may have to obtain improved tools, such as different kinds of wrenches.

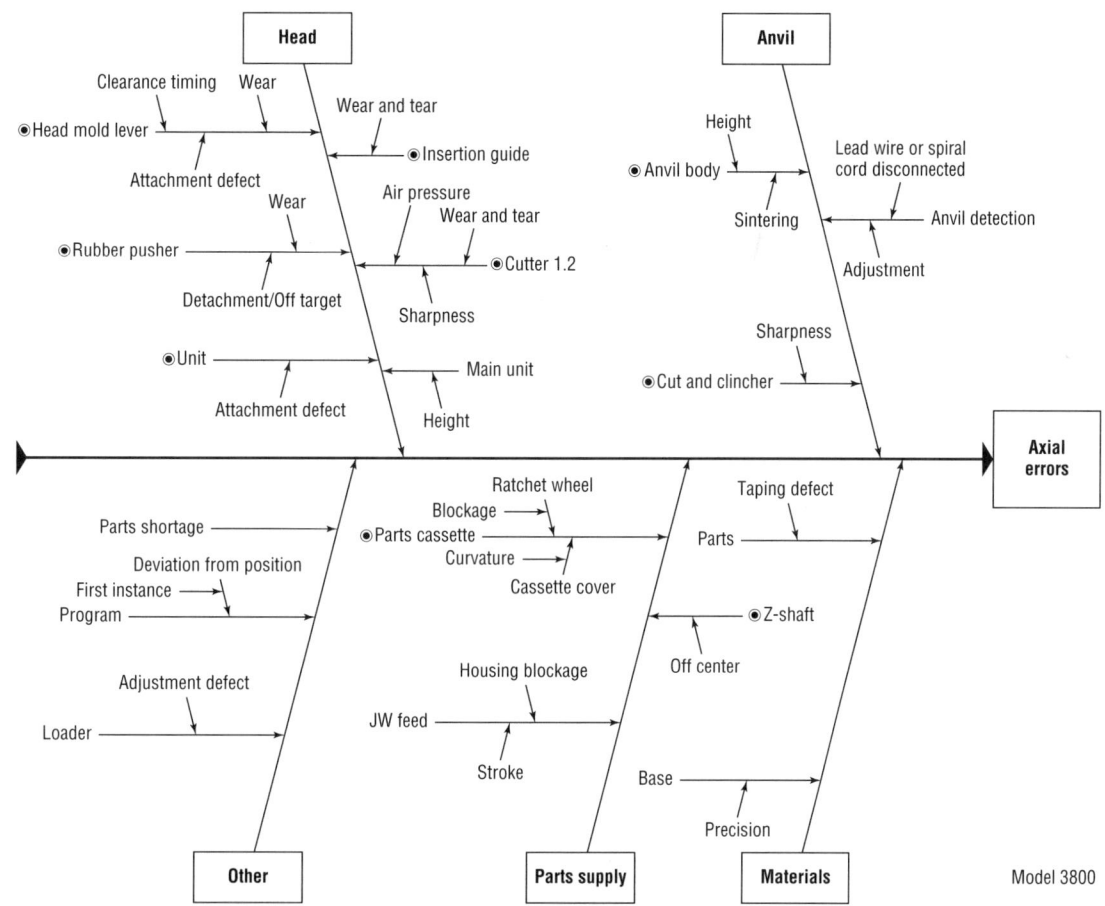

Figure 5-8. Cause and Effect Diagram for Instant Maintenance of Axial-Type Errors

Step 6: Demonstrate the Maintenance Techniques to Employees

These training sessions should be conducted with the idea that everyone is both a teacher and a pupil. For example, you can announce a mandatory training session on instant maintenance of a guide pin. Explain that you are going to demonstrate the instant maintenance techniques but that anyone is free to point out wasted time or motion in your approach, or to make other improvement suggestions. This will ensure that employee B, who has an instinct for mechanical matters, and employee C, who has a low opinion of you, will be watching intently. In addition, you will probably want to have someone videotape the session.

Then you start skillfully demonstrating each maintenance routine. Since you've been practicing, you don't get flustered. Maintenance routines that used to take half an hour now take 3 minutes. Even if you slip up, you can still reduce the amount of time needed to no more than 15 minutes.

Suppose that you don't have a lot of practical experience with machines and equipment and are all thumbs when it comes to dealing with them. In that case, you can train your trusted employee D and have her standing by. Then if you mess up, you can call on D to step in and demonstrate the routines quickly and skillfully for everyone to see.

Once the demonstration is over, call on employees B and C, who have been watching with expert eyes, and tell them it's their turn to try. If they make a mistake, you can correct them as they go through the steps.

By setting up a mutually respectful two-way learning environment, you can teach instant maintenance in an enjoyable and even humorous way. Through several repetitions of these basic steps, company K implemented instant maintenance throughout its manufacturing facilities. Figure 5-15 reviews the steps in a nutshell.

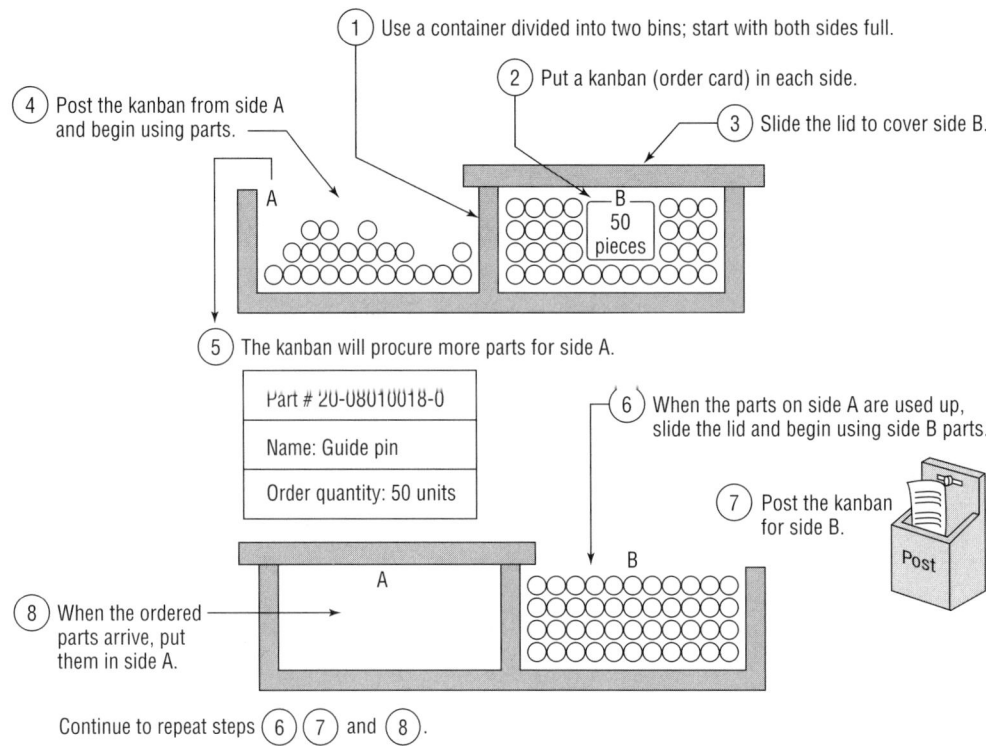

Figure 5-9. The Two-Bin System for Instant Maintenance Parts

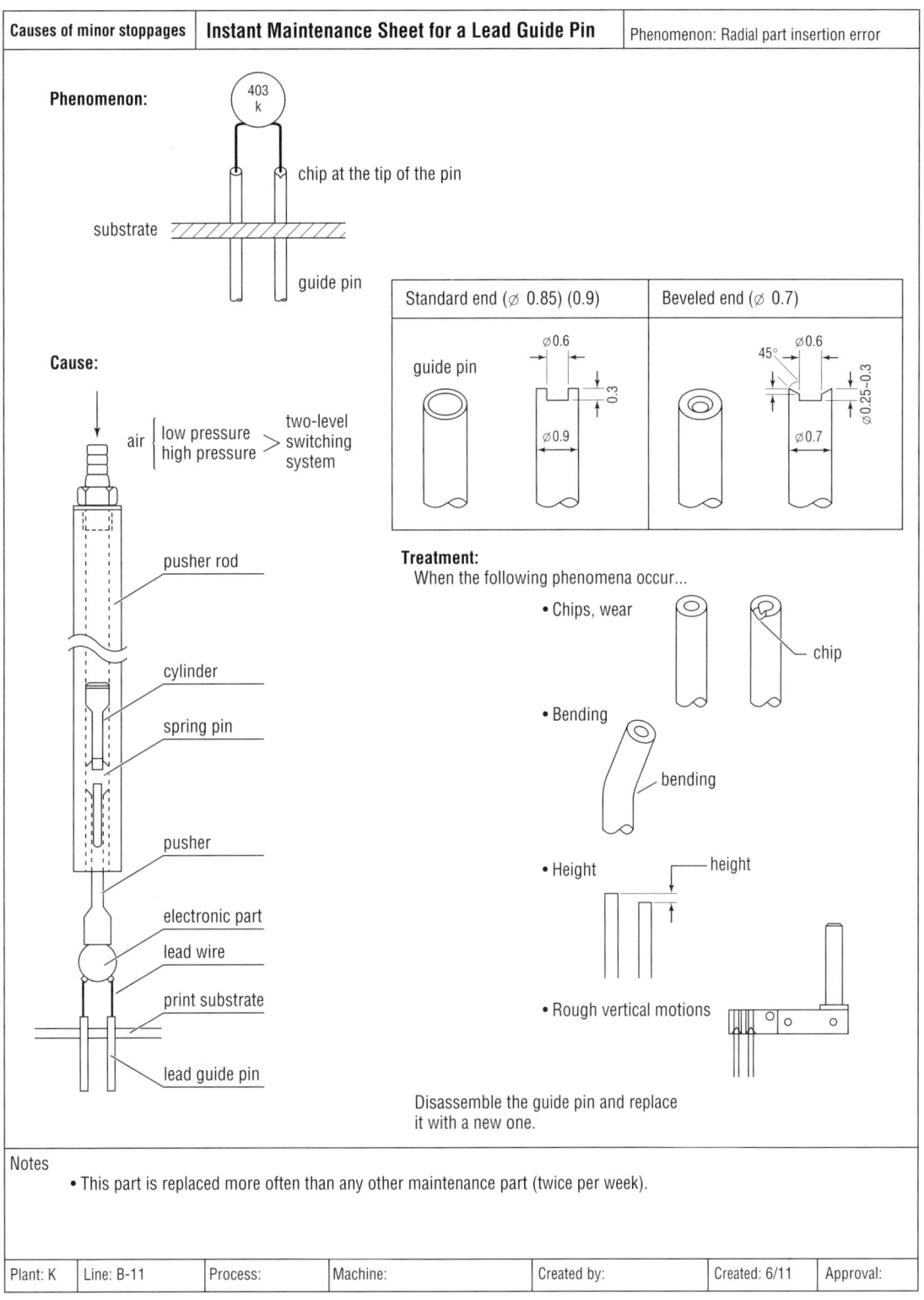

Figure 5-10. Instant Maintenance Sheet for a Lead Guide Pin

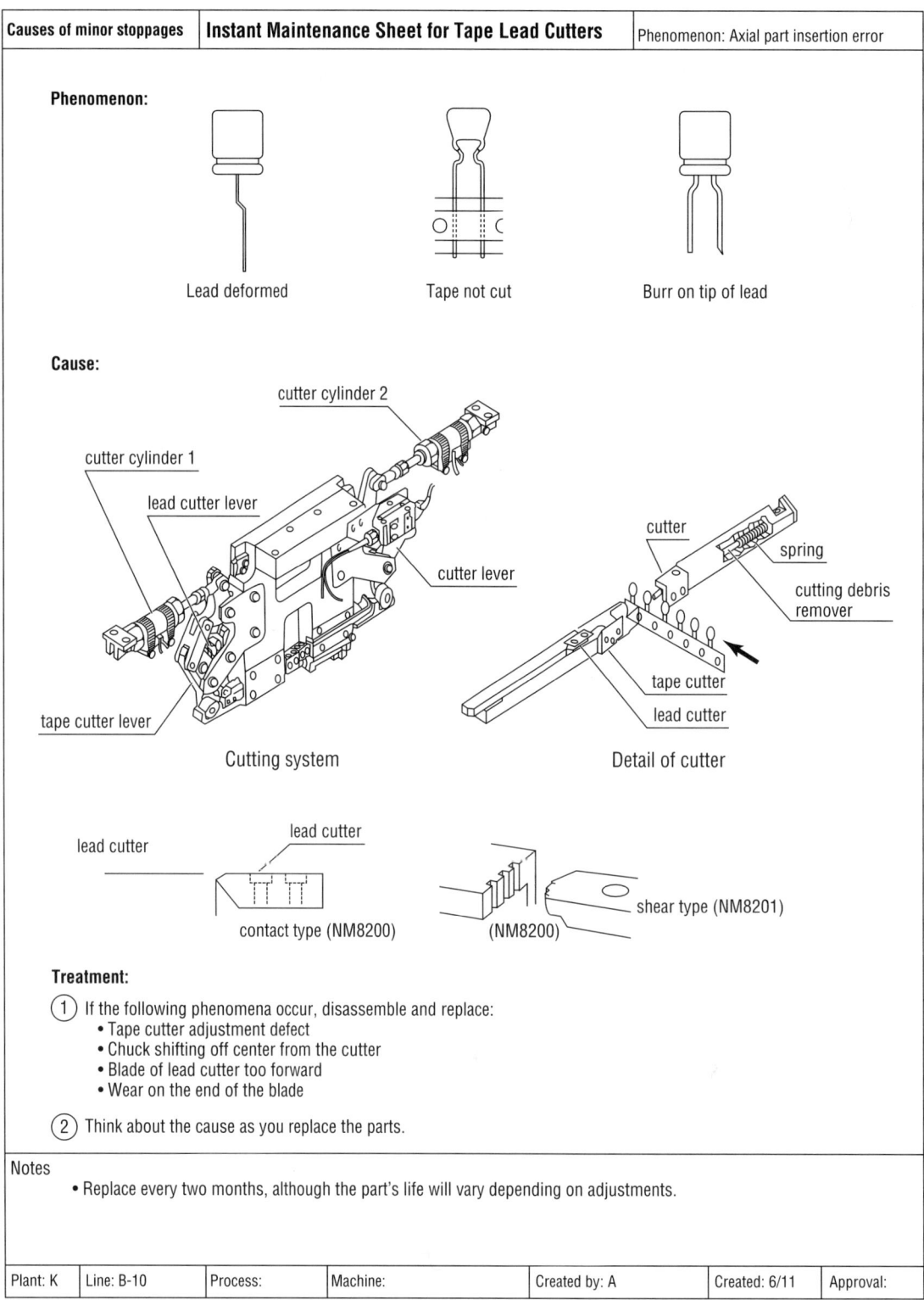

Causes of minor stoppages	**Instant Maintenance Sheet for Tape Lead Cutters**	Phenomenon: Axial part insertion error

Phenomenon:

Lead deformed   Tape not cut   Burr on tip of lead

Cause:

cutter cylinder 2

cutter cylinder 1

lead cutter lever

cutter lever

cutter

spring

cutting debris remover

tape cutter

lead cutter

tape cutter lever

Cutting system   Detail of cutter

lead cutter

lead cutter

contact type (NM8200)   (NM8200)   shear type (NM8201)

Treatment:

① If the following phenomena occur, disassemble and replace:
   • Tape cutter adjustment defect
   • Chuck shifting off center from the cutter
   • Blade of lead cutter too forward
   • Wear on the end of the blade

② Think about the cause as you replace the parts.

Notes
   • Replace every two months, although the part's life will vary depending on adjustments.

Plant: K	Line: B-10	Process:	Machine:	Created by: A	Created: 6/11	Approval:

Figure 5-11. Instant Maintenance Sheet for Tape Lead Cutter

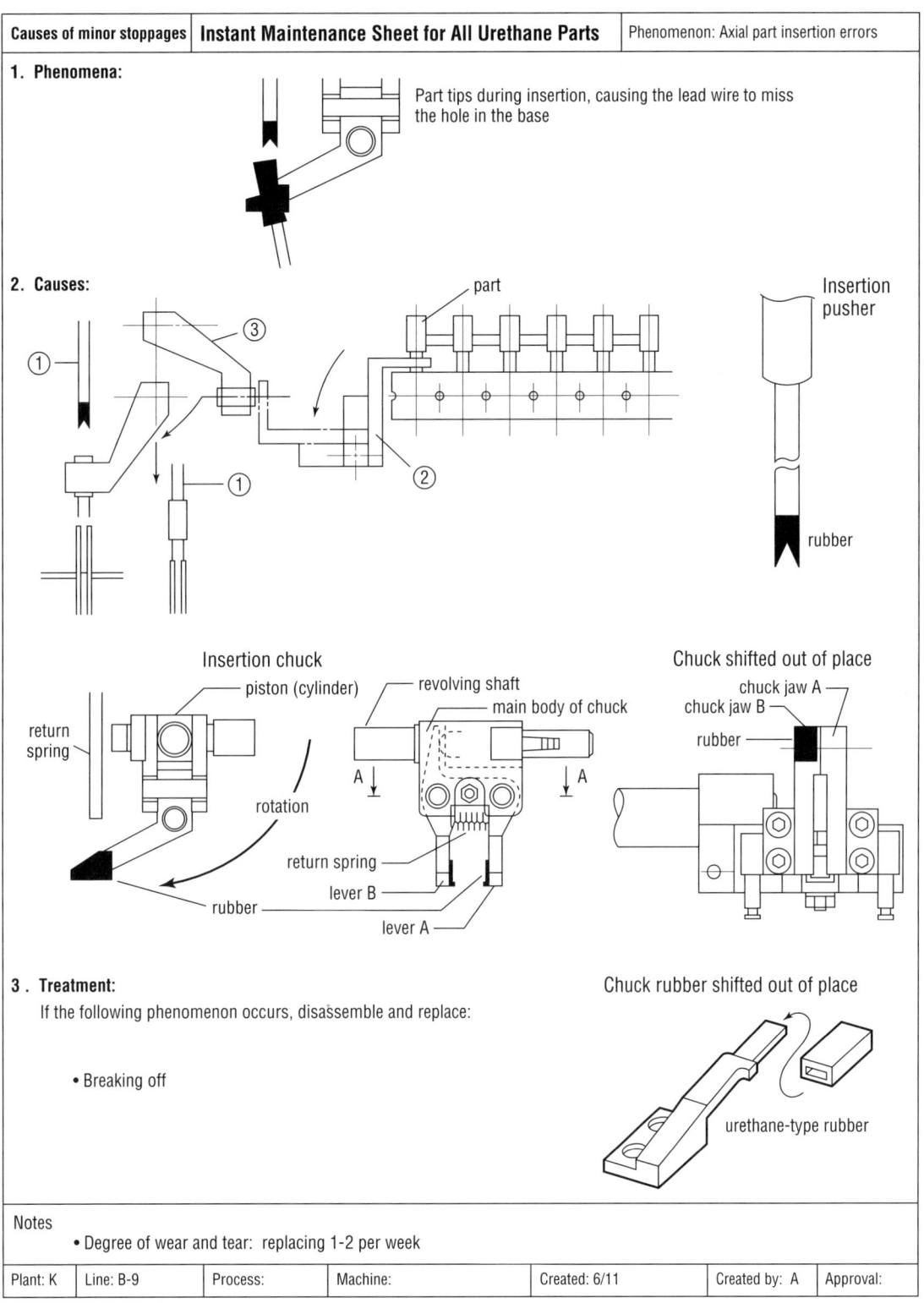

Figure 5-12. Instant Maintenance Sheet for All Urethane Parts

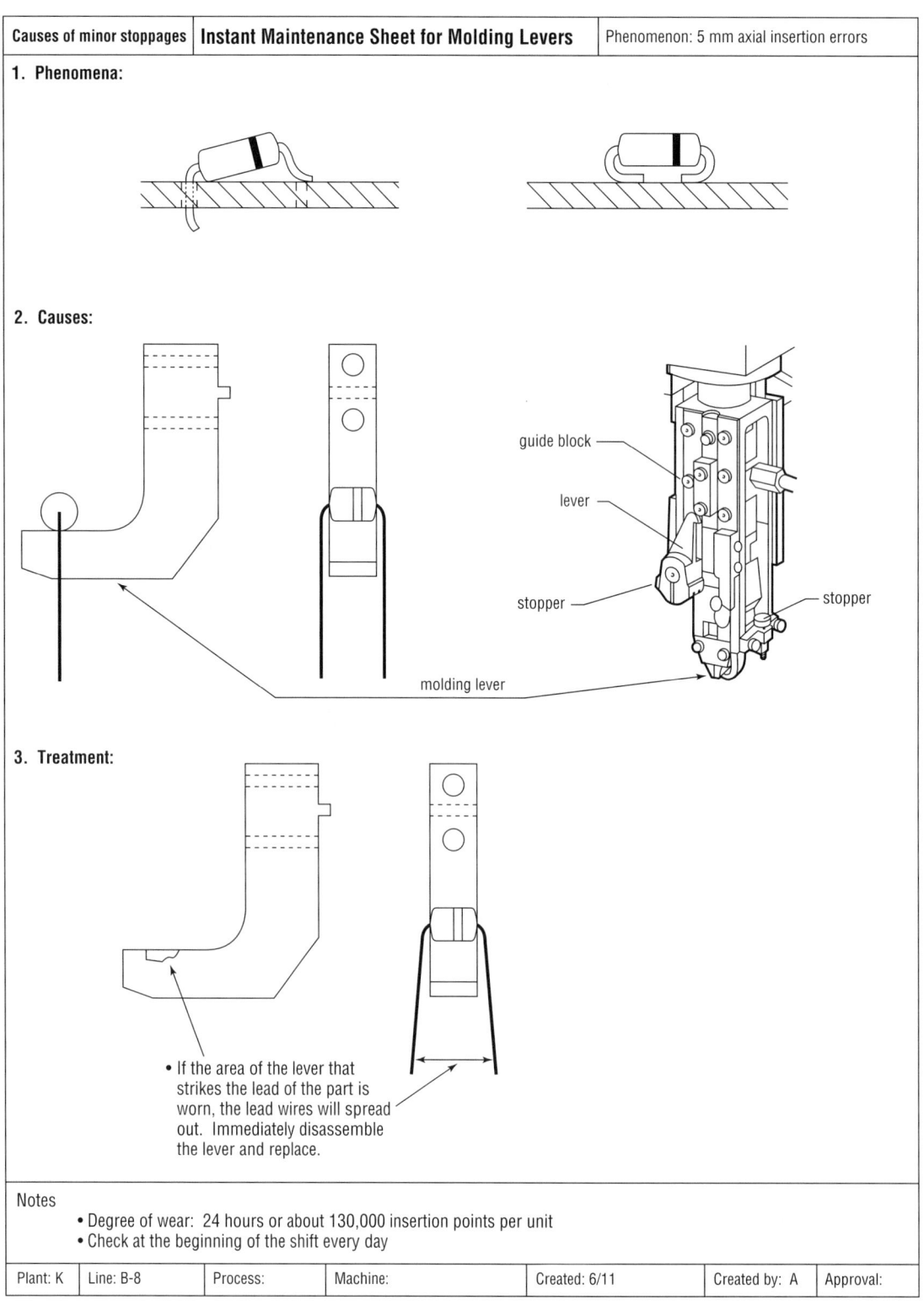

Causes of minor stoppages	Instant Maintenance Sheet for Molding Levers	Phenomenon: 5 mm axial insertion errors

1. Phenomena:

2. Causes:

guide block
lever
stopper
stopper
molding lever

3. Treatment:

• If the area of the lever that strikes the lead of the part is worn, the lead wires will spread out. Immediately disassemble the lever and replace.

Notes
• Degree of wear: 24 hours or about 130,000 insertion points per unit
• Check at the beginning of the shift every day

Plant: K	Line: B-8	Process:	Machine:	Created: 6/11	Created by: A	Approval:

Figure 5-13. Instant Maintenance Sheet for Molding Levers

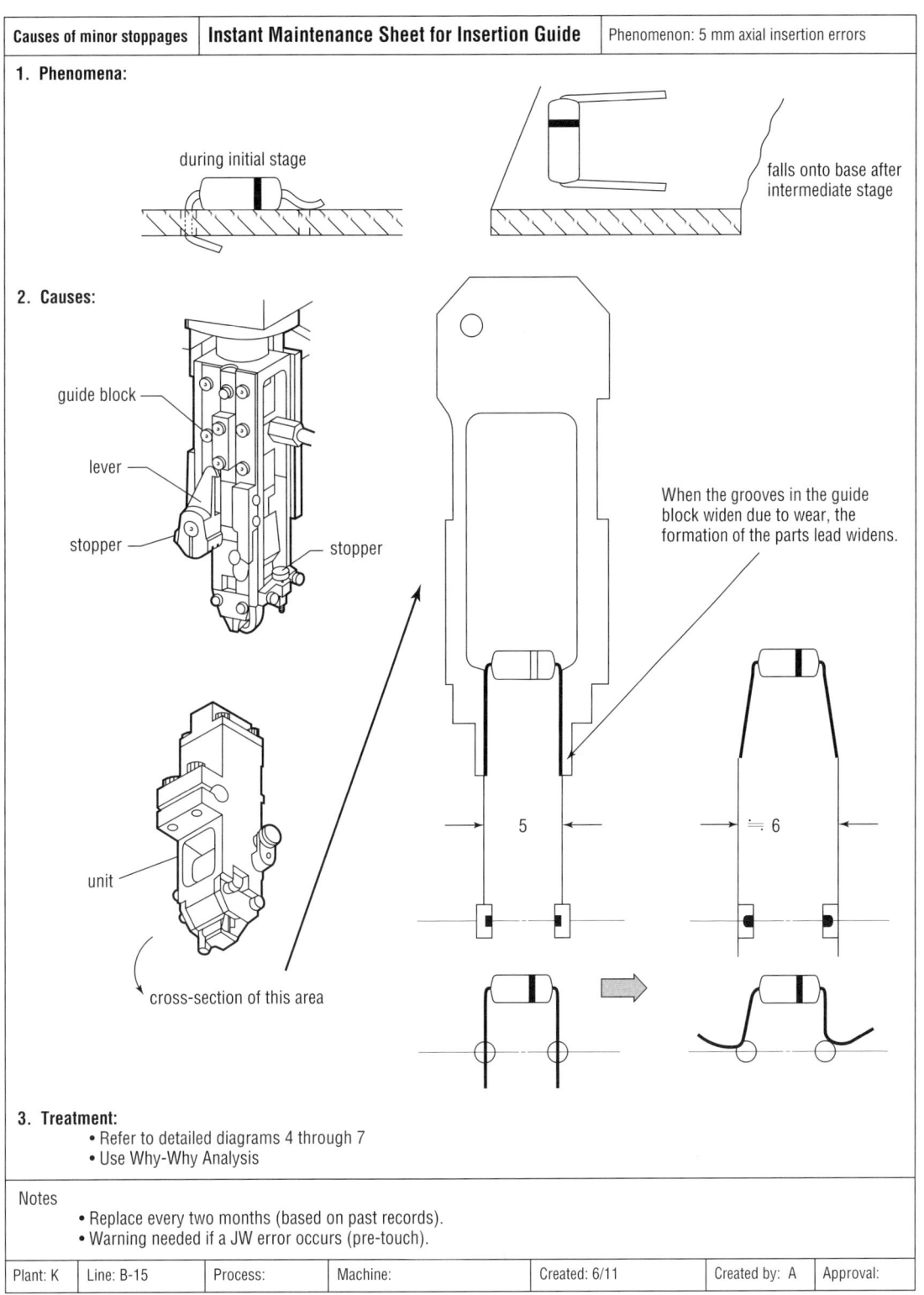

| Causes of minor stoppages | Instant Maintenance Sheet for Insertion Guide | Phenomenon: 5 mm axial insertion errors |

1. Phenomena:

during initial stage

falls onto base after intermediate stage

2. Causes:

guide block

lever

stopper   stopper

unit

cross-section of this area

When the grooves in the guide block widen due to wear, the formation of the parts lead widens.

5   ≒ 6

3. Treatment:
- Refer to detailed diagrams 4 through 7
- Use Why-Why Analysis

Notes
- Replace every two months (based on past records).
- Warning needed if a JW error occurs (pre-touch).

| Plant: K | Line: B-15 | Process: | Machine: | Created: 6/11 | Created by: A | Approval: |

Figure 5-14. Instant Maintenance Sheet for Insertion Guide

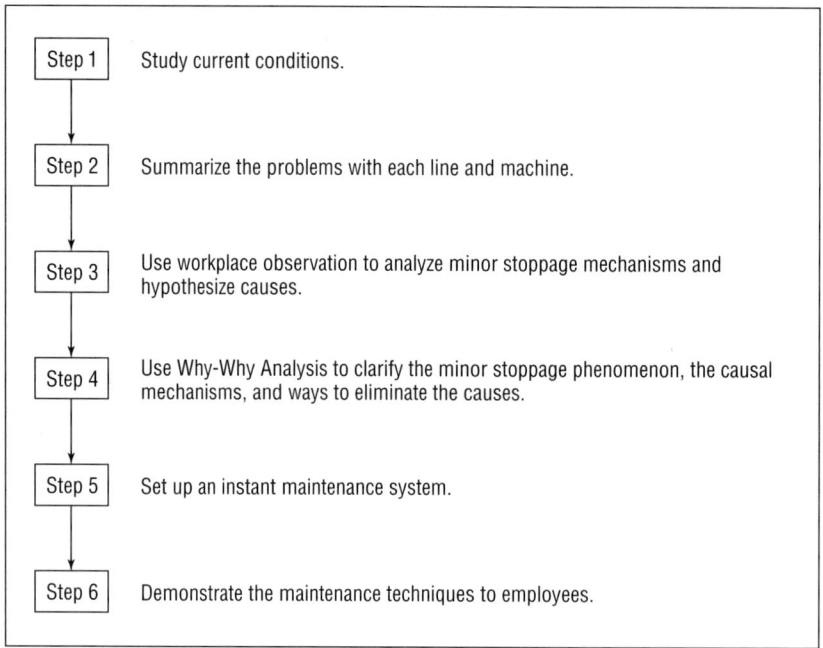

Figure 5-15. Procedure for Eliminating Minor Stoppages in an Automatic Inserter

Improving Setup Operations

6

A Well-Done Setup Can Demonstrate the Effectiveness of TPM

There's a saying in factories, "Setup is eight-tenths." In other words, if you've completed setup properly, your work is 80 percent completed. TPM follows this principle too. If you look at how a plant handles setup, you can see how powerful the effect of TPM is and determine that company's level of TPM implementation. For example, if you observe operators wandering around in confusion looking for things when replacing jigs, that is fifth-class TPM. If they deviate from the standards occasionally, but have succeeded in bringing setups down to less than ten minutes, that's first-class TPM.

Company A surveyed its machinists to find out how long setup was taking. They divided the machining process into four stages and determined the percentage of time devoted to each stage; their observations appear in Table 6-1. The data show that actual machining took up only a little over 10 percent of the time. When you consider that cleanup is actually part of setup for the next job, you could say that setup was nearly nine-tenths of the job at company A!

Why was setup taking so long? We found that the plant's setup process was aimless and disorganized. A close investigation revealed seven causes of this disorganization.

Seven Causes of Aimless and Disorganized Setup

It's no exaggeration to say that half of all setup problems are caused by disorganized setup procedures. If we look more closely, we can see that the company

Table 6-1. Setup Time at a Parts Machining Plant

Operation	Time (minutes)			Percentage of total
	Min.	Max.	Avg.	
Pre-setup	20	60	40	36
Setup and changeover	8	30	20	18
Machining	5	20	15	14
Cleanup	10	80	35	32

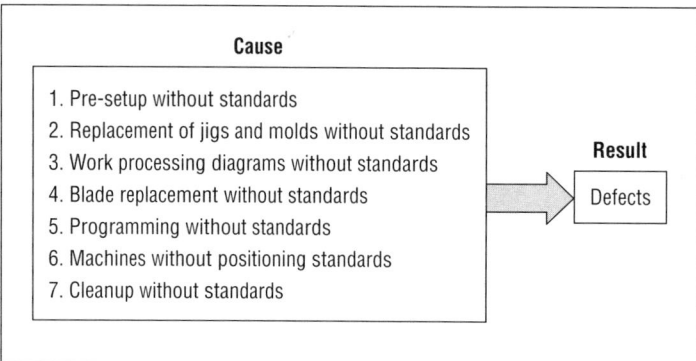

Figure 6-1. The Seven Causes of Defects

either has failed to set standards or has allowed the standards to shift in the seven areas listed in Figure 6-1.

Pre-Setup without Standards

Here's a story we heard at company A. A recent high school graduate employed there was putting in lots of overtime and coming home late nearly every evening. His father was concerned about the situation, and one night the two of them talked about it over dinner.

"What's keeping you so busy at work?" the father asked.

"Company B placed a huge order for molds with us," the young man explained, "so we're running around like crazy trying to keep up."

"So, like I asked, what's keeping you so busy?" the father persisted.

"They're always making me help look for tools and blades and stuff," the son replied.

"So everyone's too busy to keep things neat and orderly," the father surmised. "That must be what's going on there."

Actually, we would say that the real reason had nothing to do with the 2Ss. Rather, the company had not been using Why-Why Analysis to find out why so much time was being wasted. Take a look at Figure 6-2 for an analysis of the situation.

Jig and Mold Replacement without Standards

No standards, no trained people, no smarts—just a lot of adjustments and defects.

When we made an inspection tour of company A, we suggested standardization of the jigs for the machining center changeover. Since the workday was almost over, we gave the problem to the operators as homework.

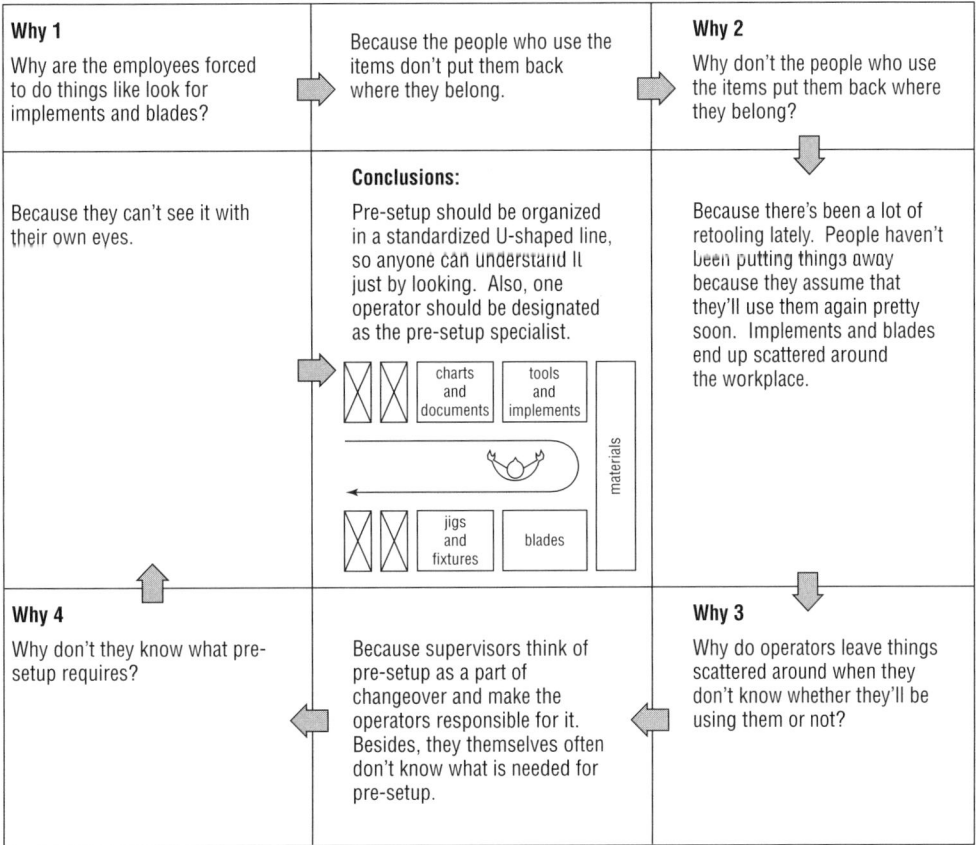

Why 1

Why are the employees forced to do things like look for implements and blades?

Because the people who use the items don't put them back where they belong.

Why 2

Why don't the people who use the items put them back where they belong?

Because they can't see it with their own eyes.

Conclusions:

Pre-setup should be organized in a standardized U-shaped line, so anyone can understand it just by looking. Also, one operator should be designated as the pre-setup specialist.

Because there's been a lot of retooling lately. People haven't been putting things away because they assume that they'll use them again pretty soon. Implements and blades end up scattered around the workplace.

Why 4

Why don't they know what pre-setup requires?

Because supervisors think of pre-setup as a part of changeover and make the operators responsible for it. Besides, they themselves often don't know what is needed for pre-setup.

Why 3

Why do operators leave things scattered around when they don't know whether they'll be using them or not?

Figure 6-2. Why-Why Analysis for Mislaid Items

At the next day's meeting, the operators reported that they had come up with compound jigs like the one pictured in Figure 6-3. They explained that the workpieces attached to these jigs were all standardized during the pre-setup phase to fit the groove in the base. When they were attached to the machine, the process went 30 percent faster. That was cause for celebration.

However, while this sort of jig is helpful when you don't deal with many types of workpieces, you can still run into some of the following problems when faced with greater variety:

1. Spending a lot of time on error adjustment during assembly.
2. Needing to stop and think during assembly. Since the taps have the same pitch, it's hard to know just what the standard is.
3. Having an excessively large jig base and leaving many unused tap holes, which increases waste during replacement and cleanup.
4. Wasting time confirming close adhesion, since the jig standards are surface standards.
5. Wasting time making adjustments if parts deviate from the standards when bolts are attached.

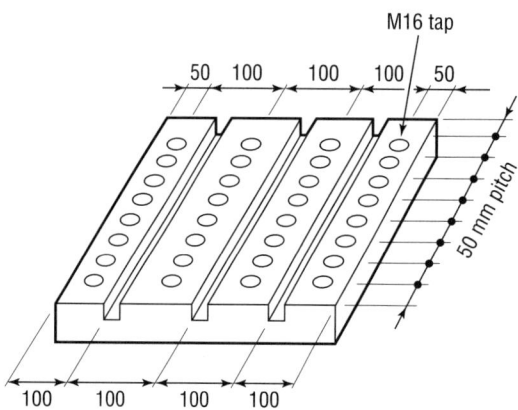

Figure 6-3. The Standard for Company A's Combined Jig

Why do these "disimprovements" occur? When we thought about it, we concluded that the people had never been taught that whenever we improve something, we have to start by studying the current situation. We've written in detail about how to do this in *Kaizen for Quick Changeover*.

Test Yourself

Assembly jigs are used in most factories because they have benefits. What are these benefits?

Our Answers

1. The jigs themselves can be reused until they break.
2. They can also be used with parts that are produced singly, saving time in making these special setups.

Work Machining Diagrams without Standards

Machining diagrams are the third cause of aimless and disorganized setup; in the worst cases, they can even lead to defects. Diagrams such as the one shown in Figure 6-4 cause the waste of having to think too much to figure things out. Here are some of the problems with this diagram:

1. So many pieces are modified that the measurements and shapes differ from the actual size.
2. The chart is not oriented toward the chucking direction.
3. The chart does not make it clear exactly which areas should be turned on the lathe.
4. The chart does not make it clear whether these are reference measurements or machining measurements.
5. The chart does not indicate the machining allowance of the surface to be finished in post-processing.

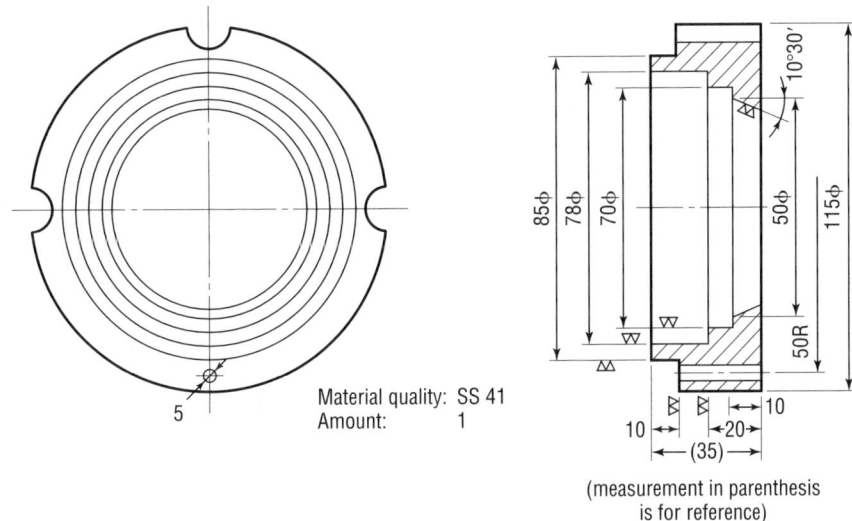

Material quality: SS 41
Amount: 1

(measurement in parenthesis
is for reference)

Figure 6-4. Lathe Processing Diagram

6. The jigs used in this process and in post-processing are not shown.

7. Most of the necessary measurements are not given, which increases the likelihood of errors by the person responsible for setup.

Looking critically at charts and diagrams, we usually find all sorts of omissions. The main reason is that the designer lays out the project before the specifications have been completely set. This situation may eventually lead to wasting time modifying the design. To help avoid this problem, we offer some suggestions for creating charts and diagrams that will support efficient setups.

Over 70 percent of defects are due to design flaws, and problems with these same 70 percent are compounded by setup methods. Designers need to keep the following points in mind:

1. The person responsible for setup needs diagrams whose main points can be understood at a glance.

    For example:

    • Make the shapes and measurements conform to the actual size. (CAD can be helpful here.)

    • Depict items with complex shapes in head-on or cutaway views.

    • When deciding the orientation of the drawing, take the machining direction into careful consideration. This is particularly true with chucking or fraise processes.

    • Show the rolling grain of boards (to prevent warping).

    • Indicate the areas to be machined with heavy, solid lines. This is helpful to the operators.

- When depicting an asymmetrical object, put the measurements in wavy brackets to make the diagram easier to understand.
- Enclose dimensions that need to be calculated in parentheses. This confirms the need for calculation and reduces the risk of errors.
- Show areas with complicated dimensions or a lot of lead lines in a separate, enlarged diagram.
- If an object requires a lot of chamfering, write the numbers into the diagram.

2. Write the standards in a clearly understandable way.

For example:

- Write the standards for attaching parts. (Raise the precision of the standard surface.)
- It's helpful to write the standards for attaching jigs and blocks.
- Measurements of long parts are easier to understand if you write them near the tip of the part.
- Write standards for centering and measuring cast metal objects.
- Make standards uniform.

For more about standards, refer to our previous book, *Kaizen for Quick Changeover*.

Blade Replacement without Standards

We happened to be watching while the operator in charge of setup at company A performed pre-setup for NC and machining center blades.

First of all, he used a presetter during setup, but he also used a measuring instrument to determine the length and diameter of each implement, entered the data into the memory of the NC apparatus, and also wrote the data onto the implement itself with a marker. Later, we saw him make further adjustments to meet the standards when he set the machine.

We got together with his supervisor and applied Why-Why Analysis to figure out why he was going to all this trouble. Here's what we came up with:

1. Since he was replacing a bare blade in the NC, he prepared a cutting tool, but this required adjustments.
2. In the machining center, he was replacing all the tools, so we considered making some of the installations permanent or semipermanent.
3. He was using worn cutting tools.
   - There were no standards for replacement.
   - There were also no standards for how much wear was acceptable.

Improving Chucks

The main thing to remember in lathe retooling is to avoid replacing the jaw element. This is in line with the fundamental rule of zero setup, "immovable standards." Figures 6-5, 6-6, and 6-7 show three examples; the details are discussed in Chapter 10.

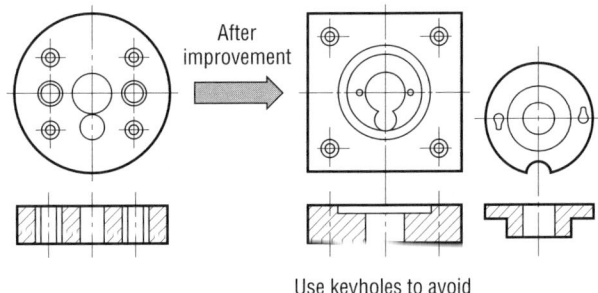

After improvement

Use keyholes to avoid entire removal of bolts.

Figure 6-5. Example of Improvement in a Chucking Plate

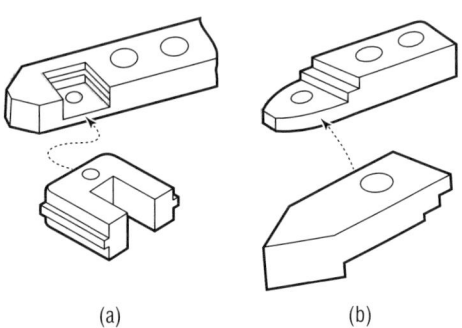

(a)　　　　　　　　(b)

Do not remove the larger jaw.
Instead, use an insert to adapt the jaw for smaller work.

Figure 6-6. Large and Small Jaws of a Chuck

nut tightens

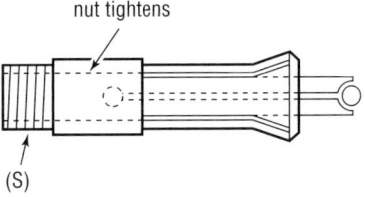

(S)

The connector shown here is of the screw type, but we can also imagine using a click connector.

　　Plan A: Use the screw type.
　　Plan B: Use the click connector type.

Which do you think is more appropriate and practical in the workplace?

Figure 6-7. Example of a Correct Chuck

Programming without Standards

The NC machining program can be input either manually or automatically, but the NC installation at company A allowed input on tape, disk, or even old-fashioned punch cards. Many different parts and components were also involved, and since they'd never been sorted or standardized, data management was a nightmare. As a result, operators not only had to look for and retrieve parts, but needed to spend a lot of time making adjustments after attaching them to the NC apparatus. In fact, this was the biggest time waster in the changeover process at company A.

Thus, after we had implemented the 2Ss and employee G had completed the main programming, we took work samples of the program correction that occurred during changeover operations. We discovered many examples of waste, including the following items:

• Looking for and counting a given number of end mills of the same size. (Cause: Wanting to use and machine end mills of the same size all at once.)

• Confirming the sequence of the program. (Cause: Employee G didn't trust the programmers who had no hands-on experience in the shop. "In particular, M sometimes makes mistakes—if I don't go over his programs, there are usually problems," he said. He evidently believed there was no way to avoid spending time correcting programs.)

• Confirming the amount of offset. (Cause: There had been programmer errors in the past, especially on the part of employee M.)

We also found waste in using an electronic calculator for computations, figuring out shapes, distinguishing rough machining from finishing, raising the grinding speed, reducing air cuts, and spending time wondering whether it was possible to increase the diameter of the end mills.

Machines without Positioning Standards
(Machines with Variable Processing Standards)

Every machine must have work positioning standards. However, even when such standards are established, they can shift during processing, causing minor stoppages. Most minor stoppages stem from this sixth form of waste.

You can begin to see the signs of this problem if you observe the work carefully. For example:

1. You have to wonder about the work positioning standards for machines where dial gauges or other measurement instruments are used during changeover. (Sometimes there is play in the machine.)

2. The positioning standards can go off if the machine vibrates a lot during operation. Even when the machine has an automatic measurement system, it will be impossible to make corrections with the automatic offset input if the standards are more than a few micrometers off, and an overhaul will be needed.

Test Yourself

Company N used to use general-purpose lathes like the one shown in Figure 6-8, but their orders no longer call for large-scale production. Instead, most of their orders have been for small lots of a wide variety of products, so they switched over to NC machines. At that time,

1. Because most of the jobs involved composite machining, process combination, and process intensification, the turning center experienced
   • a decrease in unfinished goods, and
   • a decrease in floor space, due to concentration of the machinery.
2. Because of the use of diamond, ceramic, and other modern tool materials, machining time was reduced and precision increased.

With that, the operators should have been ready for anything, but they weren't. Why not?

Our Answer

Reducing setup time especially requires reducing retooling time. It is necessary to divide the types of waste into those involving preparation, replacement, and adjustment, and to reduce the time spent on chucking, tool offset, and programming.

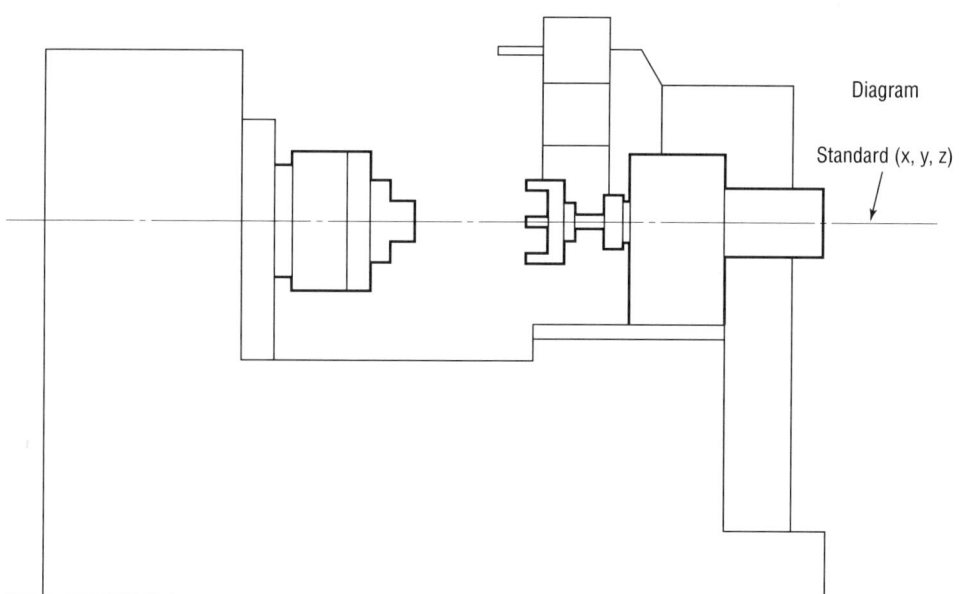

The machine's standards are as shown.
However, the standards begin to deviate if the jaw is moved during chucking.
Therefore, improvements connected with chucking should be given close attention.

Note: See question 7 in Chapter 10.

Figure 6-8. General Purpose Lathe

Cleanup without Standards

Since the plant didn't have any standards for tidying up and cleaning the tools, blades, jigs, and molds, operators put them away dirty. They didn't get around to cleaning them until several days later, so dirt adhered to the surface, and cleaning them off took a long time.

The scope of cleanup is much broader than simple cleaning. It includes disposing of shavings and taking care of problems.

Problems

In our opinion, most problems with machines are traceable to changeover. Ignoring the principle of maintaining unwavering machining standards leads to breakdowns.

When we talk about machining standards shifting, we're talking about molds, jigs, and work shifting. If any one of these shifts, the machine will begin to vibrate, and vibration leads to wear. When parts become worn, breakdowns occur, along with minor, intermediate, and major stoppages, and this causes an increase in defects and last-minute adjustments. If we implement zero setup, which presupposes unchanging machining standards, we can greatly reduce minor stoppages.

Other common causes of problems include:

1. Human factors
   - Input errors in program correction
   - Errors in selecting tools

   Some ways to prevent tool selection errors include
   - implementing pre-setup
   - standardizing hole diameters and taps at the design stage
   - color coding tools and materials

2. Problems with cutting and grinding tools

   We commonly encounter tool breakage and wear, or lose the tips of end mills. Our advice here is necessarily brief and one-sided, but the most important countermeasures begin with
   - attaching breakage sensing equipment
   - gathering data frequently, calculating the mean time between failures (MTBF), and replacing parts when 80 percent of that period has passed

Steps for Improving Setup

Multiple-product small-lot production brings with it an increase in the number of employee-hours lost to setup. According to a survey from the mid-1980s, companies affiliated with the automobile industry went through setup and changeover an average of up to ten times per day. If the same is true today, then, given a working day of 450 minutes, the average working time associated with each different product is 45 minutes, including the changeover time.

Even if we can reduce the setup and changeover time to 9 minutes, we end up with a 20 percent changeover loss:

machining time = 45 minutes − 9 minutes = 36 minutes
setup and changeover time loss = (9 minutes/45 minutes) × 100 = 20 percent

We should stop doing anything that wastes time or materials, but in reality, stopping is no simple matter. Begin with the goal of cutting the current setup and changeover time in half. Once we have managed that, we can adopt the formula of reducing that by half, and so on. Eventually we will arrive at zero setup, meaning setup that can be accomplished within three minutes. The following sections describe the important steps for achieving this goal.

Step 1. Study Current Setup Losses   **Step 2. Form Setup Improvement Promotion Team**

Actual Time Devoted to Setup and Changeover (over 7 days)

Survey item Process	Machining time (A)	Time required (B)	Setup time (C)	Setup frequency (D)		
M₁	1,633 min.	1,700 min.	1,151 min.	41		
M₂	566	1,370	1,336	56		
M₃	1,364	1,454	1,553	72		
M₄	880	1,028	1,743	54		
M₅	1,429	2,016	631	43		
M₆	1,831	2,679	352	83		
M₇	1,575	2,323	288	45		
M₈	1,959	2,528	374	39		
M₂₅	2,365	3,371	90	15		

* = machining time/other time (excluding setup time)

Studying the current situation begins with looking at current economic losses, including accurately assessing the losses that accompany that wasteful activity known as setup. The survey results displayed in the table above show that 26 percent of worker hours at a certain company were lost to setup and changeover.

Figure 6-9 shows how PQ analysis ferreted out one of the causes: The group A products were no longer produced at the plant, and only groups B and C, which are small-lot, short-term projects, remained. (PQ analysis is a specialized application of Pareto analysis; it is explained in more detail in the example in Chapter 7.)

The waste due to setup was really the responsibility of production managers who didn't try to improve the setup process, despite the need for more frequent changeovers. When the plant manager finally woke up to what was happening, he quickly formed a setup improvement team to address the situation.

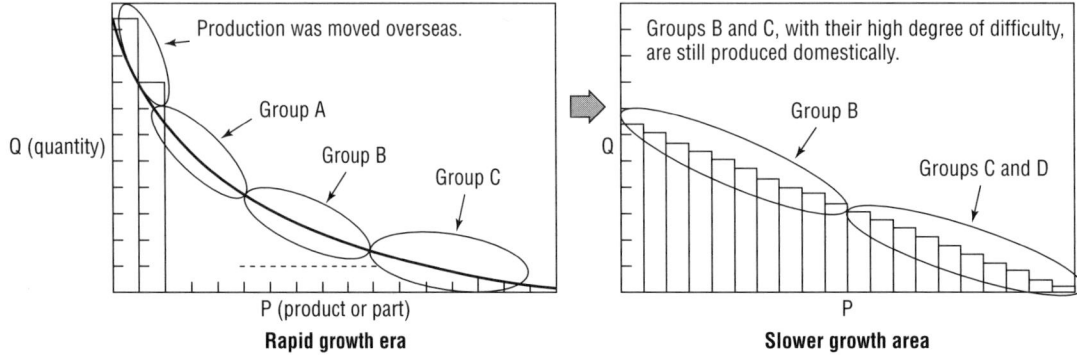

Figure 6-9. PQ Analysis

Step 3. Perform On-Site Observation and Operation Analysis

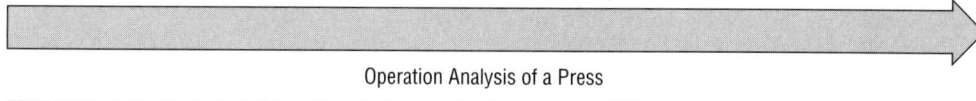

Operation Analysis of a Press

Elemental operation	Time read (min./sec.)	Net time	Operation			Improvement points
			Internal	External	Waste	
3. Turn on the switch.	24:02		○			Attach a mold.
4. Remove the air hose.	24:12		○			Make into a one-touch
5. Adjust the air hose.	24:40				○	operation.

The team leader was a factory manager, and the other members were production team leaders and specialists who actually dealt with production technology and changeover. Once a week they left their posts for about two hours to observe setups in groups of four or five.

The team defined setup time as the time during which the old molds were removed, the new molds were attached, a test run was completed, and the first acceptable goods were produced. Ideally, operation analysis reviews the working conditions of the main operator and the machine separately, but in this case they usually combined the two and took time measurements of the setup conditions.

The team classified the setup steps that can't be done without shutting down the equipment as *internal setup*. This includes activities like removing and attaching dies. Setup activities that can be done while the equipment is in operation they called *external setup*. This includes activities such as preparing molds and transferring materials. The team counted extra consultations, searching time, and adjustments as wasteful activities.

The team members did time studies of the activities of the main operators, and recorded their operation analysis on plain paper. They were also able to use video cameras to document the setup operations. When the observers were new to the evaluation process, they worked in pairs, with one person watching the time and the other thinking of ways to improve. When the team consisted of three people, the third person analyzed the equipment. Using their observations, they created operator/machine charts to give depth to their analysis and noted possible points for improvement.

The team divided the results of the time observations into time for removal, attachment, positioning, standard setting, inspection, test runs, and adjustments. Since they also standardized the method for presenting improvement plans, they easily cut their setup time in half.

Step 4. Apply Waste Elimination Concepts

During their analysis, the team identified three types of waste, as shown in Table 6-2:

• waste in preparation

• waste in replacement

• waste in adjustment

Next they set specific goals for implementing improvements—not just "Cut setup and changeover time in half," but "Cut setup and changeover time from 30 minutes to 15 minutes." They took one step at a time. After deciding on each goal, they considered techniques for achieving it, thinking up ways to counter the three types of waste.

Waste in preparation consisted mostly of looking for, finding, choosing, and transporting parts and materials. "How can we stop doing this?" they asked themselves.

Waste in replacement consisted mostly of removing and attaching bolts. "Can't we fix the bolts in place?" they wondered.

Waste in adjustment occurred at changeover because of shifting standards. For example, when setting the feed pitch, operators usually make micro adjustments by looking at a graduated scale. In this case, however, the amounts on the scale were hard to read, so the operators tended to make micro adjustments on the computer without recording what they had done.

Table 6-2. Eliminating Waste in Setup and Retooling of Presses

Macro classification	Classification	Time	Type of waste	Ease or difficulty of waste elimination plan		
				A (minor improvements)	B (medium improvements)	C (major improvements)
Waste in preparation	Preparation	10%	1. Gathering tools 2. Transferring inspection tools 3. Transferring molds 4. Layout that makes setup and changeover difficult	Getting rid of waste during preparation 1. Plating the seven tools 2. Addresses for storing molds (color coding) 3. Setup and changeover stands	1. Special revolving carts for setup and changeover	1. Make the presses a U-line operaton

Waste in replacement	Removing and attaching	20%	*Molds*
			1. Bolt tightener
			2. Mold setup and changeover
			3. Positioning molds
			4. Variation in molds
			5. Tightening tools inadequate
			6. Use of L-wrench
			7. Replacing hoop materials
			8. Hoop cord undone
			
			Materials
				Waste during removal and attachment
				4. Independent tightener (with a spring)
				5. Male and female parts snap together
				6. Survey dimensions of molds
				  a. width × length
				  b. die height
				  c. tightening height
				  d. top mold attachment method
				  e. feed height
				  f. feed width
				  g. feed center
				  h. oil supply
				  i. chute
				  j. form, etc.
				7. Band type
					2. Have two types of tighteners
					3. Don't remove bolts; C-type washer
					4. Plan for mold standardization: compatibility of machine and mold, make while referring to chart
					5. Double work on hoop materials setting (external setup)
						2. Tighten with auto clamp
						3. Standardize partial rebuilding of molds
						  a. Issue mold design standards
						  b. Determine mold for the maximum dimensions of the press
						  c. When standardization is difficult, divide them into three groups

Waste in adjustment	Positioning, setting standards, trial runs, inspection, adjustment	65%	1. Stroke adjustment 2. Feed height adjustment 3. Feed amount adjustment 4. Waste due to moving the chute 5. Waste during inspection	Waste during adjustment (waste due to shifting standards) 8. Set in a mold and attach a spacer 9. Attach a chute to the mold, allowing sliding adjustments 10. Partly remove the stroke cover and make it immediately understandable on sight (insert a graduated scale). 11. Bins with drawers for parts	6. Gauge for use in designing stroke standards (set in mold) 7. Attach an oil sprayer to the mold 8. Look into a two-person setup system	4. Floating style design (cutting off shank) 5. Determine and practice the standards for one-person setup 6. Replace only the cavitated part
	Other	5%	1. Position of oil supply 2. Air 3. Configuration of electrical wires 4. Taking out materials			
	Total	32 min.				

99

Step 5. Deploy Improvement Plans

The secret of successful improvement is getting everyone to pitch in. Once the team had formulated specific plans for waste reduction and improvement, they divided them into three categories:

1. Minor improvements: those that can be implemented immediately
2. Medium improvements: those that will require some expenditure of time and money
3. Major improvements: those that will require equipment modifications and technology studies

They developed plans to implement the improvements from 1 to 3—the simplest to the most difficult. They felt that just implementing 1 and 2 would cut changeover time losses in half.

Once again, it was important for everyone to play a role. Group discussions led to plans like those outlined in the deployment charts shown in Figure 6-10, so everyone knew who would do what and by what deadline. With the goals and the implementation plans clarified, the team wrote them on large pieces of paper and posted them in the administrative area of each work site.

Figure 6-10. Setup and Changeover Improvement Plans

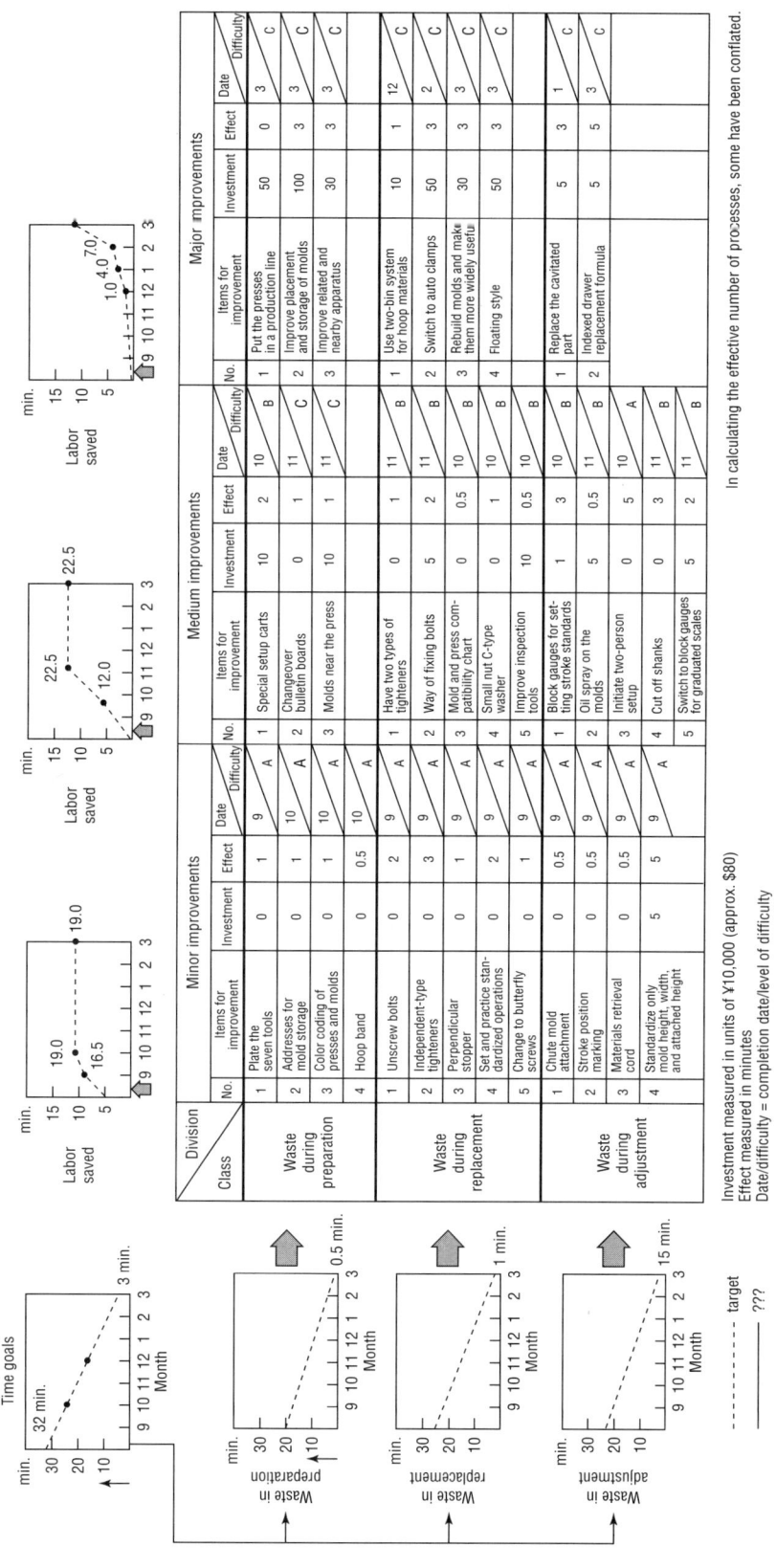

Investment measured in units of ¥10,000 (approx. $80)
Effect measured in minutes
Date/difficulty = completion date/level of difficulty

In calculating the effective number of processes, some have been conflated.

Step 6. Implement Improvements

Step 7. Evaluate and Spread Horizontally

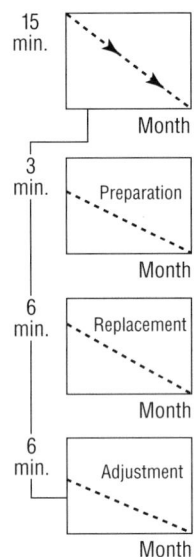

Once you have achieved zero setup
on a model production line, extend
it to other lines.

Once each improvement was achieved, the date was recorded and it was crossed off the list. The team leader offered advice on these matters. If implementation of an improvement was delayed, a special task force was delegated to take care of it. The majority of such delays are the responsibility of the production managers and plant manager. At small- and medium-sized companies and others that don't have a formal organization to direct improvements, these managers need to place orders and draw up contracts ahead of time so employees can create jigs and implements on demand.

Once zero setup has been achieved on a model production line, it should be extended to other lines. These are the basic procedures for moving toward zero setup.

Eliminating the Waste of Planned Downtime

7

Excessive Planned Downtime

Planned downtime refers to time when equipment in a processing line is not producing products because it is not needed for a particular process. This chapter presents a method for analyzing the efficiency of line layout and reconfiguring lines according to related product families so that individual equipment is used more productively. As we begin considering the subject, first take a look at Table 7-1, a monthly efficiency report broken down by production line, and Figure 7-1, an efficiency management graph. These reports use the standard work performed in any company as the standard for management. Unless setting the standard work indirectly forces you to expand your staff, these are not bad tools for providing a daily overview of each line's strengths and weaknesses in terms of efficiency.

In a market dominated by variable high-variety, small-lot production, these reports are essential tools for TPM, even if they tend to overstate costs. Setting the standard work allows you to take a good look around the work site.

The input for the monthly reports usually consists of the daily operations reports, such as Table 7-2, which classify, list, and compile employee time losses. The first aim is to determine which losses are happening at which management level and use that information to raise productivity. The second aim is to promote a reduction in operator-level time losses by increasing the operators' efficiency. An additional aim is to improve operator skills enough to allow prompt restoration of minor stoppages (lasting three minutes or less).

Still, when we look at the losses that are the responsibility of the coordinating manager, the companies that carry out these painstaking forms of management are precisely the ones most likely to neglect loss due to production plans (item 2 in the left column of Table 7-2). In other words, they ignore the very planned downtime that is the topic of this chapter.

Let's take company A as an example. If we walk around taking work samples from all the production lines there, we find more than 50 percent planned downtime. When we ask managers about the causes of planned downtime, the most common answer we hear is, "We've gone from producing mostly large-lot group A products to producing mostly small-lot group C products," while the second most common is, "Orders haven't been coming in." This is particularly true of factories that handle seasonal products.

At company A, the charge is ¥200 (approximately $1.60) per minute per production line. Since the line has already in effect been charged for the downtime, just

Table 7-1. Monthly Line Efficiency Report

Line	A Employee time on the job (min.)	B Actual employee time worked (min.)	A – B Time lost at employee level (min.)	G Test setup total (min.)	C Number of cases	Overall efficiency (C ÷ A)			Operating efficiency (C ÷ B)			Operating rate (B ÷ A)		
						This month	Last month	Highest value	This month	Last month	Highest value	This month	Last month	Highest value
1	140,880	133,870	7,010		169,943	▼120.6	127.5	127.5	▼126.9	135.5	135.5	△95.0	94.1	95.0
2	176,910	169,939	6,971		166,661	▼94.2	103.7	107.1	▶98.1	108.0	112.6	△96.1	96.0	96.1
3	161,220	155,820	5,400		130,234	▶80.8	93.7	108.0	▶83.6	100.2	119.3	△96.7	93.5	96.7
4	119,340	114,513	4,827		156,762	▼131.4	146.4	146.4	▼136.9	151.7	151.7	▶96.0	96.5	96.5
5	279,500	263,225	8,825	7,450	250,154	▶89.5	91.3	91.3	▶95.0	96.9	96.9	▶94.2	94.2	94.6
6	162,435	145,559	16,875		190,987	▼117.6	119.8	119.8	▼131.2	132.8	132.8	▶89.6	90.2	90.8
7	45,840	44,076	1,764		39,111	△85.3	84.8	99.9	▶88.7	89.0	104.2	△96.2	95.3	96.3
8	21,105	19,730	1,375		22,007	▼104.3	116.7	116.7	▼111.5	125.4	125.4	△93.5	93.1	93.9
9	173,164	158,264	6,940	7,960	149,557	▶86.4	93.0	93.0	▶94.5	100.6	100.6	▶91.4	92.5	92.5
10	161,390	156,104	5,286		150,489	▶93.2	97.0	106.0	▶96.4	100.2	110.3	▶96.7	96.8	97.1
11	189,515	183,790	5,725		211,331	△111.5	107.7	111.5	△115.5	110.6	115.0	▶97.0	97.3	97.4
12	75,480	72,687	2,793		75,002	△99.4	93.7	99.4	△103.2	96.4	103.2	▶96.3	97.2	97.5
13	102,905	99,377	3,528		117,737	△114.4	111.5	114.4	△118.5	115.8	118.5	△96.6	96.2	96.6
14	141,330	135,776	5,554		164,933	△116.7	116.3	116.7	△121.5	121.2	121.5	△96.1	96.0	96.6
15	124,045	117,297	6,748		140,625	△113.4	108.8	113.4	△119.9	115.2	119.9	△94.6	94.5	95.8
16	127,910	118,365	7,395	2,150	123,247	▶96.4	97.5	100.2	▼104.1	104.6	108.3	▶92.5	93.2	93.8
17	100,504	95,914	4,590		145,258	△144.5	128.0	144.5	△151.4	133.5	151.4	▶95.4	95.8	97.0
18	115,100	111,384	3,716		131,350	▼114.1	115.1	118.7	▼117.9	119.0	123.4	△96.8	96.7	96.8
19	84,444	81,249	3,195		96,552	▼114.3	119.2	128.3	▼118.8	123.8	133.8	▶96.2	96.3	96.7
20	83,220	76,791	6,429		96,970	▼116.5	117.8	117.8	▼126.3	131.8	131.8	△92.3	89.4	92.3
21	75,788	70,962	4,826		89,738	△118.4	113.7	118.9	△126.5	121.2	127.0	▼93.6	93.8	95.4
22	112,275	106,670	2,025	3,580	150,364	△133.9	132.2	133.9	△141.0	140.2	141.0	△95.0	94.3	95.0

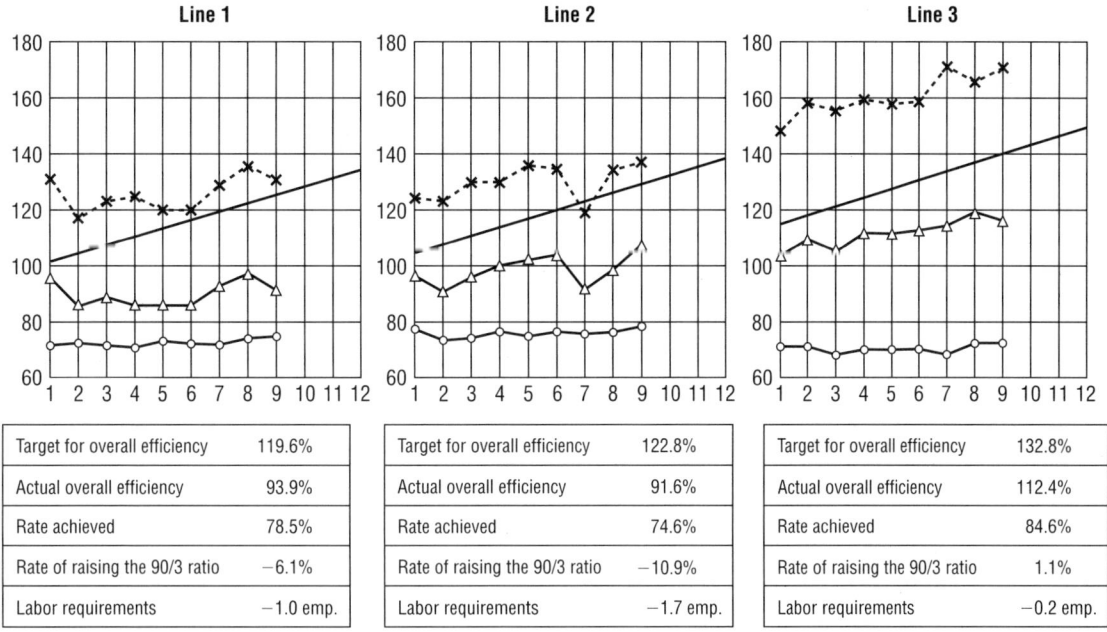

Figure 7-1. Efficiency Management Graph

Table 7-2. Classification of Employee Time Management Losses

Total losses for which management is responsible	Losses for which section managers and chief clerks are responsible	Losses for which team leader is responsible	Setup
1. Company and union events 2. Loss due to production plans 3. Other	1. Events in the workplace, meetings, consultations, morning assemblies, etc. 2. Flawed decisions for improved operations (others' responsibility) 3. Idleness due to late arrival of materials or parts 4. Other	1. Idleness due to equipment breakdowns 2. Idleness due to other causes 3. Flawed decisions for improved operations (their responsibility) 4. Remachining due to defects 5. Breaks 6. Chip cleanup 7. Measurements 8. Other	1. Setup 2. Blade replacement

ignoring the waste instead of razing the production process makes costs appear lower. The company ends up giving tacit consent to planned downtime of 50 percent or more.

Figure 7-2 shows a production line operating with 70 percent planned downtime. During normal production, one person operates about 50 machines. The takt time (daily required quantity divided by daily work time) is 720 seconds per unit, so

Number of machines: D: 28 units, M: 7 units, B: 6 units, I: 3 units, F: 3 units, W: 2 units, S: 1 unit, total: 50 units.

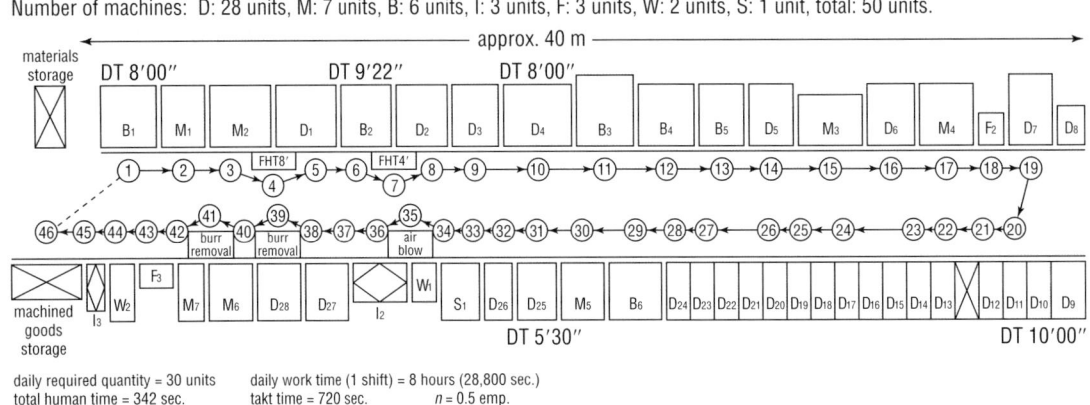

daily required quantity = 30 units   daily work time (1 shift) = 8 hours (28,800 sec.)
total human time = 342 sec.   takt time = 720 sec. n = 0.5 emp.

Figure 7-2. Production Line with 70 Percent Downtime

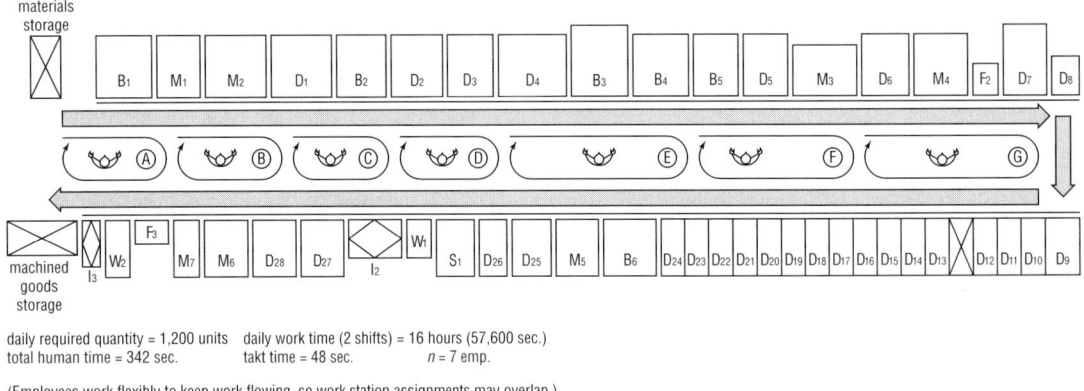

daily required quantity = 1,200 units   daily work time (2 shifts) = 16 hours (57,600 sec.)
total human time = 342 sec.   takt time = 48 sec. n = 7 emp.

(Employees work flexibly to keep work flowing, so work station assignments may overlap.)

Figure 7-3. Operating at Full Capacity with Seven People

each cycle is completed in 12 minutes, with one person running around doing all the work. At the busiest times, seven people operate the same machines, as shown in Figure 7-3, to meet a takt time of 48 seconds per unit. In this case, the line may be producing as much as it can.

Lines like this are particularly common in auto parts plants. This line was in fact being charged for its downtime, but we struggled with the question of whether it was better to leave the line the way it was or to raze the process to create more efficient lines from the same equipment.

Razing the Old Line to Build Something New

The cause of the planned downtime in Figure 7-2 is specialized production lines. There's no doubt that these are efficient when they're actually in operation, so many people believe they're efficient all the time. These days, however, you can never tell when you will end up manufacturing mostly small-lot group C products instead of large-lot group A products.

Companies often create specialized processing lines, invest millions in so-called automatic assemblers to create specialized assembly lines, install massive equipment, and institute large-scale production. Is this a good idea? Let's reconsider this question, focusing on waste due to planned downtime.

Test Yourself

Figure 7-4 is a chart analyzing the processing paths at company A. Take a look at this diagram, then answer a process design question: How many production lines should the company create?

Our Answer

Most people say between two and four lines, with three being the most common answer (see Figure 7-5).

Actually, our answer is two lines. F1 is fine with just F1 processes, but we design the process so that the F2 line makes part #26 and can flow all the way down to #22, so we create it out of the L, VM, Ke, Bo, S, M, B, Te, and inspection machines. Try to sketch out how the layout for the second line would look, then look at Figure 7-8 on page 116 to compare our solution.

If you can increase the number of common-use lines, working more of them into the budget over time, you'll find that planned downtime no longer occurs. Even if your product line changes, these general-purpose machines can be rearranged at any time.

No.	Part	Quantity	1 (L)	2 (VM)	5 (M)	10 (Ke)	20 (Bo)	50 (S)	100 (B)	200 (Te)	500 (C)	1000	2000	Process series no.*	Total operating hours
16	Cover	1												81	12.0
17	Cover	1												80	9.0
18	Sleeve	1												611	34.5
19	Spacer	1												611	12.0
20	Washer	1												1	3.0
21	Lever jig	2												501	15.0
22	Lever	1												330	57.5
23	Stopper	1												261	7.5
24	Dog	1												161	13.5
25	Block	1												332	24.0
26	Frames	1												266	13.5
27	Subtotal														201.5

* This system assigns a unique number to each machine in a line that processes different types of parts. For each type of part, the unique numbers for the machines used to make it are added together; the result is a unique process series number that identifies the part and tells in shorthand what machines are used. —Ed.

Figure 7-4. Process Path Analysis for Company A

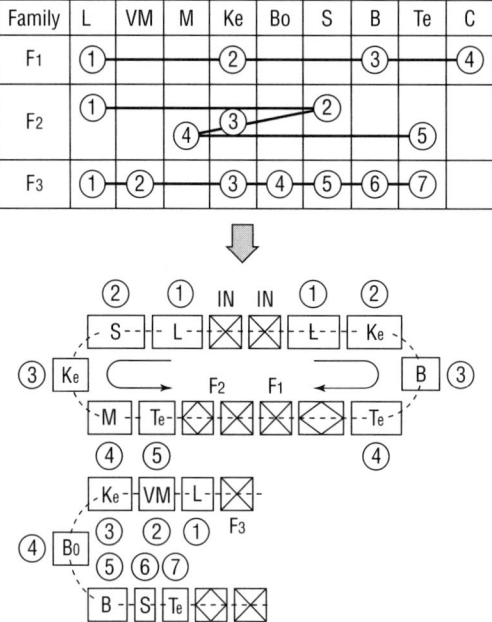

Figure 7-5. A Typical Answer

Steps for Process Design

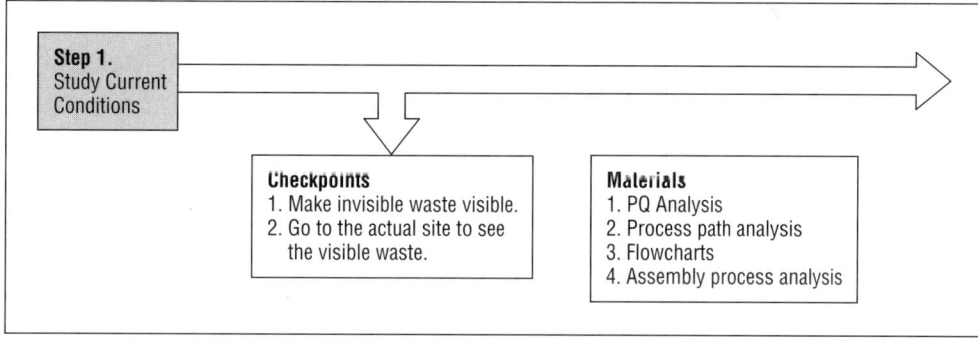

Process design is also called process creation. It consists of a series of techniques, so we've broken it down into steps. Start by going back to the beginning and taking another look at your current situation. That's because the current state of your plant and procedures is rife with fundamental forms of waste.

Step 1 asks you to identify invisible forms of waste with these instruments:

1. PQ analysis chart
2. Process path analysis
3. Flowcharts
4. Assembly process chart

Let's start with simple explanations of PQ analysis and processing path analysis.

PQ Analysis

A PQ analysis like the one shown in Figure 7-6 allows you to see the current status of wide-variety, small-lot production. The "P" refers to "parts"—the different kinds of products made. The "Q" refers to the quantity of each different product type. Think of it as a Pareto-chart ranking of the amounts produced of each type of product. It's based on the information in Table 7-3.

To use this chart, we take the top 20 percent of the number of product types and see whether it corresponds to 80 percent of the amount. This is the 2:8 line. We also draw 3:7, 4:6, and 5:5 lines in the same way.

> 2:8 reflects large-scale production and Pareto's Law (a few types account for most of the quantity).
>
> 3:7 is close to Pareto's Law and still falls under the classification of large-scale production.
>
> 4:6 signifies a move toward wide-variety, small-lot production (the most common form of manufacturing in Japan).
>
> 5:5 indicates production based entirely on wide variety and small lots.

Table 7-3. Product/Quantity Table

Line	Type of product	Quantity	Cumulative total	Cumulative total percent	Line	Type of product	Quantity	Cumulative total	Cumulative total percent
1	P1	2,180	2,180	8.3	17	P17	680	22,850	86.7
2	P2	2,100	4,280	16.2	18	P18	680	23,530	89.3
3	P3	2,060	6,340	24.1	19	P19	410	23,940	90.8
4	P4	1,780	8,120	30.8	20	P20	310	24,250	92.0
5	P5	1,460	9,580	36.3	21	P21	300	24,550	93.1
6	P6	1,320	10,900	41.4	22	P22	300	24,850	94.3
7	P7	1,310	12,210	46.3	23	P23	300	25,150	95.4
8	P8	1,300	13,510	51.3	24	P24	280	25,430	96.5
9	P9	1,290	14,800	56.1	25	P25	200	25,630	97.2
10	P10	1,280	16,080	61.6	26	P26	200	25,830	98.0
11	P11	1,200	17,280	65.6	27	P27	150	25,980	98.6
12	P12	1,200	18,480	70.1	28	P28	130	26,110	99.1
13	P13	1,030	19,510	74.0	29	P29	100	26,210	99.4
14	P14	1,000	20,510	77.8	30	P30	100	26,310	99.8
15	P15	970	21,480	81.5	31	P31	50	26,360	100.0
16	P16	690	22,170	84.1					

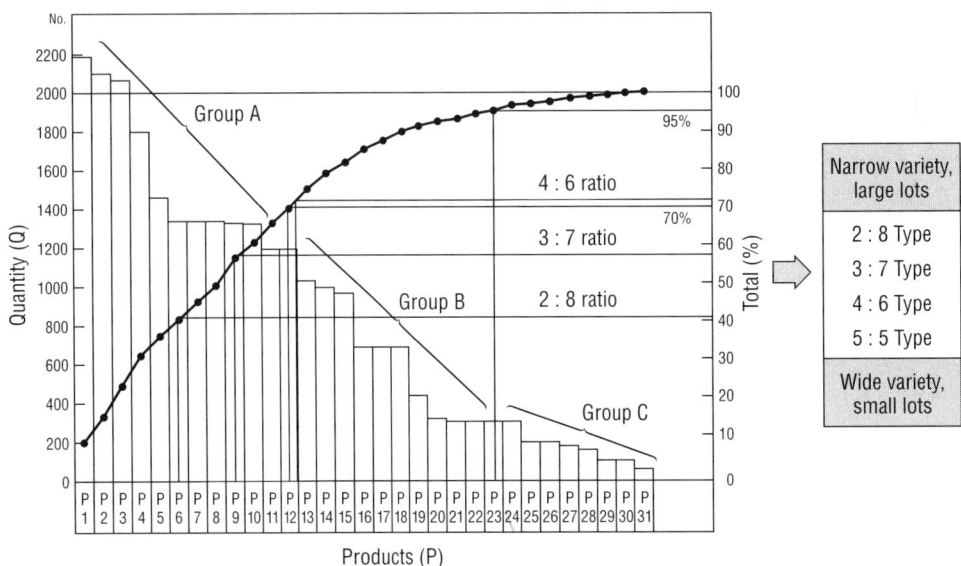

Figure 7-6. PQ Analysis

Process Path Analysis

Take another look at Figures 7-4 and 7-5. Companies that make a lot of products may end up creating a separate production line for each product, and there never seem to be enough machines to handle everything. That's why we need to use process path analysis to group similar products into families. This allows us to treat the manufacture of each family of products as large-scale production, and we end up with fewer lines. It's no exaggeration to say that the quality of process design is reflected in the designer's success at dividing the products into families.

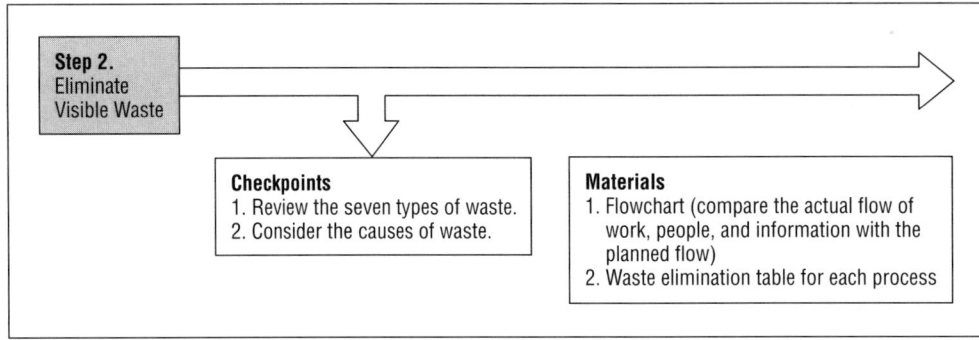

Step 2 amounts to a rejection of the current situation and an elimination of the waste that is easily visible. It requires you to visit the plant, mentally discard the current way of doing things, and rethink everything from scratch. (Of course, you don't discard your goals.) Table 7-4 summarizes the seven most important types of waste to look for as you use this approach.

Table 7-4. The Seven Types of Waste and Their Causes

Waste (ranked by importance)	Cause
1. Planned downtime (idle 50 to 70 percent)	Using specialized lines for group C products (very low quantity)
2. Minor, medium, and major stoppages	Ignoring minor stoppages and not seeking out their causes
3. Long setup time	Making setup improvements that do not last or do not go far enough
4. Manual rework, defects, poorly manufactured goods	Performing changeover with shifting standards; continuing production without fully understanding the causes of minor adjustments and defects
5. Time wasted wandering around searching for things	Failing to set standard positions and standard labels; performing insufficient pre-setup
6. Overproduction	Using large equipment, which often breaks down and takes a long time to repair
7. Idled workers	Accepting processing problems (defects, missing parts, changeover); unmotivated attitude

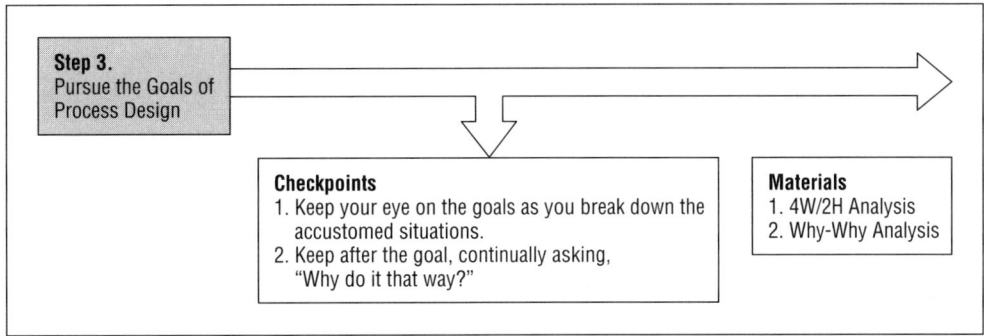

Ask yourself why you've let things remain as they are, despite 50 to 70 percent planned downtime. The best method for getting to the bottom of this phenomenon is to ask the Four Ws and Two Hs as in Table 7-5, and then to apply Why-Why Analysis as in Figure 7-7.

Table 7-5. Four Ws and Two Hs Analysis

Who?	Operator M alone
What?	Operating 50 machines
When?	Often finishes by noon; not enough for a full day's work.
Where?	On the A line, with specialized jigs for use with group C products
How?	Working alone in a U-shaped line
How much does it cost?	The machines are out of service at least 70 percent of the time, but since everything is charged against the line, the "cost" is zero. However, since setup takes about an hour, it costs almost $100 extra per day ($1.60 per minute x 60 minutes). Even more waste is traceable to the failure to use equipment that could be profitably producing.

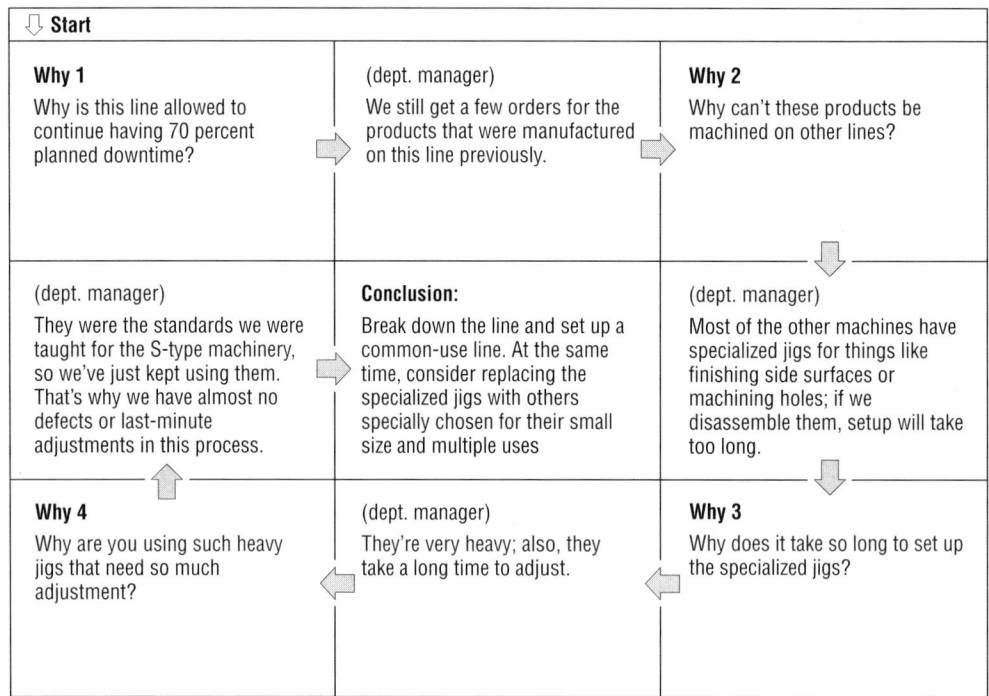

Figure 7-7. Why-Why Analysis of Planned Downtime

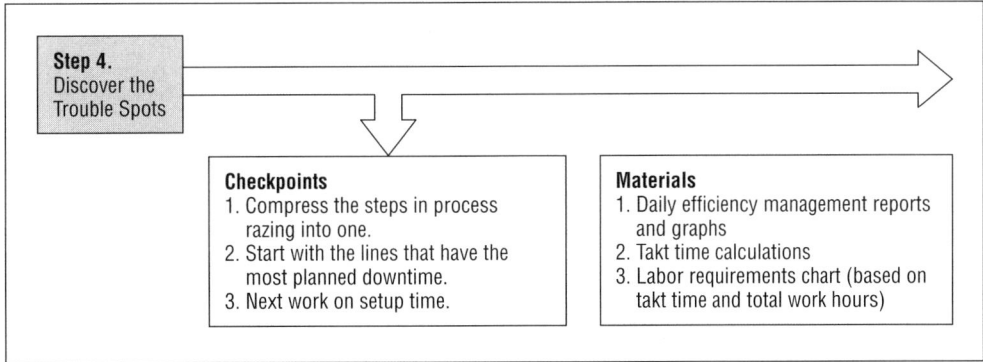

In Step 4, we find the trouble spots among the seven types of waste. The first trouble spot is planned downtime. The second trouble spot is setup. The third trouble spot is minor stoppages.

Defects, manufacturing errors, and the constant need for minor adjustments are unmistakable signs of trouble. Start with the phenomena that are most readily visible in the workplace.

1. Take another look at Table 7-1 and Figure 7-1 to learn more about efficiency management.

2. Calculate takt time (number of minutes or seconds in which one unit must be completed):

$$\text{takt time} = \frac{\text{daily work time}}{\text{daily required quantity}}$$

3. Use takt time and the total human work time per piece to determine the number of personnel needed:

$$\frac{\text{total human work time per piece}}{\text{takt time}} = n \text{ employees}$$

The purpose is to determine the most appropriate number of operators per line.

4. Sometimes you end up with a fraction for the number of employees (taking company A as an example):

$$\frac{110 \text{ sec.}}{60 \text{ sec.}} = 1.8 \text{ people}$$

In this case you tentatively end up with two people. If you promote simple automation, however, you can reduce the operator time required.

$$\frac{80 \text{ sec.}}{60 \text{ sec.}} = 1.3 \text{ people}$$

By further eliminating wasted motions and the like, you can reduce this to one person.

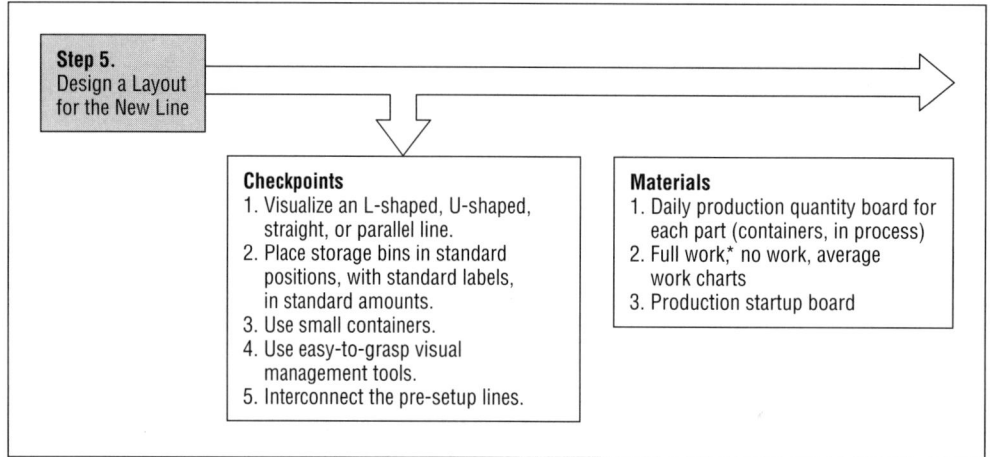

Step 5. Design a Layout for the New Line

Checkpoints
1. Visualize an L-shaped, U-shaped, straight, or parallel line.
2. Place storage bins in standard positions, with standard labels, in standard amounts.
3. Use small containers.
4. Use easy-to-grasp visual management tools.
5. Interconnect the pre-setup lines.

Materials
1. Daily production quantity board for each part (containers, in process)
2. Full work,* no work, average work charts
3. Production startup board

* "Full work" refers to a situation in which two processes are interconnected so that process 1 shuts down when a predetermined quantity of its WIP output is lined up waiting for process 2. "No work" refers to the reactivation of process 1 when the WIP waiting for process 2 falls below a predetermined quantity.—Ed.

The lines remade through process design can take any shape: L-shaped, U- shaped, straight-line, or parallel lines. You can create a graphic image of the entire process, including the constraints on the layout, work, or machines.

To optimize machine operation, it's best not to create a layout in which the machines are just lined up in row. Furthermore, if you expect to move soon from group A products (narrow variety, large lots) to group C products (wide variety, small lots), avoid creating specialized lines and concentrate on common-use lines.

Include the storage bins in your drawing, making sure they are in standard positions with standard labels and with standard amounts. Also include production startup boards, signposts, and other easy-to-grasp visual management tools. Of course, be sure to set up the machines in accordance with the families created during process path analysis.

When we gave the process design quiz, earlier in this chapter, we omitted the details of our answer. Figure 7-8 shows how we would set up the F2 line. How does this design compare with yours?

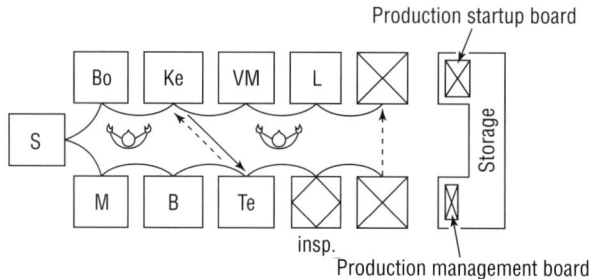

Figure 7-8. The New F2 Line

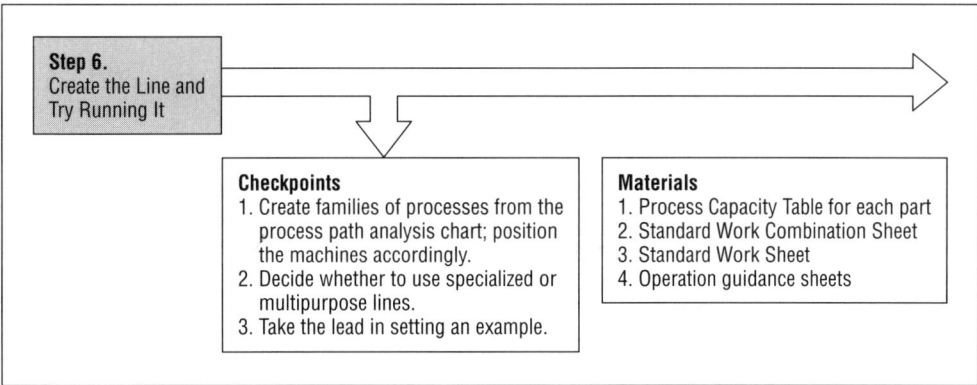

Once the line is in place, the supervisors should take the initiative to demonstrate the new system to the operators. They may find it difficult, having been off the production line for a while, but they can practice after hours and learn to pull off a demonstration that will leave the operators exclaiming in admiration.

Unless you have put together an effective combination of people, things, and machines, you'll end up with a wasteful production line. To avoid this, you need to standardize the work sequence, cycle time, and amount of inventory into what is called "standard work." To develop standard work, you will use three important instruments:

• A process capacity table for each part or component

• A standard work combination sheet that shows time required for manual and machine work

• A standard work sheet that diagrams the flow of the process from station to station

Table 7-6 shows how these charts were filled in for a pinion machining line.

For a new production line to function smoothly, the operating directions also should be summarized on an operation guide sheet before being publicly implemented.

Table 7-6. Process Capacity Table, Standard Work Combination Sheet, and Standard Work Sheet

Revised___/___/___   Page___of___

Approvals			Process Capacity Table	Part no.	41211-20092	Model	RY	Position	Name
								432	HS
				Part name	8-inch pinion	Quantity	1	442	NK

	Process	Machine no.	Basic time (min./sec.)			Cutting tools		Machining capacity (1,294)	Time display; notes
			Manual operation time	Automatic feed time	Completion time	Number replaced	Replacement time		Key: —— = manual work
									- - - = automatic feed
	Taking raw materials	—	1		1			—	
1	Gear cutter, rough cutting and polishing	CC614	5	38	43	300	2'30''	1,324	5'' 38'' —— - - - - -
2	Gear cutter, small edge chamfering	CH228	6	7	13	2,000	1'00''	4,330	6'' 7'' —— -
3	Gear cutter, lead surface finishing	GC1444	6	38	44	300	2'30''	1,294	6'' 38'' —— - - - - - .
4	Gear cutter, rear surface finishing	GC1445	6	30	36	300	2'30''	1,578	6'' 30'' —— - - - -
5	Pin diameter measurement	TS1100	7	3	10			5,760	7'' 3'' —— - -
	Completed item		1		1			—	

Part number and name	41211-20092	Standard Work Combination Sheet	Date created		Daily quota	1,252 units	Key:
Process			Position		Number needed	___min.46 sec.	—— = manual work / - - - = automatic feed / ∿∿ = walking around

	Name of operation	Time (min./sec.)		Operation time (in seconds)
				6'' 12'' 18'' 24'' 30'' 36'' 42'' 48'' 54'' 1' 1'06'' 1'12'' 1'18'' 1'24'' 1'30'' 1'36'' 1'42'' 1'48'' 1'54'' 2'
1	Retrieve raw materials	1		
2	Gear cutter, rough cutting and polishing	5	38	
3	Gear cutter, small edge chamfering	6	7	
4	Gear cutter, lead surface finishing	6	38	
5	Gear cutter, rear surface finishing	6	30	
6	Pin diameter measurement	7	3	
	Completed item	1		

CT

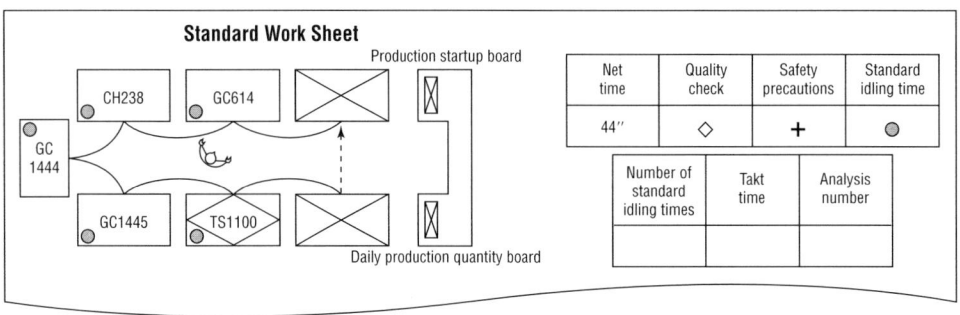

Standard Work Sheet

Production startup board

Net time	Quality check	Safety precautions	Standard idling time
44''	◇	+	●

Number of standard idling times	Takt time	Analysis number

Daily production quantity board

Eliminating Abnormalities within the Process

Many minor stoppages and product quality defects can be traced to abnormalities within the manufacturing process. If we use standard materials and work under standard conditions, our machines won't break down and cause defects. When we use materials of varying quality and work under inconsistent conditions, however, our machines run into trouble.

The brief definition of an abnormality is something that deviates from the standard. If the work, molds, jigs, and other machine-related aspects of the process deviate from the established standards, you'll get increasing numbers of minor stoppages and product defects. Thus, if you can reduce the machine-related abnormalities within the process, the machine will run under favorable conditions, breakdowns and defects will decrease, and the useful life of the machine will be extended.

The top graph in Figure 8-1 shows how defects within the process gradually decreased machine productivity and hastened the end of their useful life. By contrast, the bottom graph shows the result when machines are improved and used in the most efficient way. This reflects the usual situation in Japan's larger companies and in the best of the small- and medium-sized companies.

This chapter focuses on machine-related processing abnormalities and ways to determine the optimal processing conditions for avoiding failures and defects. The examples in the following sections come from the printed circuit board mounting line at company A.* Most of the defects in printed circuit board processing occur during the solder dip stage. The cause of these defects is failure to adjust the conditions in the massive solder tanks according to the size or density of the circuit board being dipped. We find up to 100 percent defects in nearly any plant we visit.

Example: Decreasing Soldering Defects in Printed Circuit Boards

Printed circuit boards fall into three general categories based on size. For purposes of manual processing, we further classify the boards according to the density of their parts (another set of three categories). Crossing these categories makes nine possible families of boards. We then determine the optimal dip conditions for each family; by maintaining these conditions, we can cut the current level of defects in half. The following sections outline the steps we have developed for determining optimal conditions.

---

* For another look at this example from a cell design perspective, see Ken'ichi Sekine, *One-Piece Flow* (Productivity Press, 1992).

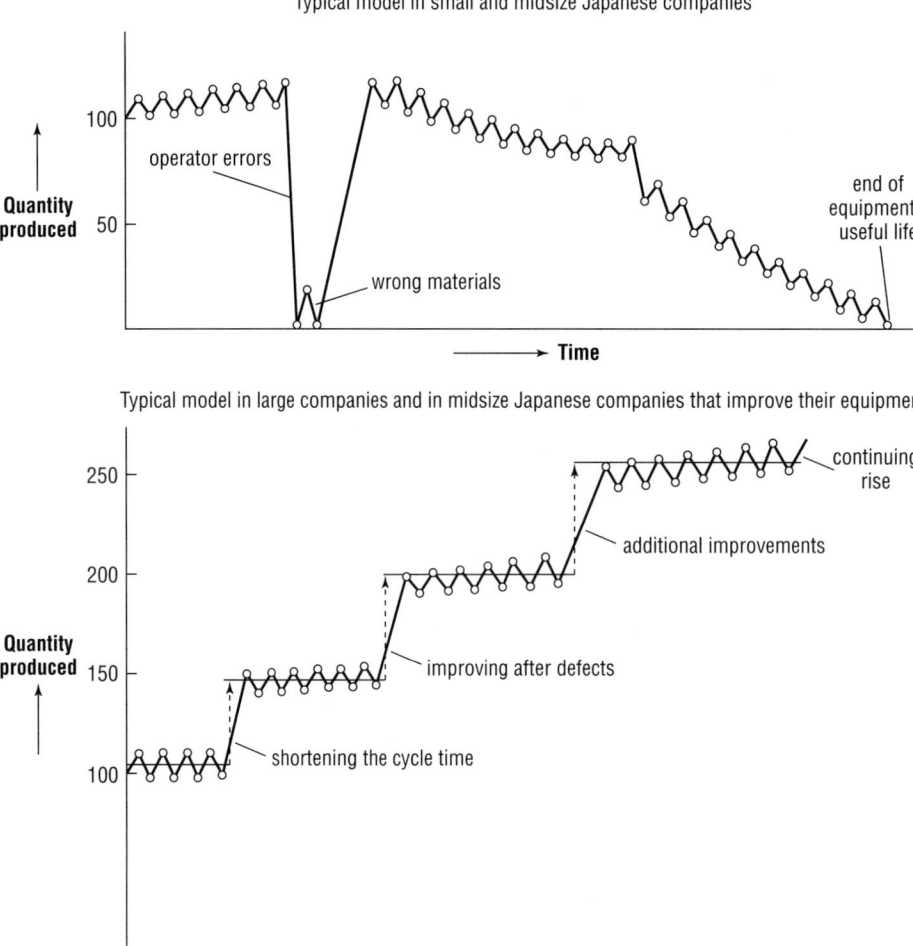

Figure 8-1. Changes in Quantities Produced by Machines and Equipment

Step 1. Study Current Conditions

First, make a chart like Table 8-1, which summarizes the defects. This way you can see what is actually going on with the different kinds of defects in the three size categories. Table 8-1 shows that

• The densest circuit boards have the most defects.

• Tunnels in the solder are the most common defect.

Next, with this data in hand, we head for the work site to observe the printed circuit boards themselves. As a result, we learn that

• Lifting of parts is most common with connectors and rare with integrated circuits and capacitors.

• Almost all the parts that exhibit tunneling are parts that are inserted by hand.

Table 8-1. Summary of Defects by Circuit Board Model and Location

Circuit board size	Number produced	Average insertion points per part	Defects / Area	Bridge Uniform	Bridge Random	Tunnel Uniform	Tunnel Random	Lifting Uniform	Lifting Random	Insufficient solder Uniform	Insufficient solder Random	Total Uniform	Total Random
A (small)	50	50	Top left		1								1
			Top right				1						1
			Lower left	3								3	
			Lower right										
B (medium)	30	80	Top left			10				2		12	
			Top right			20						20	
			Lower left			5						5	
			Lower right			20						20	
C (large)	60	160	Top left		4	40	10					40	14
			Top right			5	15	2				7	15
			Lower left				5	5				5	5
			Lower right			30	30	5				35	30
Total	140			3	5	130	61	12	0	2	0	147	66

Step 2. Create Families of Circuit Boards

Step 1 involved looking at defects based on circuit board size. In Step 2 we classify the circuit boards into nine families based on the possible combinations of size (a) and density of their components (b), as in Table 8-2. Then we perform PQ analysis to rank the parts within each family in the order of the quantity manufactured (see letters noting product groups A, B, and C in Table 8-2). We then circle the items in each family that display the fewest defects.

In Table 8-2, we looked at the density of all the parts mounted on the circuit board, but since most of the defects found at company A were in the form of tunnels, we probably should have used only the densities of the parts inserted manually. As you can see, there is some trial and error in this approach.

Step 3. Investigate the Primary Factors for Solder Dip Defects

The best way to deal with the primary factors is to find blind spots in the way things are done in the workplace. One method is to create a causal factors diagram of distinguishing factors, by making a hierarchical chart of primary factors like Table 8-3, then using it to draw the main "ribs" and smaller branching bones of the diagram. Finally, we determine the smallest branches by asking the operators for their input. Based on these results, we circle what we think are the most important causes of defects. Table 8-4 is a control factor chart that summarizes the factors we have found.

Table 8-2. Families of Circuit Boards

Circuit board size \ Density of parts	Low (β_1)	Medium (β_2)	High (β_3)
A (small − α_1)	A 12845 150 A 12643 120 (A 12342 70) A 13651 40 A 13655 40	A 12963 80 (A 12865 70) A 13642 60 A 13688 60 A 13901 30	A 14621 120 A 14930 100 (A 14650 60) A 14989 20 A 14993 20
B (medium − α_2)	B 02561 80 B 10541 60 B 11231 40 (B 10623 30) B 11986 20	B 10356 70 B 10457 50 B 03563 30 (B 11245 20) B 12003 20	B 13653 40 (B 13352 20) B 13251 20
C (large − α_3)	C 10051 120 C 10063 100 C 10541 30 (C 10627 20) C 10932 20	C 08765 30 (C 10161 20)	(C 12602 20)

Circled items = boards with fewest solder defects

Table 8-3. Causal Factors Table

Primary factors	Secondary factors	Tertiary factors
Line N, from hand insertion to sight inspection	(divide into elemental operations) Circuit board attachment Transport Flux Dipping once Cutting	(method for each elemental operation) E.g., attaching parts by hand, sliding them to the left with the right hand

Table 8-4. Control Factors

Symbol	Factor	Levels
A	Brand of flux	Current A_1 : Company A_2
B	Amount of flux (time)	Current amount B_1 : Amount B_2
C	Flux consistency	Current consistency C_1 : C_2
D	Flux temperature	Current temperature D_1 : D_2
E	Brand of solder	Current E_1 : Brand E_2
F	Amount of solder	Current F_1 : F_2
G	Solder temperature	Current G_1 : G_2
H	Soldering method (Two-stage formula: solder is jet-sprayed twice to get rid of the air that accumulates at the end of the part in the direction of the spraying.)	Jet single-stage H_1 : Two-stage formula H_2
J	Solder replacement frequency	Once a day J_1 : Twice a day J_2
K	Circuit board attachment method (This factor may be a block factor, but depending on the circuit board attachment method, the solder sometimes flows in the working direction.)	Uncontrolled flow attachment K_1 : Controlled flow attachment K_2
L	Preheater temperature	Current L_1 : New temperature L_2

Step 4. Look for Optimal Conditions While Carrying Out Daily Production

There are two methods of looking for the optimal conditions:

• The ordered factor elimination method

• The orthogonal method

Table 8-5. Ordered Factor Elimination Method

Factor	Brand of flux	Amount of flux	Temperature	Height of liquid surface	Defect data
No. Symbol	A	B	C	D	
1	1	1	1	1	y_1
2	2	1 (F)	1 (F)	1 (F)	y_2
3	2	2	1 (F)	1 (F)	y_3
4	2	1 (restore)	2	1 (F)	y_4
5	2	1 (F)	1 (restore)	2	y_5

The ordered factor elimination method begins with finding the optimal conditions for each of the nine families individually in the course of daily operations. We focus first on finding the principal factors for the most common defect, tunneling. Table 8-5 shows how this method is applied. In row 1, we gather defect data (y_1) under current conditions in the process.

In row 2, we change the first factor (brand of flux) to level 2, keeping the other factors the same, then take defect data again (y_2). If y_2 is better, in row 3 we then change the second factor (amount of flux) to its level 2 and obtain data y_3 in a similar fashion. We compare y_2 and y_3. From there on, we continue comparing combinations of factors to obtain the optimal values.

We present only a brief outline of the orthogonal method. Again, we think of the current situation in terms of two levels for each factor. We then combine the factors in an orthogonal array, as shown in Table 8-6 (note that you may need to use different arrays for different types of factors). For example, we create two-level images of soldering by the wave soldering method (level 1 in the first column) and by the reflow method (level 2 in the first column).

Furthermore, we seek the optimal conditions for levels 1 and 2 of the amount of flux and the conveyor speed (time) in the course of carrying out daily manufacturing operations. If things aren't clear after using an L_8 array, we can use a larger array. Design of experiment methods such as this use orthogonal arrays to greatly reduce the number of experiments required to test different combinations of factors. Through this approach, we can understand most of the causes of the defects and determine the optimal conditions.

Company A was processing all printed circuit boards in their large-scale solder dipping tanks under one set of fixed operating conditions, so it's no wonder they were getting up to 100 percent defects. The first task in improving this situation was to reduce defects by 50 percent in the course of daily manufacturing operations. To achieve this, we applied the steps described in the next section, which are easy to understand and have proved their value over time. They allow you to reduce the number of defects in the process and eliminate minor stoppages. The useful life of your equipment will increase, and productivity will rise dramatically.

Table 8-6. Orthogonal Array

Date	Test No.	Solder method	Amount of flux	Conveyor speed (time)	Amount of flux	Conveyor speed (time)
		A	B	C	D	E
5/11	1	1 Wave method	1	1	1	2
12	2	1	1	2	2	1
13	3	1	2	1	2	1
14	4	1	2	2	1	2
5/17	5	2 Reflow method	1	1	2	2
18	6	2	1	2	1	1
19	7	2	2	1	1	1
20	8	2	2	2	2	2

Testing for Optimal Conditions for Circuit Board Soldering

Step 1. Perform Experiments for Finding Optimal Conditions

Company A produces a wide variety (60 types) of printed circuit boards in small lots (average size 20 units); Figure 8-2 shows the PQ analysis for these products. If we define a defective circuit board as one that needs even one touchup operation, company A's products are coming out almost 100 percent defective. That's why we decide to analyze the effects of different levels for several factors, as shown in Table 8-7.

Step 2. Summarize the Data

We look for the optimal conditions to determine which of the factors in Table 8-7 has the greatest effect. We do this by adding the total defects counted for the two levels of each factor. Our findings are summarized in Table 8-8.

Step 3. Put the Findings into Graphic Form

After summarizing the findings, we graph the results, as in Figure 8-3.

Step 4. Identify the Effective Factors and Levels

As Figure 8-3 shows, factor *D*, the preheater temperature, has the greatest effect (the largest difference between high and low defect counts for the two levels). Focusing on the lower values (fewer defects) for each factor, we see the results shown in Table 8-9.

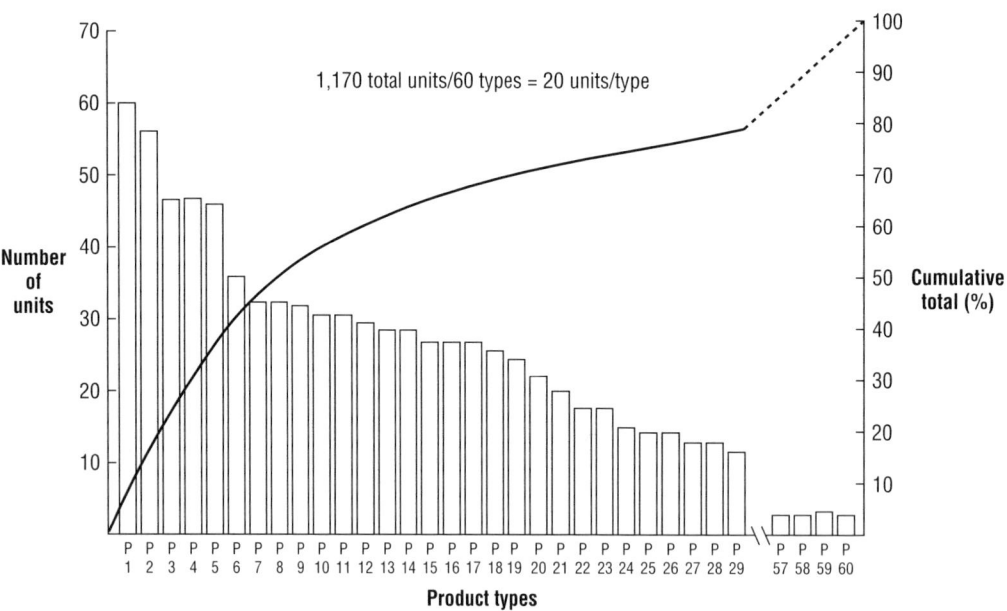

Figure 8-2. PQ Analysis

Step 5. Estimate the Defect Rate under Optimal Conditions

As we saw in Figure 8-3, the optimal conditions are A_1, B_2, C_1, and D_1; the difference between levels for factor E was not as great, but E_2 was slightly better than E_1. For these factors levels, the estimated value ($\hat{p}$) of implementing A_1, B_2, C_1, D_1, and E_2 is

$$\hat{p} = \frac{\left(\frac{1}{A_1}-1\right)\left(\frac{1}{B_2}-1\right)\left(\frac{1}{C_1}-1\right)\left(\frac{1}{D_1}-1\right)\left(\frac{1}{E_1}-1\right)}{\left(\frac{1}{T}-1\right)^{n-1}}$$

If we take the computed values and the data obtained from Steps 1 and 2 and substitute the defect rates of A_1, B_2, C_1, D_1, E_2, and T, we get

$$\hat{p} \doteqdot 0.041$$

In other words, we should be able to get solder dip defects down to 4.1 percent by using the lower levels of the five factors. After that, we repeat the testing process a number of times to bring the defect rate down to 0.2 percent.

Step 6. Test with a Confirmation Experiment

Even if we can't achieve the value we have estimated is optimal for the conditions, we can adjust the levels, confirm them, and then try actually working under those conditions. If we don't even get close to the estimated value ($\hat{p} = 0.041$), then some factor or level other than the ones we tested must be contributing to the defects. The thing to do then is to plan a second set of on-site experiments.

We believe that if you find the optimal conditions through unhesitating application of these steps, you will be able to cut your defect rate in half. Once you have

Table 8-7. Experimental Data for Finding Optimal Dip Conditions

Sample: Number of defective points on 50 printed circuit boards

Factor / Test no.	A Solder formula sp gr	B Flux temp.	C Conveyor speed	D Preheater temp.	E Solder temp.	P1 K1 S	P1 K1 U	P1 K1 I	P1 K2 S	P1 K2 U	P1 K2 I	P1 K3 S	P1 K3 U	P1 K3 I	P2 K1 S	P2 K1 U	P2 K1 I	P2 K2 S	P2 K2 U	P2 K2 I	P2 K3 S	P2 K3 U	P2 K3 I	P3 K1 S	P3 K1 U	P3 K1 I	P3 K2 S	P3 K2 U	P3 K2 I	P3 K3 S	P3 K3 U	P3 K3 I	Total
1	new	current	current	current	current																18									8			26
2	new	current	−α	+α	+α						22										17		32						14				85
3	new	+α	−α	current	+α																25								7				32
4	new	+α	current	+α	current									6							16									10			32
5	old	current	current	current	+α				5			8									23									12			48
6	old	current	−α	+α	current				7			24						6	28		19		22			18	4		21	14	25		188
7	old	+α	−α	current	current																21									28			49
8	old	+α	current	+α	+α							6			9			7			23						6			7			58
Total									12		22	38		6	9			13	28		162		54			18	10		42	79	25		518

Defective items: **S** = short **U** = unsoldered **I** = insufficient solder

Locations of defects on circuit board

	P1	P2	P3
K1			
K2			
K3			

Table 8-8. Summary of Findings

A: Solder formula specific gravity
1. New $(1 + 2 + 3 + 4) = 26 + 85 + 32 + 32 = 175$ 2. Old $(5 + 6 + 7 + 8) = 48 + 188 + 49 + 58 = 343$
B: Flux temperature
1. Current $(1 + 2 + 5 + 6) = 26 + 85 + 48 + 188 = 347$ 2. $+\alpha$ $(3 + 4 + 7 + 8) = 32 + 32 + 49 + 58 = 171$
C: Conveyor speed
1. Current $(1 + 4 + 5 + 8) = 26 + 32 + 48 + 58 = 164$ 2. $-\alpha$ $(2 + 3 + 6 + 7) = 85 + 32 + 188 + 49 = 354$
D: Preheater temperature
1. Current $(1 + 3 + 5 + 7) = 26 + 32 + 48 + 49 = 155$ 2. $-\alpha$ $(2 + 4 + 6 + 8) = 85 + 32 + 188 + 58 = 363$
E: Solder temperature
1. Current $(1 + 4 + 6 + 7) = 26 + 32 + 188 + 49 = 295$ 2. $+\alpha$ $(2 + 3 + 5 + 8) = 85 + 32 + 48 + 58 = 223$

cut your process-related defects in half, your minor stoppages rate will also fall by half.

Unless your machines and equipment are truly wrecks ready for the scrap heap, operating them with uniform materials under uniform operating conditions should help you avoid defects and stoppages. Eliminating processing abnormalities is the first step in improvement.

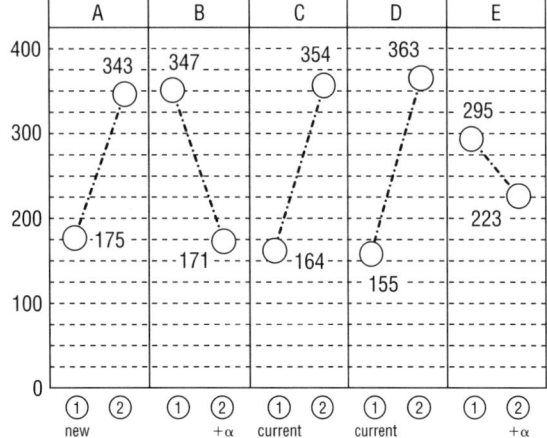

Figure 8-3. Graph for Each Factor

Table 8-9. Optimal Conditions

A: Solder formula specific gravity	new	(A_1)
B: Flux temperature	$+\alpha$	(B_2)
C: Conveyor speed	current	(C_1)
D: Preheater temperature	current	(D_1)
E: Solder temperature	$+\alpha$	(E_2)

Promoting Daily Equipment Inspections

9

The remarkable increases in automation and acceleration of factory equipment in recent years have underscored the importance of increasing the operating rates of these machines. Yet when we go out to the factory floor and observe the activities for a day, we often see that operating rates actually have decreased. We can easily fix the minor stoppages, but it's not unusual for machines to exhibit first medium-length stoppages and then major stoppages before breaking down completely.

These problems arise because we use the equipment until it breaks down, without ever conducting daily spot checks. If minor stoppages occur, we make a slight adjustment without really investigating the causes, or else we just run the machine at a slower speed. When we do this, the apparent operating rate rises, but the amount produced decreases. Since we use the machines in a partly broken down state, the standards are out of whack, defects occur, and the yield keeps decreasing.

Even if we decide that we can't use a certain machine anymore, we can't always scrap it, because its depreciation period may not be up yet. At many plants, inoperative machines just stand around like extra furniture, getting in the way of daily work. One way to introduce the concept of daily inspections into such a plant is to start with minor stoppages, implement minor improvements, and then go after the causes of their causes.

If all we do is conduct spot checks under current conditions, we will keep discovering previously undetected mechanical problems and keep trying quick and dirty ways to fix them. With this approach, the efficiency of our spot checks and the effectiveness of our machines only worsen. That's why we introduce the following method for promoting daily inspections.

Introducing Daily Inspections and Implementing a Program

Step 1. Summarize the Problems at Each Station

1. Review the daily operations report for the past six months, looking especially for mechanical breakdowns lasting five minutes or more.

2. Create a summary list, using the format shown in Table 9-1.

Display the results for each station. Figure 9-1 shows a way to arrange the summary of results according to processing stations. You can list the types of problems by frequency of occurrence or by total downtime (frequency of occurrence multiplied by downtime for each incident). Listing by frequency of occurrence is more common.

Table 9-1. Format for Summary of Problems

Item	Date of occurrence
G part LS disconnected	5/6, 6/7, 6/9, 6/13
C part lot broken	6/30, 7/13
V belt replaced	7/12

Step 2. Analyze the Breakdown Mechanisms

1. Use Pareto analysis to determine the most significant problems to address.
2. Look carefully at the parts and components that have broken down.
3. Conduct on-site observations of workstations where breakdowns have occurred.
4. Use Why-Why Analysis to get at the true causes and figure out solutions (see Figure 9-2).

Step 3. Implement Measures to Eliminate the Causes

Step 4. Attach Daily Inspection Labels to Equipment

1. Once minor stoppages have decreased, attach daily inspection labels in the sequence of inspection.
2. Be sure operators can perform instant maintenance if some abnormality turns up during the daily inspections.

Step 5: Divide Parts and Components into Three Groups

Create a Pareto chart showing the frequency with which parts and components break down. Use the scores to divide them into three groups:

Group A: parts to always have on hand

Group B: parts contracted for immediate delivery

Group C: parts for general purchase

Step 6. Manage the Spare Parts Inventory

Frequently used spare parts are easiest to manage using a two-bin system for small parts with multiple uses (see Chapter 5, p. 73). The order dates for all other parts should be determined by the time required for them to be delivered. The formula for calculating the order point is

Breakdowns per day × delivery time × safety margin (buffer)

Step 7. Improve Parts Replacement Procedures

Eliminate waste through on-site observation; create an improvement plan, and implement it immediately. For example, the three-pronged connector in Figure 9-3, which was formerly attached with screws, can be redesigned so that it simply plugs in. Store tools and replacement parts near the machines.

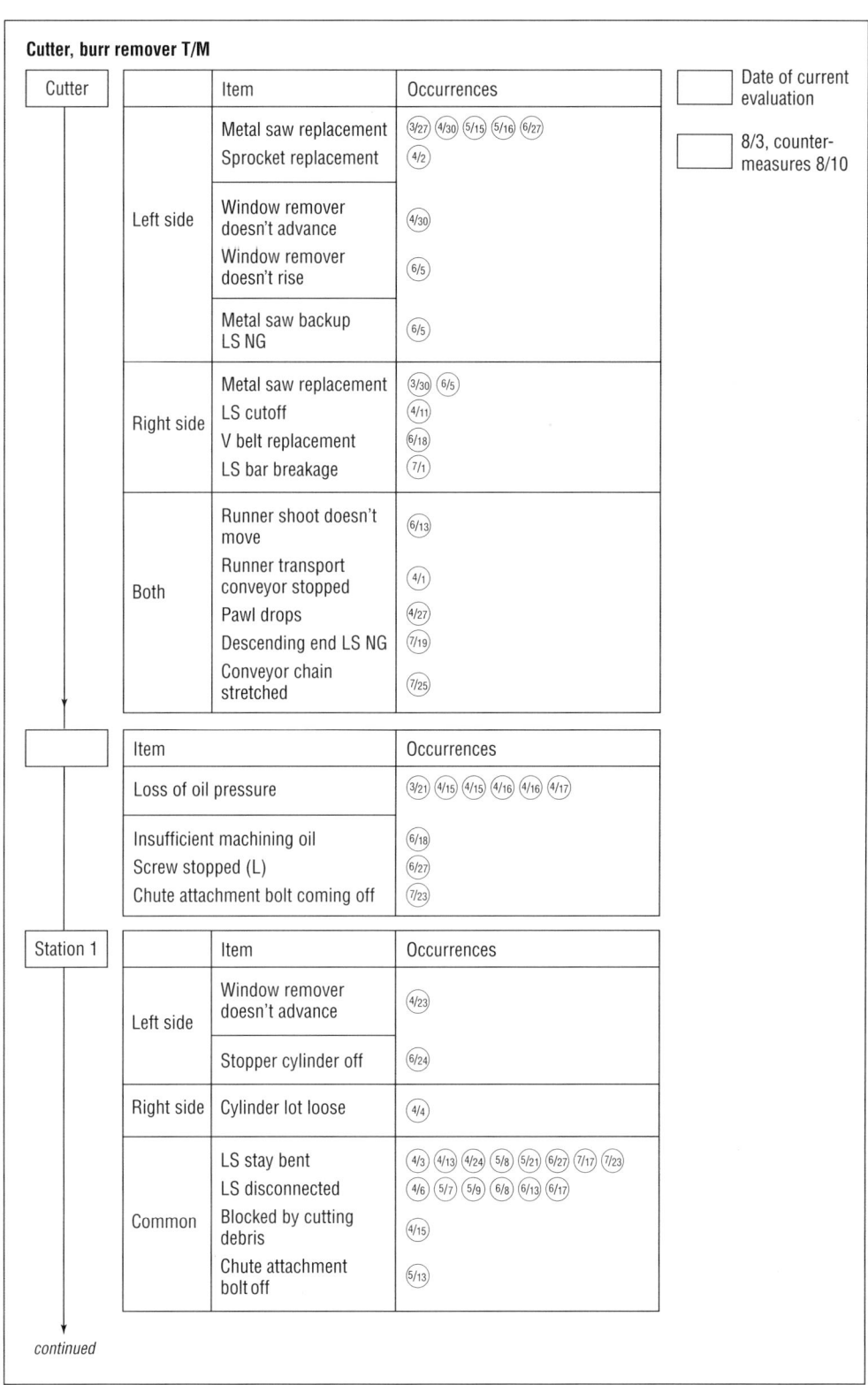

Cutter, burr remover T/M

Cutter		Item	Occurrences
	Left side	Metal saw replacement	(3/27) (4/30) (5/15) (5/16) (6/27)
		Sprocket replacement	(4/2)
		Window remover doesn't advance	(4/30)
		Window remover doesn't rise	(6/5)
		Metal saw backup LS NG	(6/5)
	Right side	Metal saw replacement	(3/30) (6/5)
		LS cutoff	(4/11)
		V belt replacement	(6/18)
		LS bar breakage	(7/1)
	Both	Runner shoot doesn't move	(6/13)
		Runner transport conveyor stopped	(4/1)
		Pawl drops	(4/27)
		Descending end LS NG	(7/19)
		Conveyor chain stretched	(7/25)

	Item	Occurrences
	Loss of oil pressure	(3/21) (4/15) (4/15) (4/16) (4/16) (4/17)
	Insufficient machining oil	(6/18)
	Screw stopped (L)	(6/27)
	Chute attachment bolt coming off	(7/23)

Station 1		Item	Occurrences
	Left side	Window remover doesn't advance	(4/23)
		Stopper cylinder off	(6/24)
	Right side	Cylinder lot loose	(4/4)
	Common	LS stay bent	(4/3) (4/13) (4/24) (5/8) (5/21) (6/27) (7/17) (7/23)
		LS disconnected	(4/6) (5/7) (5/9) (6/8) (6/13) (6/17)
		Blocked by cutting debris	(4/15)
		Chute attachment bolt off	(5/13)

Date of current evaluation

8/3, counter-measures 8/10

continued

Figure 9-1. Summary of Machine Breakdowns by Station

Cutter, burr remover T/M, *continued*

Station 2		Item	Occurrences			
	Left side	Cylinder lot broken	(6/17) (5/31) (7/12)			
		Disconnection	(3/26)			
		V belt replacement	(6/22)			
		Cutter seized up	(4/18)			
		Terminal disconnection	(5/11) (6/11)			
		Motor replacement	(7/18)			
	Right side	Cylinder lot broken	(4/8) (4/23) (5/30)			
		V belt replacement	(4/9)			
		LS disconnection	(5/9)			
	Common	Product out of position	(4/18)			

Date of current evaluation

8/3, counter-measures 8/10

Station 3		Item	Occurrences
	Left side	Blade spill	(3/21)
		Bearing replacement	(6/20)
		Short circuit in motor	(7/2)
	Right side	V belt replacement	(4/13)
		Cutter chipped	(7/17)
	Common	LS breakdown	(4/17)
		LS cutting dust blockage	(4/5)

Station 4		Item	Occurrences
	Left side	Unit LS adjustment	(7/22)
	Right side		

Attachment

Assembly

Discharge	Item	Occurrences
	LS adjustment	(3/22)
	Discharge conveyor stopped	(4/18) (4/25)

Figure 9-1. Summary of Machine Breakdowns by Station, *continued*

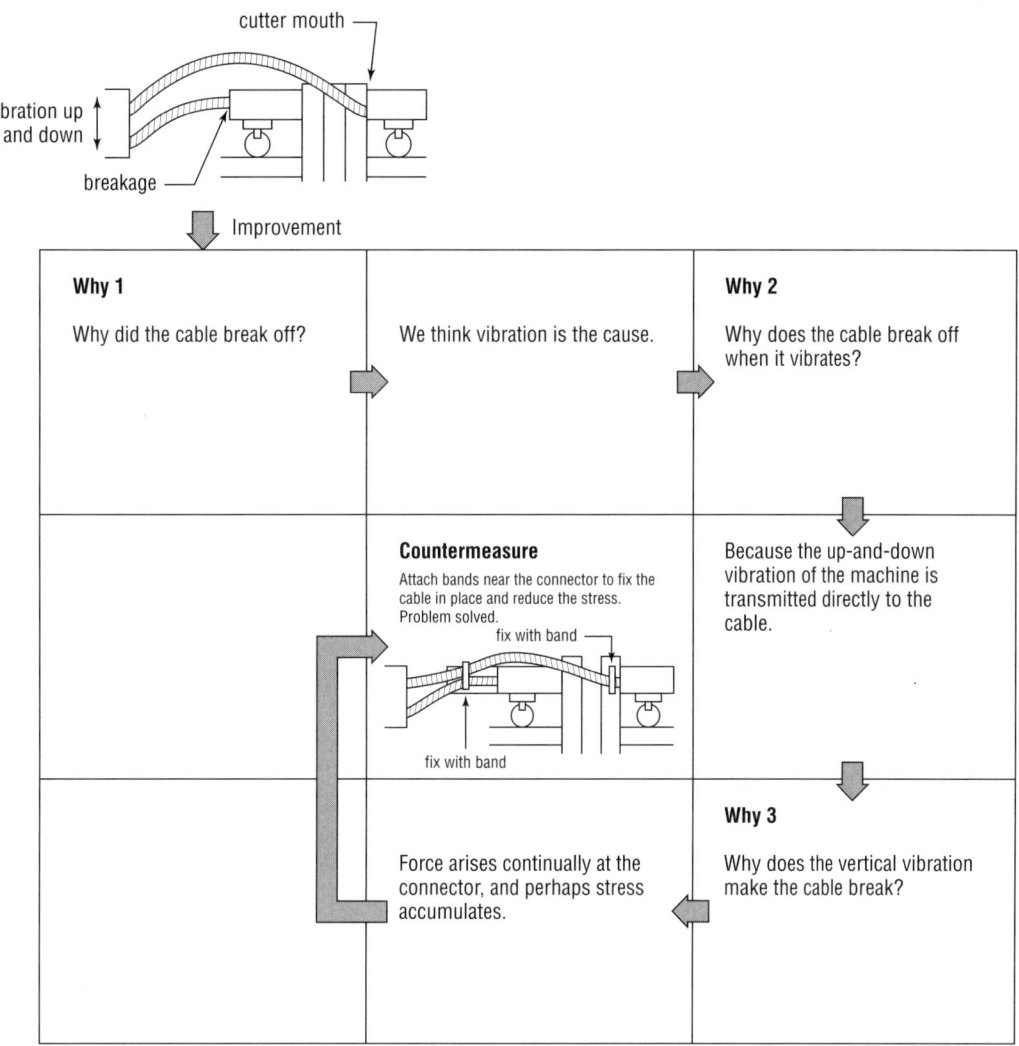

Figure 9-2. Why-Why Analysis for a Machine Part Failure

Step 8. Investigate Broken Parts

Implementing the first seven steps is not the end. You will continually have to inspect all broken parts and use techniques like Why-Why Analysis to get to the bottom of the causes and devise countermeasures.

Step 9. Thoroughly Train Employees in Daily Maintenance

The supervisor should conduct a daily equipment inspection to teach operators the procedures. Operators learn to use their senses of sight, hearing, smell, touch, and (only in safe situations) taste during daily inspections.

• Looking carefully at the machine can help operators detect oil leaks, drops in oil level, dirt, nicks, corrosion, rust, loosening, and pressure changes.

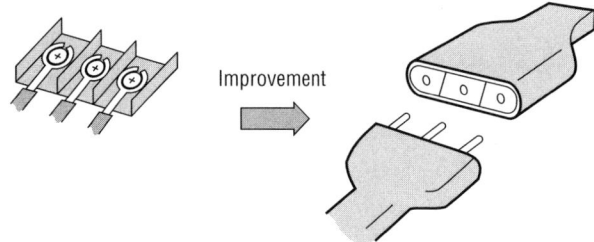

Figure 9-3. Improvement Example: Replacing Screws with a Plug

- Listening for different sounds such as groaning or grinding in a press or motor can help identify abnormal operation.

- Smelling a scorched odor indicates that overheating is occurring.

- Touching the machine gives an indication whether it is too hot or vibrating too much, and whether fluid is someplace it shouldn't be.

- Tasting is occasionally used to identify salt and other foreign elements that can be detected by taste. Of course, you should never do this in situations where there is any risk of ingesting a poisonous or harmful substance.

Detection through the senses is described in more detail in "Using Sensory Inspection to Detect Machine and Equipment Abnormalities," beginning on page 138.

Eliminating Minor Stoppages

Minor stoppages are minor breakdowns in machines and equipment, those caused by variations in workpieces or associated equipment. This section discusses the basics of how to eliminate them.

Classifying Minor Stoppages

Minor stoppages can be classified into two categories:

- *Automatic shutdown.* This occurs when a machine or piece of equipment detects trouble due to overloading or abnormal readings and turns itself off.

- *Idling.* Due to some problem or other, the flow of work stops, and the machine or piece of equipment runs idly for a while. When this is not discovered in time, the machine shuts down.

Why Do Minor Stoppages Occur?

Design flaws reportedly cause 70 percent of all minor stoppages. Design problems include product designs with no machining or assembly standards as well as weak points in the specified materials, shapes, and construction of the equipment mechanisms, parts, or tools.

Manufacturing management accounts for another 20 percent of minor stoppages, largely due to managers' failure to spend time at the work site to improve and standardize the situation. Suppose the plant contains a piece of equipment that has tiny flaws. Although previous "standards" for attachment and machining were based on instinct, experience, and a try-anything attitude, they may remain fossilized in the current procedures if managers do not support a systematic approach.

The final 10 percent of minor stoppages results from operators' failure to follow established operating procedures. The percentage of operator-caused minor stoppages will be higher in plants where operators are involved in maintenance and support tasks, but ignore standards for cleaning, lubricating, tightening, setup, and adjustment, and just follow their own inclinations.

Why Minor Stoppages Lead to Significant Waste

Minor stoppages are a form of waste in themselves, but they are linked to even more significant waste. The following factors are the principal causes of minor stoppages:

1. "Fixing" the minor stoppages is so simple that people never implement improvements that will eliminate them forever.
2. Problems are hard to observe or quantify, so improvements are postponed or left undone.
3. There is variation between machines and equipment.
4. Minor stoppages occur under different circumstances with different parts and products.
5. The stoppages don't always occur in the same place.
6. The minor stoppages have become so chronic that people don't connect product defects with the reduced performance rate.

These factors trigger minor stoppages, with the following bad results:

1. The performance efficiency of the machines and equipment drops, leading to a decrease in productivity and overall effectiveness.
2. Since the operators can never tell when a minor stoppage will occur and need fixing, they have to watch each machine all the time, which means that every machine requires its own operator.
3. When a minor stoppage occurs on a production line, operators rarely turn off all the other machines and equipment, leaving them to idle, so that consumption rises on the counters, but productivity does not go up. The cumulative effect is a big energy drain.
4. Even though the company has gone to the trouble of installing new machines, they use them in the same way as the old ones, and minor stoppages reduce their efficiency.
5. Blockages and hang-ups on the production lines require extra employee hours, not only to restore the equipment but also to inspect for nicks, deformations, and other flaws in the resulting products. These hours burden the payroll.
6. Mistakes in parts insertion or attachment and the presence of foreign debris lead to increased defects.

How Minor Stoppages Differ from Breakdowns

Minor stoppages and breakdowns are similar in that they cause machines and equipment to shut down, but they differ on the following four points:

1. Minor stoppages usually arise due to variability in employees, raw materials, machines, or methods. Examples include blocked chutes, malfunctioning sensors, parts dropped and bent, and materials damaged by embedded foreign materials.
2. Increased variety of parts and number of parts are a huge factor in minor stoppages. Higher variety and number of parts leads to variations in precision; factories often try to address these variations with adjustment operations, but these do not lead to stability. Many factories are trying to combine parts or decrease the number of parts to decrease the possible variation during processing; the automobile industry in Japan has drastically reduced the number of different parts used.
3. Minor stoppages take the form of slowdowns when parts are dropped or lost during manual assembly.
4. Breakdowns mean a loss of functionality and useful life, and restoring them requires extra labor, time, parts, and money.

Steps for Eliminating Minor Stoppages

Figure 9-4 shows the basic steps in eliminating minor stoppages. Worksheets to support several of these steps appear in the final section of this book.

Step 1: Study Current Conditions

Minor stoppages that come to the surface are only the tip of the iceberg. That's why we begin by getting a more complete picture of the situation.

1. Record the daily direct operating time (DT) of each machine and piece of equipment.
2. Record the total daily time on the job (TT).
3. Determine the operating rate (DT/TT) for the machines and equipment.

   When the operating rate is low, investigate how much downtime the machines and equipment have each day. This includes setup, adjustment, fixing defects, minor stoppages, medium stoppages, major stoppages, and planned stoppages.
4. Form a clear picture of the stoppages, their places of occurrence, and their frequency for each machine, piece of equipment, and unit.

Step 2: Rethink the Process from the Ground Up

Observe the stoppages carefully, rank them in order of importance, and investigate their frequency and the time between occurrences. Analyze this information to create an image of the way things ought to be.

Step 3: Pursue the Goal

The goal is zero minor stoppages, so the items for improvement should include the following:

1. Make sure you don't have to shut down because you're out of parts or materials or because the materials vary too much. (Don't run out of or misplace parts.)
2. Make sure there are no defects in parts or materials. (The parts manufacturer needs to conduct exhaustive quality checks.)

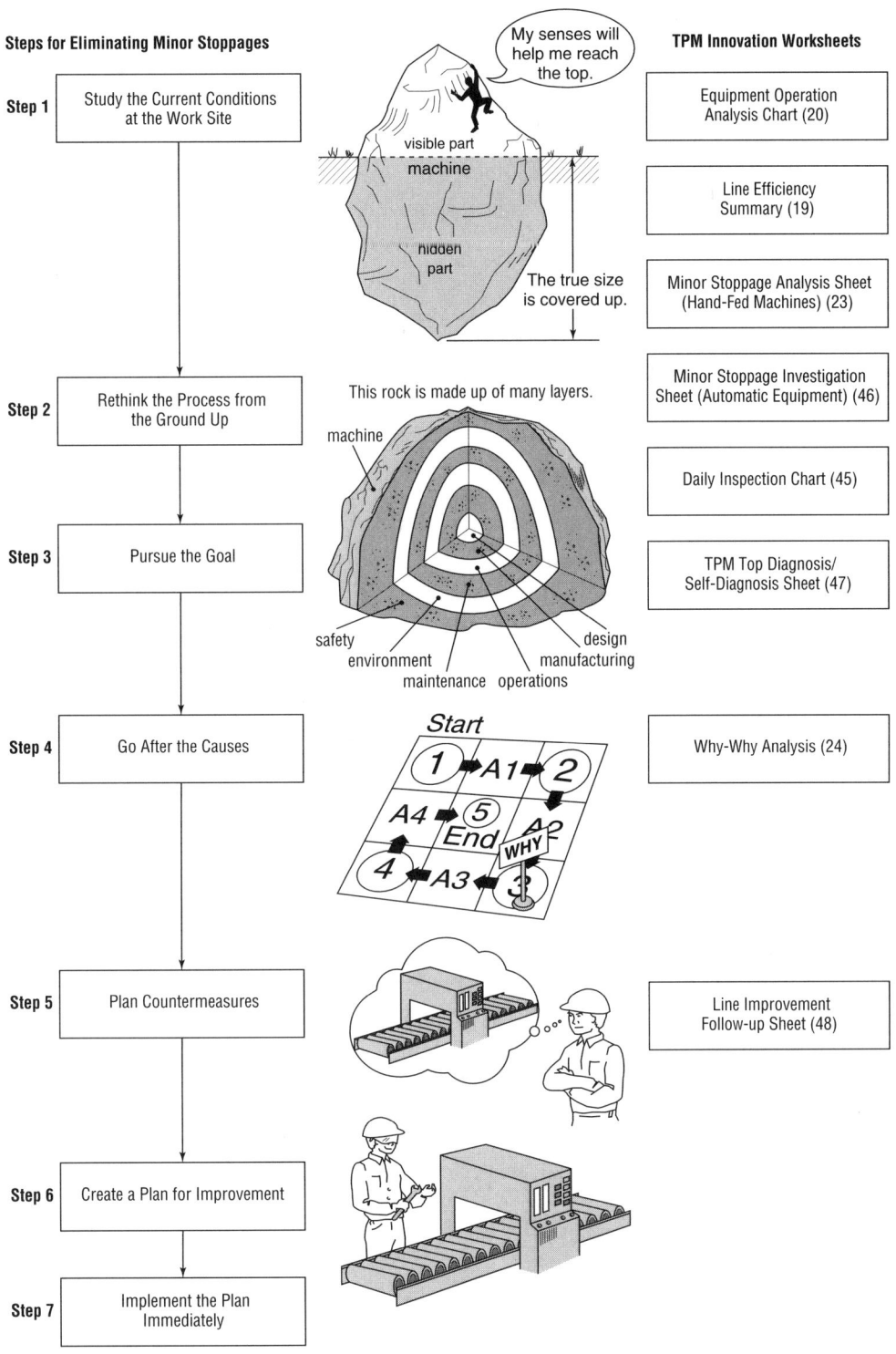

Figure 9-4. Steps for Eliminating Minor Stoppages

3. Make sure all the parts in each unit that are likely to get dirty or wear out quickly can be easily replaced.

4. Make sure debris and other foreign material don't contaminate the parts and materials.

5. Make sure low-speed, medium-speed, and high-speed changeover are synchronized in each unit.

6. Make sure your procedures apply mistake-proofing and poka-yoke systems.

7. Make sure your formulas for setting standards mesh smoothly with one another and require no adjustment.

With these measures in place, you can move further in your quest for zero stoppages by comparing troublesome equipment with well-functioning equipment and figuring out where they differ.

Step 4: Root out the Causes

Use Why-Why Analysis to root out the true causes of the minor stoppages.

Step 5: Plan Countermeasures

At company Y, the team first got rid of all debris. Then they eliminated the causes of variation in each equipment unit. Next, they consulted with the designer to improve the weak points, reevaluating the designs of the equipment mechanisms, the shapes of the parts, the quality of the materials, the shapes and structures of the jigs, and the quality of the sensors. Then they made appropriate adjustments in any areas that needed improvement.

When you plan improvements, make sure to include the following elements.

1. Fixed methods for setting standards
   • Standard methods for attaching and removing workpieces
   • Standard positions for parts
   • Standards for machining parts
   • Standards for inspection and other aspects of production
2. Optimal conditions in place for machining (tightening pressure, tightening force, amplitude, number of revolutions, etc.)

Step 6: Create an Improvement Plan

Divide the proposals into small-scale, mid-scale, and large-scale improvements, according to the amount of time and effort you expect them to require.

Step 7: Implement the Plan Immediately

Don't be resigned to minor stoppages; grapple seriously with the task of eliminating them. Ignoring minor stoppages simply adds to the kind of waste attributable to inattentive management.

Using Sensory Inspection to Detect Machine and Equipment Abnormalities

Only the on-site operators can detect the symptoms of abnormalities early, prevent breakdowns before they happen or discover them immediately after they've

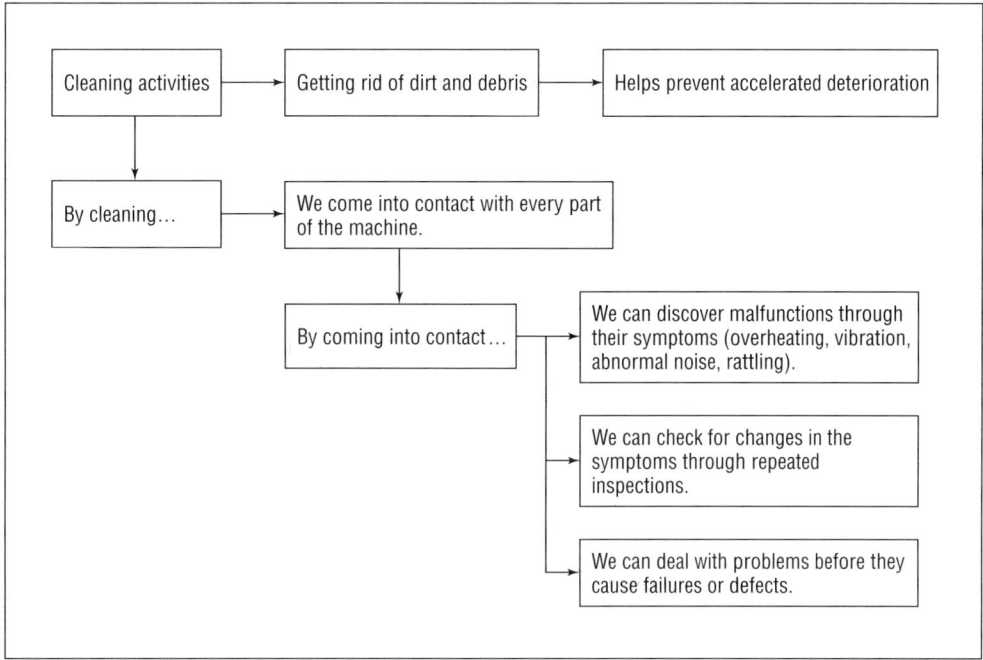

Figure 9-5. Benefits of Cleaning and Inspection

occurred, and fix them before they turn into massive problems. For this reason, the next few pages discuss how operators can use their senses to find abnormalities.

The Operator's Role

Operators are in a good position to detect the symptoms of abnormalities as soon as possible and either fix them or inform the person responsible for fixing them. A maintenance technician is like a doctor who analyzes the operators' information about vibration, noise, pressure, oil level, and other factors. He or she then applies specialized knowledge and hands-on experience to fix the problem.

To help the maintenance technicians assess problems correctly, it is important for operators to conduct daily inspections. Daily cleaning is a way for operators to discover abnormal conditions in the smallest, most out-of-the-way parts of the equipment. Figure 9-5 summarizes the benefits of cleaning and inspection.

Detecting Abnormalities and Their Symptoms

In detecting abnormalities, operators use their senses: sight, hearing, touch, smell, and sometimes even taste. Test yourself by trying to match these senses with the lubrication inspection items listed in Table 9-2. (Our own answers appear in Table 9-11 on page 155.)

Detection by Sight

In equipment inspection, nothing is more significant than human vision. For example, here are some abnormalities you can discover by looking at the oil in a machine:

Table 9-2. Lubrication Inspection Items

The Operator's Role

| 1. Discover abnormal symptoms early. Train the senses.
2. Carry out daily inspections punctually.
• Oil supply
• Cleaning
• Operation according to standards
3. Perform instant maintenance when possible.
4. Inform maintenance and related departments immediately when major problems are discovered.
5. Work with other functions for concurrent engineering. | Guarantee delivery times.
• Zero setup
• Zero minor stoppages
• Zero breakdowns
Guarantee amount produced.
• Zero inventory
• Reduced staff
Bring down costs.
Maintain quality.
• Zero defects
Maintain safety.
• Zero injuries |

No.	Nature of abnormality	Standards for determination	Inspection methods (senses)	Notes
1	Overheating	No overheating at lubricated points		
2	Vibration	No abnormal vibration		
3	Abnormal noise	No abnormal noise at lubricated points		
4	Leak in oil supply	No leakage from element, lid, or oil seal		
5	Abnormality in operation of oil supply	Oil supply system operating normally		
6	Sight glass dirt	No breakage or dirt on level gauge or eyeball gauge		
7	Oil seal leakage	No leakage from oil seal		
8	Grease splattering	No splattering of coupling grease		
9	Filter blockage	No abnormal filter blockage		
10	Running out of oil	Level gauge, oil surface visible		
11	Level too low	Display beyond MIN		
12	Level too high	Display beyond MAX		
13	Contamination of oil in use	No change in drain oil color, abnormal moisture content, or metal shavings		
14	Cloudiness of oil in use	No opacity or whiteness in drain oil		
15	Rust in oil	No rust in drain oil		
16	Metal shavings in oil	No wear particles in drain oil		
17	Foreign bodies in oil	No debris in drain oil		
18	Presence of drain water	No separated water in drain oil		
19	Change in oil color	No change in color of drain oil		
20	Oiler pipe blockage	No blockage		

• When the oil darkens, oxidation or deterioration has occurred.

• When the oil turns black, it is contaminated with wear debris.

• When the oil is cloudy, it is contaminated with water.

This is just one tiny example; others include external observation, instrument readings, and observing the behavior of machines and equipment. Table 9-3 shows some points to watch for.

Detection by Sound

The abnormality most often detected by ear is noise, and there are many factors involved. Try evaluating your own plant and see how the machines and equipment rate on the scale shown in Table 9-4. Table 9-5 shows several inspection points that include sound.

Detection by Touch

The phenomenon most often detected by touch is vibration. Moreover, since there is a strong link between vibration and noise, using both senses to detect these phenomena makes it easier to pinpoint the location of the problem. Table 9-6 describes six levels of vibration.

Even when dealing with something like machine oil, where you determine the temperature with a thermometer, it's also convenient to use the sense of touch to get a subjective, "ballpark" reading. Table 9-7 shares a handy chart that shows temperature ranges for a specific type of oil, and also includes "touch" tests for gauging temperature by feel.

Finally, Table 9-8 shows a number of inspection points that rely on touch.

Detection by Smell

We can detect bad odors with our sense of smell. The human sense of smell is inferior to that of dogs and other animals, but it's still an important sense.

Bad smells often indicate the presence of harmful substances, which are not only injurious to the environment and human health but can also lead to life-threatening accidents within the plant. Whenever we read newspaper articles about chemical explosions in factories, we almost always find an account of someone noticing an abnormal odor just before the explosion. That shows how important it is to pay attention to our sense of smell. Table 9-9 shows several inspection items involving smell.

Inspection Pointers

The flow chart in Figure 9-6 shows the elements of a sensory inspection of the inside of a tank. To determine an appropriate cycle for inspection, look at the average rate of breakdowns. Table 9-10 suggests the frequency for several types of periodic inspections for reference.

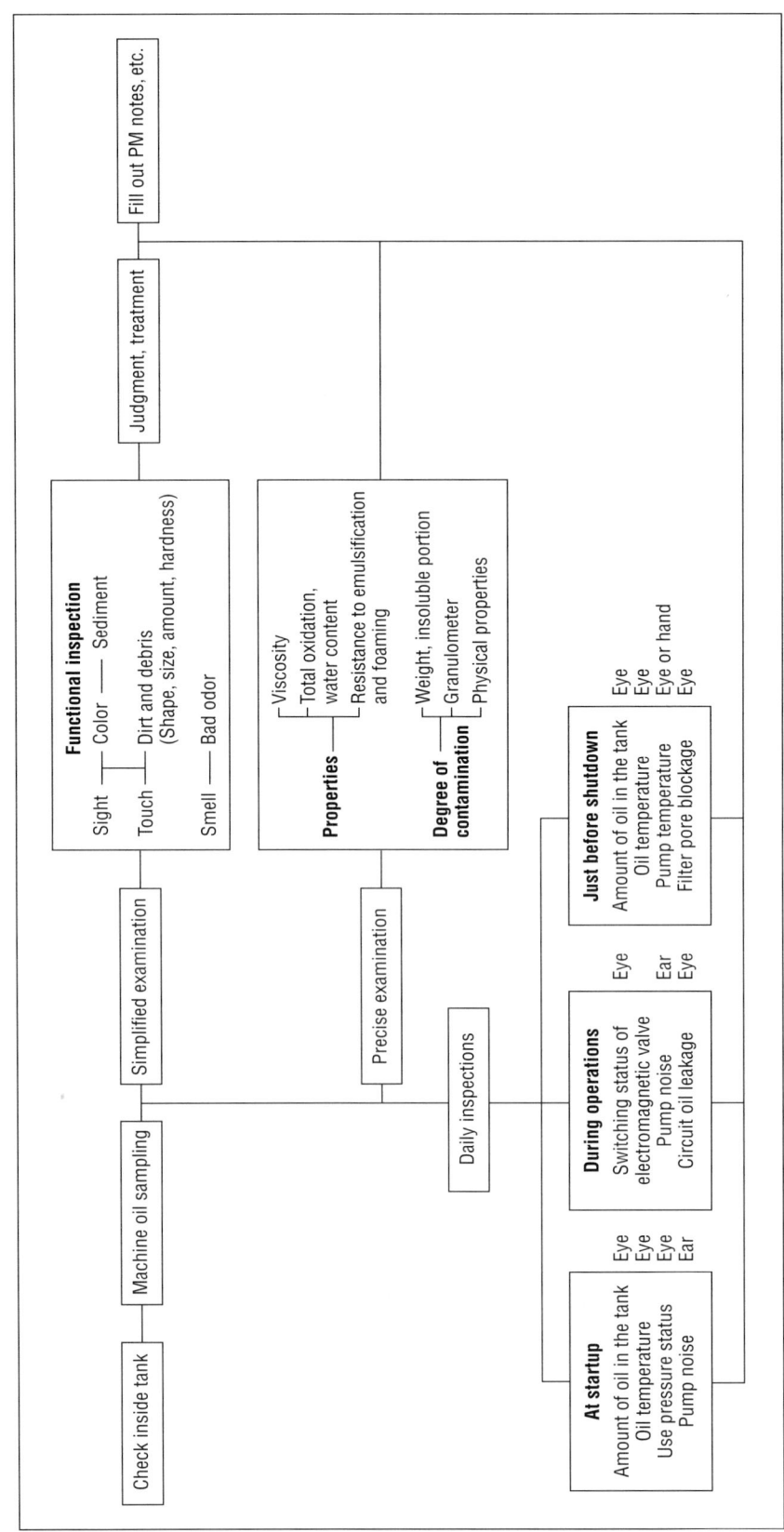

Source: Seijiro Hanawa, "Manage Machining Oil on Site," *TPM Age 3* (6).

Figure 9-6. Sensory Inspection of a Tank

Table 9-3. Visual Inspection Items

Symptoms of abnormality, phenomena	Place	Important inspection items and methods	Operator's counter-measures	Urgency	Notes
Oil leakage		1. Correct adjustment of flow amount?	⊙		A drop in the amount in the tank within a short time is a sign of trouble.
		2. No bolts loose?	⊙		Oil leaking from the sleeve surface is a supply abnormality.
		3. No deterioration in shape of seal packing?	○		
		4. No wear in the shaft?	○		The gasket problem is due to
		5. Pipes and joints firmly attached?	○	▣	• Strain in the gasket contact surface
		6. Valves and other oil pressure devices normal?	○		• Wrong choice of gasket
		7. No oil leaking from gasket contact surface?	○		• Wrong gasket size • Poor installation
		8. Places to be oiled, type of oil, cycle, and amount all confirmed?	○		• Bolts only half tightened upon attachment or improperly attached
Change in oil color • Cloudiness • Air absorption • Black cloudiness • White cloudiness during operations		1. Confirm change in color through the inspection window.	○		A change in oil color is a symptom of abnormality.
		2. Any contamination by foreign bodies, water, air, or metal particles?	○	▣	It turns white if water gets in.
		3. Try comparing it to fresh oil.	○		It may develop white clouds if mixed with another type of oil.
					It may also turn white if air gets in.
					Contamination with wear debris causes it to turn black.
					See if there's a difference between the color during operations and the color at rest.
					Don't change the type of oil.

admixture of air — max — min

▣ = Emergency   ○ = Inspection   ⊙ = Inspection process step

(continued)

Table 9-3. Visual Inspection Items, continued

Symptoms of abnormality, phenomena	Place	Important inspection items and methods	Operator's counter-measures	Urgency	Notes
Gauge Abnormalities in metering instruments		1. No loosening of the lock nut? 2. Measurement instruments and meters not broken? 3. Alarm set value confirmed? 4. Doesn't reach the limits in the normal range of operations?	⊙ ○ ⊙ ○	▣	Confirm the gauge's zero point. Does the needle indicator rise and fall smoothly? Item 4 requires classification by color. Is it oscillating? How are the numerical values? Is it fluctuating? How are the numerical values?
Wear Corrosion		In machine tools, for example, wear and play are most common in the sliding surfaces of the heads, table, or saddle, or in the lead or feed screws. 1. Lubricant present in the proper amount? 2. No breakage, collapse, or blockage in the pipe system? 3. No foreign matter or cutting debris embedded in the sliding surfaces? 4. No burrs or nicks in the dog standard pads? 5. No abnormality in the attachment points for the main bodies of the devices and the nozzles? (Use hearing and touch.) 6. If devices, pipes, or measuring instruments are subjected to such severe conditions as wear from granular particles or acid corrosion, a daily inspection may not be sufficient.	⊙ ○ ⊙ ⊙ ○ ○	▣ ▣	Cleaning and maintenance of sliding surfaces is important. Replace wipers. Depending on their uses, they may be made of felt, rubber, hardened plastic, or brass. In any case, you must check for deformation and wear. When load is applied during cutting, the blade hold and the unit must not move from side to side or up and down. The majority of the problems in item 5 arise from corrosion (crevice corrosion, stress corrosion) during manufacturing.
Play and rattling		1. No play in screws? 2. Spaces in thrust and inlays not too large? 3. Sliding parts move smoothly? 4. Taper strikes normally, causing no bruises or burrs? 5. No looseness in the leveling or the anchor bolt? 6. Pin, cotter, key not broken off?	○ ○ ○ ⊙ ○ ○	▣ ▣	Abnormal noises or shocks at startup are a sign of deterioration. Abnormal vibration caused by the cutting load hasten wear and turn into major rattling. If you ignore minor rattling, it will turn into something major!

▣ = Emergency ○ = Inspection ⊙ = Inspection process step

(continued)

Table 9-3. Visual Inspection Items, continued

Symptoms of abnormality, phenomena	Place	Important inspection items and methods	Operator's counter-measures	Urgency	Notes
CPU screen		Everything displayed correctly?	⊙		
Abnormalities in products Burr marks Difference in level Biting Small cracks Striking off center Variable dimensions		1. Rattling in the main shaft or bearings? 2. Wear in the slide surface? 3. Blade worn or improperly fitted? 4. Sufficient oil on the sliding surfaces? 5. Installation bolts of machines loose?	○ ⊙ ⊙ ⊙ ○	▪ ▪ ▪	Abnormalities in the finished surface of a product or variations in the measurements are signs of trouble.
Equipment malfunctions		1. Correct valve on the oil pressure devices? 2. Limit switch working correctly? 3. Wear in the dogs? 4. Correct oil temperature? Confirm with instruments. 5. Valves opening and closing normally? 6. Easily understandable display of functions of seal valves and emergency operation valves? 7. Oil film normal?	○ ○ ○ ⊙ ○ ⊙ ○	▪	If the oil temperature in an NC machine exceeds 41° C, malfunctions are more likely. When a machine malfunctions, shut it down immediately and report the situation to the appropriate department.

| Brake slippage | | 1. Stops immediately after brakes are applied?
2. No abnormal noise when brakes are applied?
3. Normal startup and stopping? | ○
○
○ | ▪ | Oil: penetration of foreign matter
Asymmetrical wear on drum?
Good brake adjustment? |

▪ = Emergency   ○ = Inspection   ⊙ = Inspection process step

(continued)

Table 9-3. Visual Inspection Items, continued

Symptoms of abnormality, phenomena	Place	Important inspection items and methods	Operator's counter-measures	Urgency	Notes
Leakage from pipes (valves)		1. Oozing or leakage from the welded parts of the pipes (those influenced by heat)?	○	▣	Once a leak occurs, it only intensifies, so operators should check for them periodically, using their ears and hands as well as their eyes.
		2. Bolts tightened all the way?	⊙		
		3. Leakage from gaskets?	⊙		
		4. Leakage or oozing from gland packing?	○		The way of checking for item 5 during operation differs according to the fluid and brand used.
		5. Debris embedded in valves or valve stands (nicks developing or cutoff with excessive force)?	○		
		6. Erosion of bending parts?	○		You can check for item 6 by touch.
Smoke		1. Motor overheated?	○	▣	If you see smoke, it's too late!
		2. Belts at proper tightness? (Also check outward appearance of belts.)	⊙		All the V belts should be replaced.
		A. Cracks in the underside and side surfaces?			The flexion value should be 10 to 15 mm or more.
		B. Wear in the cover cloth on the side surface?			If the belt is deeper than the outer diameter, wear on the pulley should be repaired.
		C. Belt twisted?			
		• Is the bottom of the V-pulley groove shiny?	○		
Thermal action		1. Does it occur in the work to be machined? Has overloading occurred?	⊙		If you let this go, you may burn out your motor!
		2. Confirm the ammeter.	○		If you don't repair cracks as soon as they're discovered, they could cause major breakdowns.
		3. Is the motor moving without effort?	○	▣	
Cracks		4. Cracks in the attachments and parts to be tightened?	○	▣	
		5. Cracks in the lever handles?	○	▣	If you suspect cracks in vital areas, use color checks or magnetic inspection.
		6. Cracks in the dovetail grooves?	○	▣	
		7. Cracks in the gears?	○	▣	
		8. Cracks in the welded parts?	○	▣	

▣ = Emergency   ○ = Inspection   ⊙ = Inspection process step

(continued)

Table 9-3. Visual Inspection Items, continued

Symptoms of abnormality, phenomena	Place	Important inspection items and methods	Operator's counter-measures	Urgency	Notes
Loosening		1. Loose bolts? 2. Loose rivets? 3. Loosening in the taper? 4. Loose cotter? 5. Loose terminals or screws?	⊙ ○ ⊙ ⊙ ○	 ▪ ▪	Periodic inspections Tightening is one of the important aspects of autonomous maintenance.
		Tighten it. Are the nuts and bolts marked so that you can tell immediately whether they're loose?			

▪ = Emergency   ○ = Inspection   ⊙ = Inspection process step

Table 9-4. Abnormal Noise Levels and Factors

Level	Sensory perception	Factors to consider
5	Sound pressure and tone are normal.	Can be considered normal or in proper working order.
4	There is an unusually high-pitched sound in addition to normal noise.	Something is rubbing against something else.
3	Sound pressure fluctuates on a regular cycle.	The shaft is off-center or wobbling.
2	Sound pressure is high, and tone becomes irregular or is punctuated by odd noises at regular intervals.	There is foreign matter embedded or local damage.
1	Sound pressure is high, and tone becomes irregular at irregular intervals but abnormal sounds are continuous.	There is pitching or scoring.

Source: Adapted from Seijiro Hanawa, "Manage Machining Oil on Site," *TPM Age* 3 (6).

Table 9-5. Auditory Inspection Items

Symptoms of abnormality, phenomena	Place	Important inspection items and methods	Operator's counter-measures	Urgency	Notes
Sound of vibration Sound of cracking	Motor bearing shaft	1. Confirm balance.	○		Loosening of plinth attachment bolt When the plinth starts to vibrate at the characteristic vibration of the electric motor, this can cause abnormal vibration.
	Chucks	2. Flaws in work attachment or tightening?	⊙		
	Bed column gearbox	3. Loosening in setting bolt leveling?	○	▣	The flapping caused by a loose belt can be transmitted to the machine and cause vibration.
		4. Difference in shaft core? Wear in joint parts?	○	▣	
	Pulley	5. Swelling in the shaft?	○		When the screws on electrical parts come loose, you can get a vibrating sound in the control panel.
		6. Rattling due to key wear?			
	Various valves	7. Insufficient amount of oil?	⊙		
	Oil pressure pipe	8. Oil seepage normal?	○		
	Cover	9. Loose attachment bolts?	⊙		
	Overall	Spots that are susceptible to vibration should be periodically checked with appropriate portable instruments.	○		Are all these positions and numbers marked?

▣ = Emergency   ○ = Inspection   ⊙ = Inspection process step

(continued)

Table 9-5. Auditory Inspection Items, continued

Symptoms of abnormality, phenomena	Place	Important inspection items and methods	Operator's counter-measures	Urgency	Notes
Abnormal noise	Rotor	1. Wear, damage, or lack of oil in the bearings?	⊙		If a rotor undergoes continued wear, vibration may occur.
		2. Loosening, sliding, or wear of the belt?	⊙		It is necessary to make occasional inspections using a stethoscope.
		3. Any cavitation of the pump?	○	▣	
		4. Defect in the cap ring? Characteristic defect in the bolts?	⊙		In the case of a humming sound in the relay, take note of the correct value.
	Pipes	1. Water hammer?	○	▣	
		2. Leakage of gas or liquid?	○	▣	
		3. Vapor leakage due to a trap defect?	⊙		
	Electrical instrumentation	1. Loosening of the terminal or wear or dirt on the contact points?	⊙		
		2. Operating defect in the device?	○	▣	
		3. Air leakage in electromagnetic valve or adjustment valve?	○	▣	
Noise	Pump	1. Intake pipe blocked?	○		Rethink the pipes and the arrangement of the equipment to see if you can make the intake pressure greater than the improved value.
		2. Suction filter pores blocked?	⊙		
		3. Taking in air through the intake pipe or joints?	⊙		
		4. Air bubbles inside the tank?	⊙		Check the pipes and joints.
		5. Oil surface too low?	⊙		
		6. Attachment panel not hard enough?	⊙		
		7. Centering defect in the coupling?	⊙		
		8. Oil temperature too high or too low?	⊙		
		9. Wear in the pump sleeve?	⊙		Check to see how the gears inside the pump mesh.
		10. Wear or damage to bearings?	⊙		
		11. Assembly defect?	⊙		

▣ = Emergency   ○ = Inspection   ⊙ = Inspection process step

Table 9-6. Abnormal Vibration Levels and Factors

Level	Sensory perception	Factors to consider
6	The nature and degree of vibration are as usual.	Normal or in proper working order.
5	No change in the nature of the vibration, but it periodically grows stronger and weaker.	Wobbling of shaft or gears, shaft off center or shaking in the shaft direction.
4	Vibration seems to be slightly above normal, and at regular intervals there seems to be slight shock vibration.	Foreign matter embedded, local damage, or play in the drive system.
3	Vibration seems to be more than normal, and at regular intervals there is strong shock vibration.	Scoring, pitching, or play in the drive system.
2	Vibration seems quite strong and there is light shock vibration at irregular intervals.	Scoring, pitching, or play in the drive system.
1	Vibration seems very strong, and there is strong shock vibration at irregular intervals.	Scoring, pitching, or play in the drive system.

Source: Seijiro Hanawa, "Manage Machining Oil on Site," *TPM Age* 3 (6).

Table 9-7. A "Touch" Guide to Machine Oil Temperatures

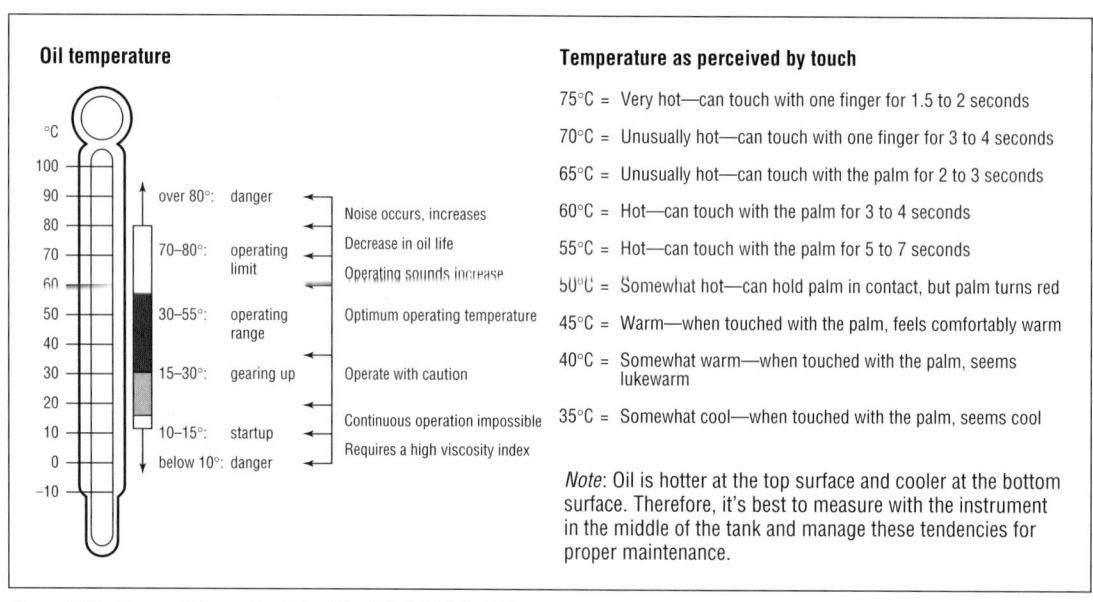

Source: Seijiro Hanawa, "Manage Machining Oil on Site," *TPM Age* 3 (6).

Table 9-8. Touch Inspection Items

Symptoms of abnormality, phenomena	Place	Important inspection items and methods	Operator's counter-measures	Urgency	Notes
Overheating	Bearings	1. Good oil supply?	⊙	▣	Overheating caused by wear or loss of gears or sliding parts.
	Conduction equipment	2. Good oil surface?			Learn and remember the normal temperatures by touch.
	Motor	3. Ventilation window blocked with dust or other debris?	⊙		The optimal temperatures may vary between 35° and 55°C depending on season, but make sure you know them more or less.
	Decelerator Gear shift	4. Touch gearbox to check temperature.	○		
	Machining oil tank	5. Check oil temperature.	○		
	Pump	6. Compare pump and tank temperatures.	○		Somewhat warm — About 32°C
	Packing	7. Packing too tight or only partly tightened?	○		Warm — About 38°C; Can feel heat (can touch for about a minute) — About 48°C
	Brakes	8. Gaps in the lining?	○		Quite hot (can touch for 15 seconds) — About 56°C

	Range	Interpretation
Oil temperature	°C 100, 90 Dangerous temperature range	Do not use under any circumstances.
	80, 70 Limit temperature range / 65, 60 Caution temperature range	1. Oil life short. 2. Oil cooler equipment needed. 3. Oil life cut in half for each 8°C rise in temperature.
	55, 50 Safe temperature range / 45, 40 Ideal temperature range	Optimal temperatures
	30, Normal temperature range / 20	No special treatment needed, but as viscosity rises, effectiveness drops.
	10 Low temperature range / 0	No-load operation, heating needed.

Note: Pump intake side should be 55°C or less.

(continued)

Note: The position of the lubrication oil surface for bearings and the like is determined by the lubrication formula.

Take a good look at the standards on the oil gauge and don't add more than that. You need to understand why the standards are in place. For example, if you add too much oil, oil agitation becomes more active, splattering increases, air bubbles are generated, lubrication performance decreases, and oil deterioration is accelerated.

Table 9-8. Touch Inspection Items, continued

Symptoms of abnormality, phenomena	Place	Important inspection items and methods	Operator's counter-measures	Urgency	Notes
Vibration	Bearings	1. Worn bearings?	◯	▣	Vibration that occurs suddenly is dangerous.
	Conducting shaft	2. Revolving parts in balance?	◯	▣	Stop the machine and check it immediately.
	Coupling Pulley	3. Differences in shaft and core?	◯		Vibration is linked to malfunction.
	Fixed bolts and nuts	4. Loose nuts and bolts?	⦿		Vibration accelerates equipment deterioration.
	Motor	5. Attachment positions wrong? Fitting bolts loose?	◯	▣	Try holding your hand on it.
	Saddle	6. Compare pump and tank temperatures.	◯		
	Bed Column Gear box	7. Give adjustment normal?	◯		
	Cover	8. Gear key loose?	⦿		
Oil deterioration	Lubrication oil tank	1. Density and viscosity normal?	◯		Preventive maintenance begins with lubrication management.
	Machining oil tank Automatic hot water supply tank	2. No contamination by foreign matter, air, water, or cutting debris?	◯		Deterioration accelerates if oil temperature exceeds 60°C.
	Oil temperature	3. Oil at appropriate temperature?	◯		Judge oil temperature by touch. Also use heat-sensitive labels.
	Grease	4. Hardening or change of color?	◯		

motor

vibration

This wheel is turning really hard all of a sudden!

▣ = Emergency   ◯ = Inspection   ⦿ = Inspection process step

(continued)

Table 9-8. Touch Inspection Items, continued

Symptoms of abnormality, phenomena	Place	Important inspection items and methods	Operator's counter-measures	Urgency	Notes
Is the wheel hard to turn? Too easy to turn?	Wheel	1. Good oil supply to sliding surfaces and bearings?	⊙	▣	If the wheel suddenly becomes hard to turn or too easy to turn, watch out!
	Cutting in	2. No foreign matter embedded?	○		
	Feed				
	Table	3. Loose bolt or missing key pin?	○		
	Other	4. Shaft twisted or off center?	○		
Rattling	Bearings	1. Good supply of lubrication oil?	⊙	▣	Adjustment and maintenance of the oil film have a huge influence on performance and quality.
	Fitted parts	2. Wear, adjustment, vibration, or overheating in any part?	○		Check for rattling or play in every moving part (left–right or up–down) during operations.
	Thrust				
	Propeller				
	Pin	3. Pin key cotter broken off?	○		If you ignore even minor rattling, it will cause deterioration and can lead to major accidents.
	Key				
	Cotter				
	Bolts	4. Any loosening?	⊙		
	Nuts				
	Gears	5. Wear on the gear surfaces? Backlash?	○		

▣ = Emergency   ○ = Inspection   ⊙ = Inspection process step

Table 9-9. Smell Inspection Items

Symptoms of abnormality, phenomena	Place	Important inspection items and methods	Operator's counter-measures	Urgency	Notes
Odor of burning coil insulation	Motor	1. Abnormal odor?	⊙	▣	If you smell burning rubber or plastic, suspect a defect in the electrical system (short circuit or contact defect). Turn off the power and fix the problem.
	Plastic cord	2. Check cord connection terminal contact point.	○		
	Bakelite	3. Check for overheating by touch.	○	▣	
	Coil	4. Check for overheating by touch.	○	▣	When a fuse blows, the current becomes single phase and surges.
	Solenoid valve	5. Check for overheating by touch.	○	▣	Overloaded, overheated operation
	Clutch	6. Check gaps in the lining.	○		
	Lining	7. Fuse blown?			Too hot!
	Fuse box	8. Deterioration due to debris or water?	○	▣	

Smell of burning rubber	V belt	1. How the belt is stretched
		2. Confirm the standards for stretching belts.
		  Since the deflection load also changes depending on the shape of the belt and the diameter of the small pulley, be sure to confirm it.
		  In general, the amount of flexion in millimeters is a uniform 1.6 mm for each 100 mm of the span length. The flexion load is measured with a spring. Check the flexion load by holding a carpenter's square against it and seeing the change from the standard value.
		3. Widen the distance between shafts and apply tension to the belts. (Occasionally turn the belt by hand to equalize the tension.)
		4. After several hours, stretch the belts again.
		5. Check the tension 2 to 7 days after that.
		6. After that, check the belts regularly every 1 to 3 months. In any case, do not allow the flexion load to fall below the minimum value.
			⊙		There should be absolutely no oil adhering.
					
					Belt replacement. Replace all of them.
					
					Flexion value should be 10 to 15 mm or more.

Chemical plant	Leaks in machines or pipes	The first thing you need to do is practice sharpening your sense of smell.	○	▣	When there is a leak, you must shut down the system and take care of it.

▣ = Emergency ○ = Inspection ⊙ = Inspection process step

Table 9-10. Periodic Inspection Standards

Frequency	Mechanical parts	Electronic parts
Daily	1. Air three-point settings: Dirt, amount of oil, pressure 2. Pressure, temperature, amounts in equipment 3. Components that can influence quality defects 4. Components that can affect safety 5. Bolts on vibrating parts 6. Parts where accumulation of dirt can't be prevented 7. High-speed spindles: Amount of oil drip, etc. 8. Dirt on dogs, etc., their position	1. Dirt in or on the positioning sensor, and the limit and proximity switch
Weekly	1. Air three-point settings: Dirt in drains, filter, etc. 2. Work holders:Centering, pawls, noise in main shaft 3. Parts that affect quality	1. Limit switch attachment 2. Proximity switch attachment 3. Moving parts in external wiring, etc. 4. Overheating, noise, or dirt in the motor 5. Noise, overheating, or dirt in the electromagnetic valves
Monthly	1. Worn parts in the drive and conduction systems 2. Amount of flexion in the drive belts and chains 3. Amount of forced discharge of lubrication oil, etc. 4. Parts influencing quality and breakdowns 5. Leakage in the Sealex joints	1. Abnormalities in the control and operations panels 2. Damage or dirt on fixed measurement devices 3. Connectors
Quarterly	1. Shaft joints 2. Deterioration in the lubricant or oil pressure 3. All nuts and bolts	1. Dirt or abnormalities in motors that have clutches or brakes 2. Wiring fixtures

Source: "Creating Written Standards for Cleaning and Lubrication," *TPM Age* 2 (4).

Table 9-11. Authors' Answers for "Senses Used" in Table 9-2

No.	Inspection method	Comments
1	Touch	
2	"	
3	Hearing	
4	Sight	
5	"	
6	"	
7	"	
8	"	
9	"	
10	"	

No.	Inspection method	Comments
11	"	
12	"	
13	"	Need to take a sampling
14	"	"
15	"	"
16	"	"
17	"	"
18	"	"
19	"	"
20	"	"

Examples of Small-Scale Improvements

This section shares some examples of small-scale improvements from the experiences of company K.

Figure 9-7 shows a method for preventing disconnection of limit switches. The improvement team used stays to fix the wiring in place so vibration from the machine wouldn't shake the cable connector.

Figure 9-8 shows a method for preventing oil pressure hose breakage. Machine vibration made the hoses pinch and break, causing oil leakage. The simple solution was to wind a spring around the entire hose, which kept a gentle curve that avoided pinching and breakage.

Figure 9-9 shows a way to prevent disconnection of a crimping terminal from the cable attachment of a rapidly turning motor. The problem was solved by attaching a stay to the terminal, fixing it as close to the base as possible.

Figure 9-10 is an example of instant maintenance. Tightening the bolt covers was an irksome step in replacing the V-belt, so the staff carried out the minor improvement of eliminating the bolt, making the hole into a U-slot, and fastening it with a butterfly nut. You can always find a better method for achieving your goal!

Figure 9-11 shows a solution for problems with the location where the stay got bent. When that happened, they had to shut down the machine to fix the bend, and what should have been a minor stoppage often ended up as a major stoppage. This problem led initially to a quick-release connector to make instant maintenance possible.

Figure 9-12 shows how investigating the bent stay problem led to the discovery of a link between it and deviations in product positioning. Detection of position deviations was improved through use of a photoelectric tube.

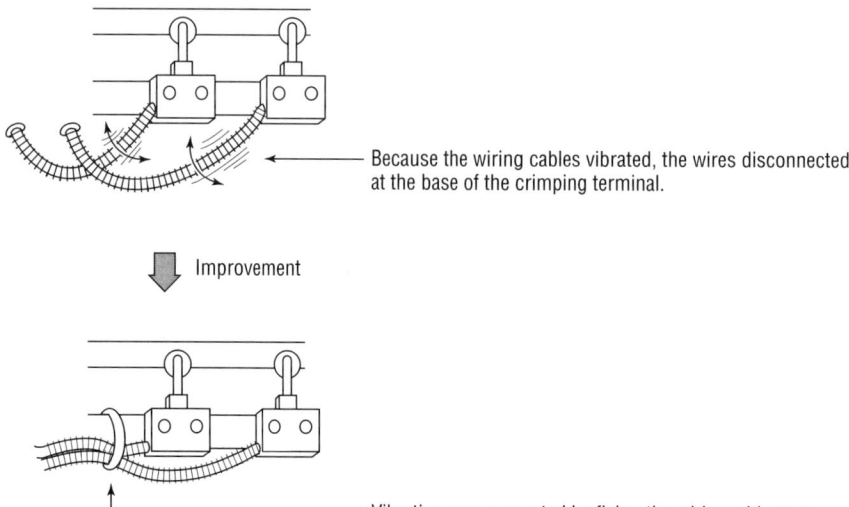

Because the wiring cables vibrated, the wires disconnected at the base of the crimping terminal.

Improvement

Vibration was prevented by fixing the wiring with stays.

Figure 9-7. Plan for Preventing Limit Switch Disconnection

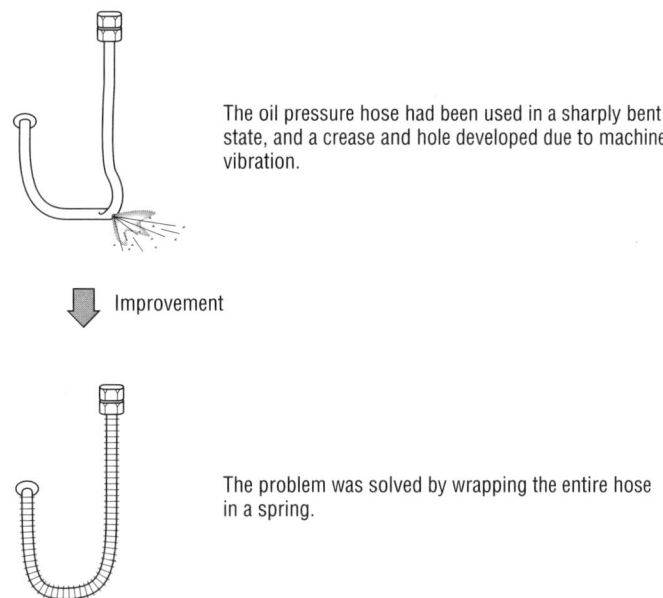

The oil pressure hose had been used in a sharply bent state, and a crease and hole developed due to machine vibration.

Improvement

The problem was solved by wrapping the entire hose in a spring.

Figure 9-8. Preventing Breakage of an Oil Pressure Hose

Figure 9-13 shows how the runner of a cutter was getting caught on the chute and causing minor and medium stoppages. The staff prevented that by installing a photoelectric tube for detecting deviations from the proper position. Figure 9-14 on pages 163 through 165 summarizes a number of related examples and improvement steps.

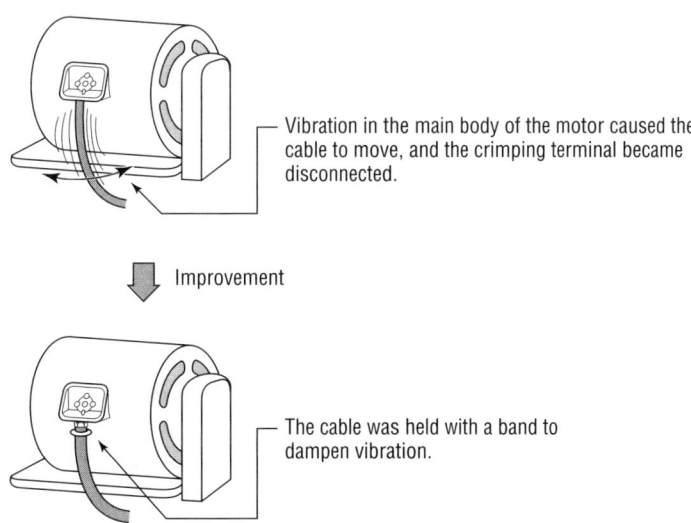

Vibration in the main body of the motor caused the cable to move, and the crimping terminal became disconnected.

Improvement

The cable was held with a band to dampen vibration.

Figure 9-9. Improvement from Stopping Vibration in a Motor Terminal

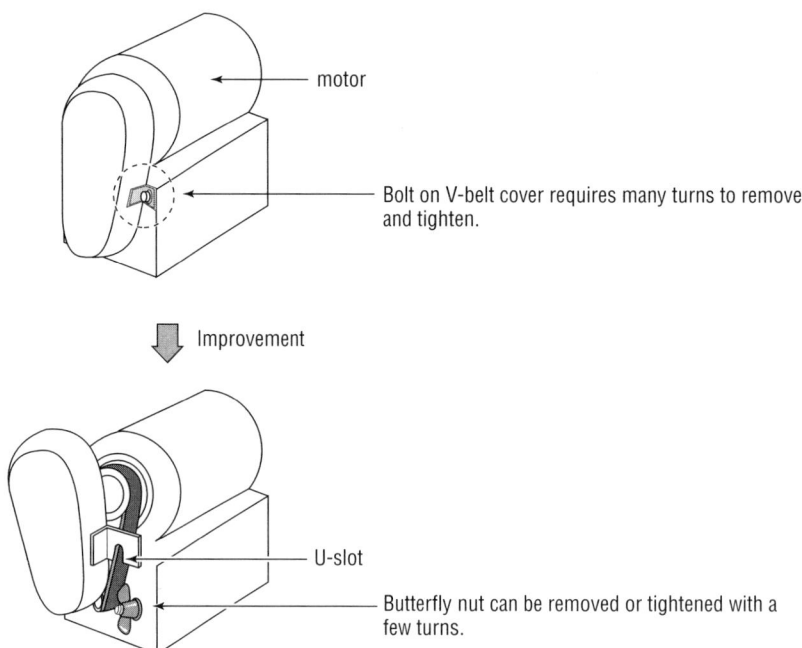

Figure 9-10. Simplifying Replacement of a V-Belt

Daily Inspection Charts

This section shares examples of actual daily inspection charts used in various companies. Companies that have not created their own daily inspection charts can use these for reference, adapting them to their own plants and processes. Even if your plant already uses daily inspection charts, you may want to take this opportunity to reconsider your routine.

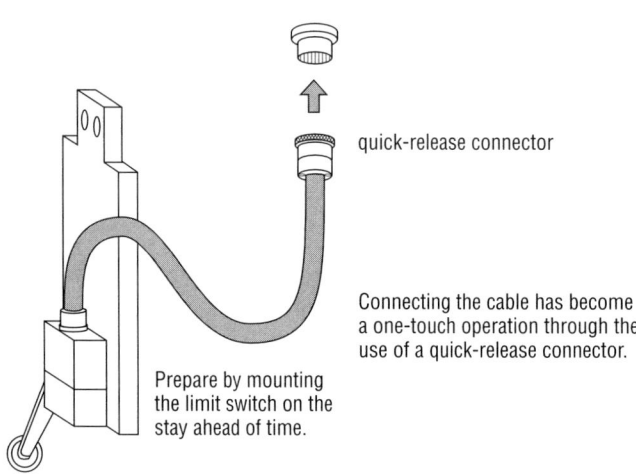

Figure 9-11. Improving Bending of a Limit Switch Stay

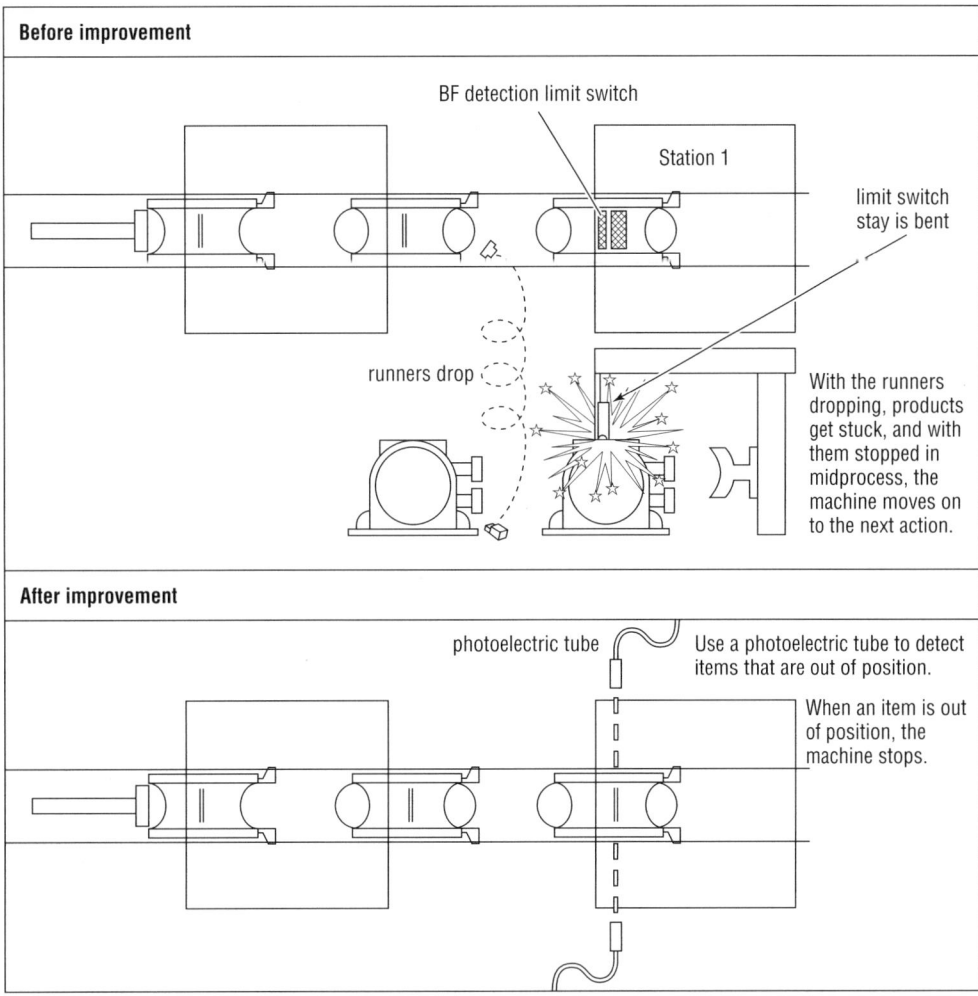

Figure 9-12. Improving Detection of Out-of-Position Workpieces

Here's an important point to remember: You need to predict the areas where minor, medium, and major stoppages are most likely to occur and the places where defects are most likely to occur. For example, you can figure out which parts of which machine tools are most likely to suffer oil stains and leaks, rust, corrosion, contamination by cutting debris, wear, and nicks.

The likely trouble spots may include the gear box, the sliding surface, the tail stock (of a lathe), the motor box, the lead screw, the feed screw, the spindle, the carriage, the table, the cutting oil tank, or the pressure gauge. These points should be checked thoroughly and included on your chart.

Try using some of the following charts to summarize your daily inspection routine.

Daily Inspection Standards for an Electric Motor

Table 9-12 is an example of inspection standards taken from some of our reference materials. If you read carefully, you'll see how it is constructed around the 4Ws and 1H.

Who: Who uses it?

What: What is the object?

Where: Location

How: Inspection standards, inspection spots

When: Weekly, monthly

Once you have determined which locations are to be checked, all you have to do is attach labels and carry out your inspections in the order indicated by the labels.

Instant Maintenance Manual for an Assembler

Figure 9-15 and Table 9-13 are taken from the instant maintenance manual and a check sheet used at the start of each shift at company K. Since diagrams are attached to the information, even new employees can start participating in inspections immediately.

Daily Inspection Chart for a Mold

Figure 9-16 shows the daily inspection chart for a mold from company D, a plastics manufacturer. This is a powerful and comprehensive instrument, but it takes time to complete, which cuts heavily into operating time. That's why company N has adopted inspection charts like the one in Figure 9-17, which condense the inspection items into a more manageable form.

Inspection Items for a Solder Tank

Figure 9-18 shows the inspection items for a solder tank at company H. The numbers written on this chart are transferred to small labels and posted on the solder tank. This indicates the order in which the operators carry out inspections at the appointed time.

An Inspection Sheet for an Electronic Device

Figure 9-19 is a check sheet used at company C. This device is several meters long, and since it is impossible to walk up and down both sides, they have separate check sheets for each side. This is a unique check sheet, incorporating the wisdom of the workplace.

Other Inspection Charts

As Figures 9-20 and 9-21 suggest, there as many kinds of check sheets as there are factories, and they may be the best summary of how inspections are conducted at the work site. We're going to avoid the question of which type of check sheet is best; each has its advantages and disadvantages. The best check sheet is the one that's easiest to use for your plant and the processes that take place in it.

Test Yourself

The cause and effect diagram shown in Figure 9-22 concerns oil leaks in an oil pressure device. See if you can fill in the blank numbered boxes with problems to check for.

Our Answers

1. Oil temperature too high
2. Wrong type of oil
3. Joint defect
4. Contact
5. Attachment bolt loose

6. Useful life
7. Dirt
8. Dirt
9. Bad attachment
10. Defective seal material

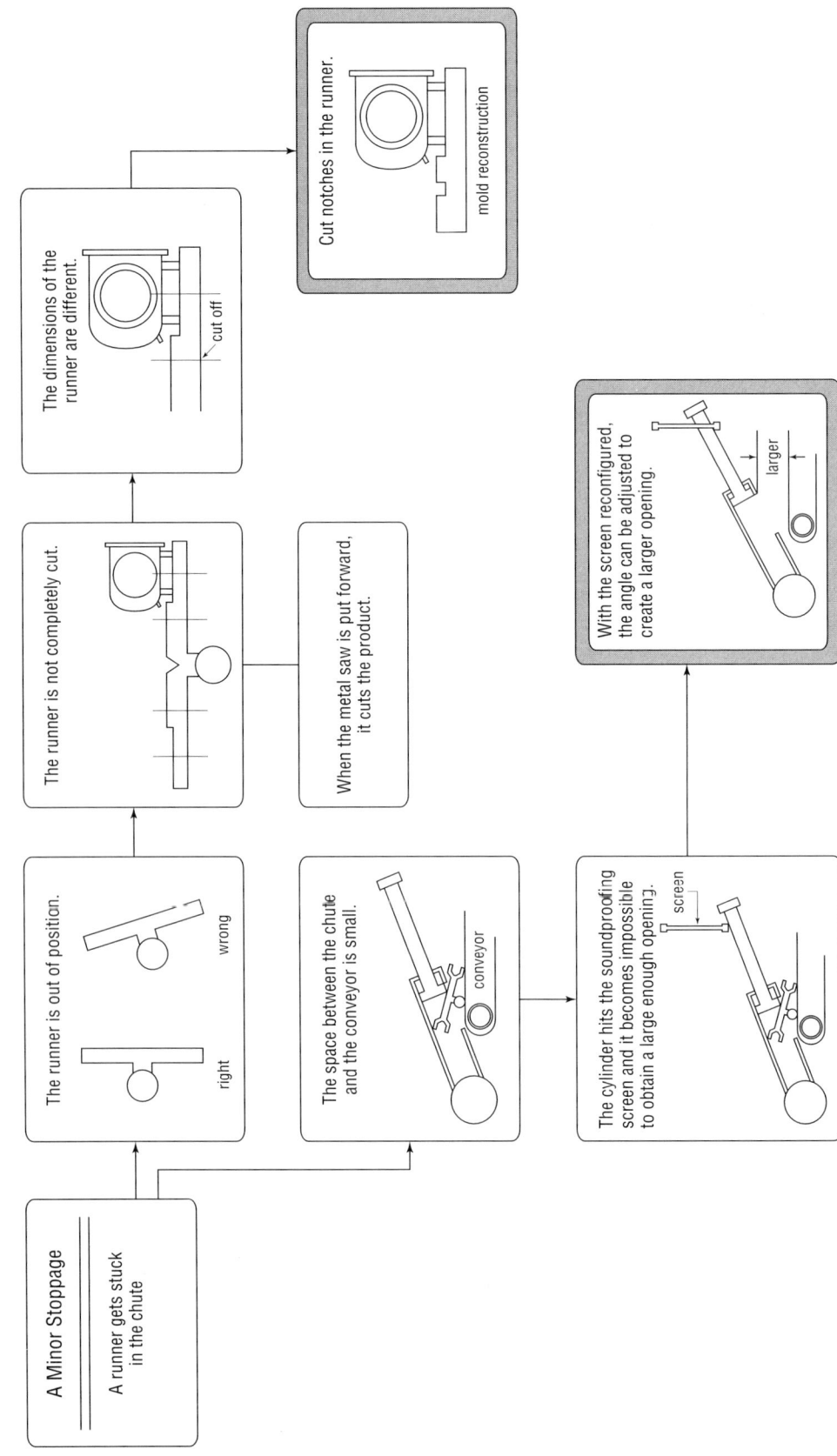

A Minor Stoppage

A runner gets stuck in the chute

The runner is out of position.

right

wrong

The runner is not completely cut.

When the metal saw is put forward, it cuts the product.

The dimensions of the runner are different.

cut off

Cut notches in the runner.

mold reconstruction

The space between the chute and the conveyor is small.

conveyor

The cylinder hits the soundproofing screen and it becomes impossible to obtain a large enough opening.

screen

With the screen reconfigured, the angle can be adjusted to create a larger opening.

larger

Figure 9-13. Improvement of a Y₂ Body Cutter

162

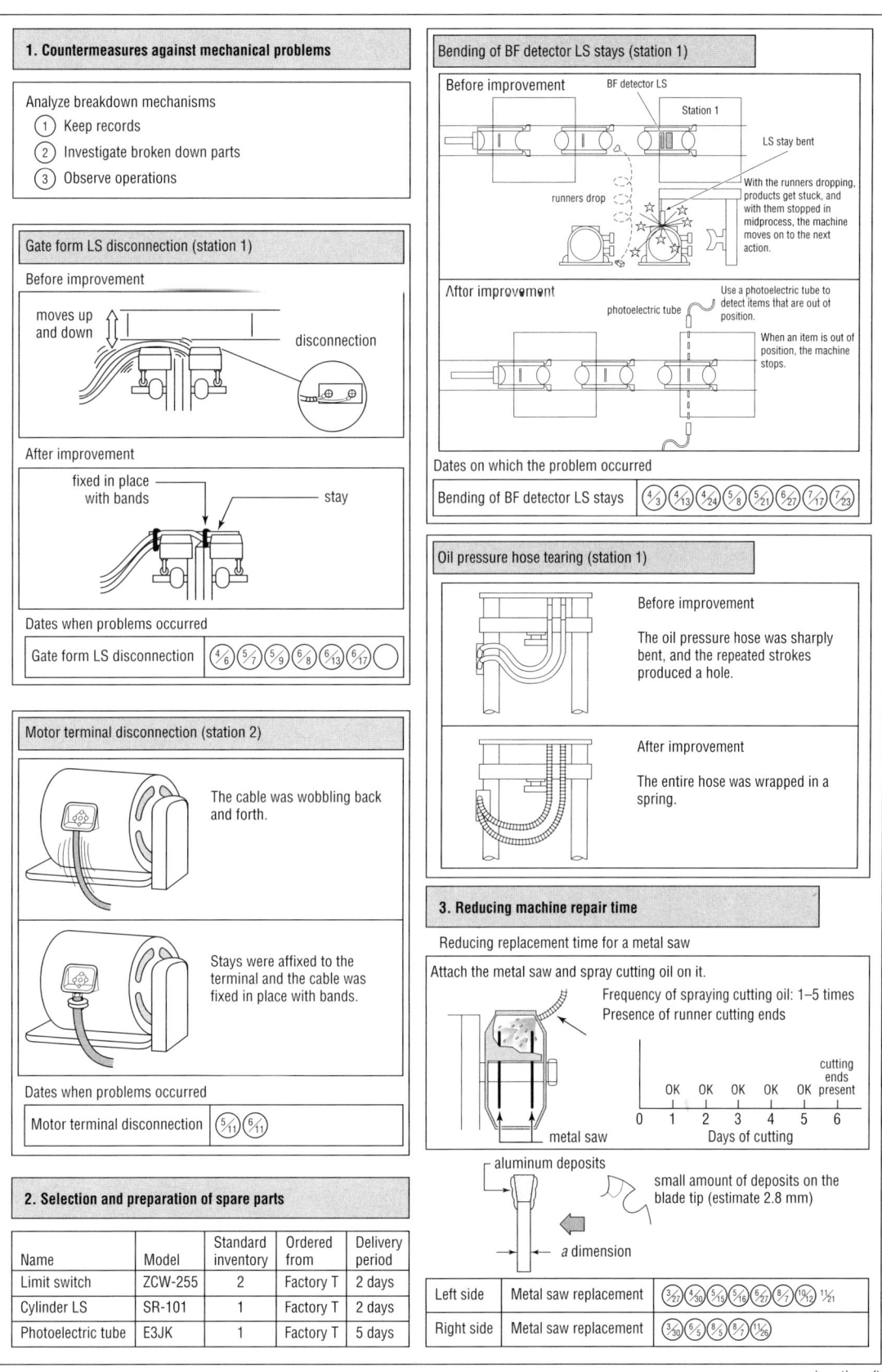

1. Countermeasures against mechanical problems

Analyze breakdown mechanisms
①　Keep records
②　Investigate broken down parts
③　Observe operations

Gate form LS disconnection (station 1)

Before improvement

moves up and down

disconnection

After improvement

fixed in place with bands

stay

Dates when problems occurred

| Gate form LS disconnection | 4/6 5/7 5/9 6/8 6/13 6/17 |

Motor terminal disconnection (station 2)

The cable was wobbling back and forth.

Stays were affixed to the terminal and the cable was fixed in place with bands.

Dates when problems occurred

| Motor terminal disconnection | 5/11 6/11 |

2. Selection and preparation of spare parts

Name	Model	Standard inventory	Ordered from	Delivery period
Limit switch	ZCW-255	2	Factory T	2 days
Cylinder LS	SR-101	1	Factory T	2 days
Photoelectric tube	E3JK	1	Factory T	5 days

Bending of BF detector LS stays (station 1)

Before improvement

BF detector LS

Station 1

LS stay bent

runners drop

With the runners dropping, products get stuck, and with them stopped in midprocess, the machine moves on to the next action.

After improvement

photoelectric tube

Use a photoelectric tube to detect items that are out of position.

When an item is out of position, the machine stops.

Dates on which the problem occurred

| Bending of BF detector LS stays | 4/3 4/13 4/24 5/8 5/21 6/27 7/17 7/23 |

Oil pressure hose tearing (station 1)

Before improvement

The oil pressure hose was sharply bent, and the repeated strokes produced a hole.

After improvement

The entire hose was wrapped in a spring.

3. Reducing machine repair time

Reducing replacement time for a metal saw

Attach the metal saw and spray cutting oil on it.

Frequency of spraying cutting oil: 1–5 times
Presence of runner cutting ends

metal saw

OK　OK　OK　OK　OK　cutting ends present

0　1　2　3　4　5　6　Days of cutting

aluminum deposits

small amount of deposits on the blade tip (estimate 2.8 mm)

a dimension

Left side	Metal saw replacement	3/27 4/30 5/19 5/16 6/27 8/1 10/12 11/21
Right side	Metal saw replacement	3/30 6/5 8/5 8/1 11/26

(continued)

Figure 9-14. Collected Improvement Examples, continued

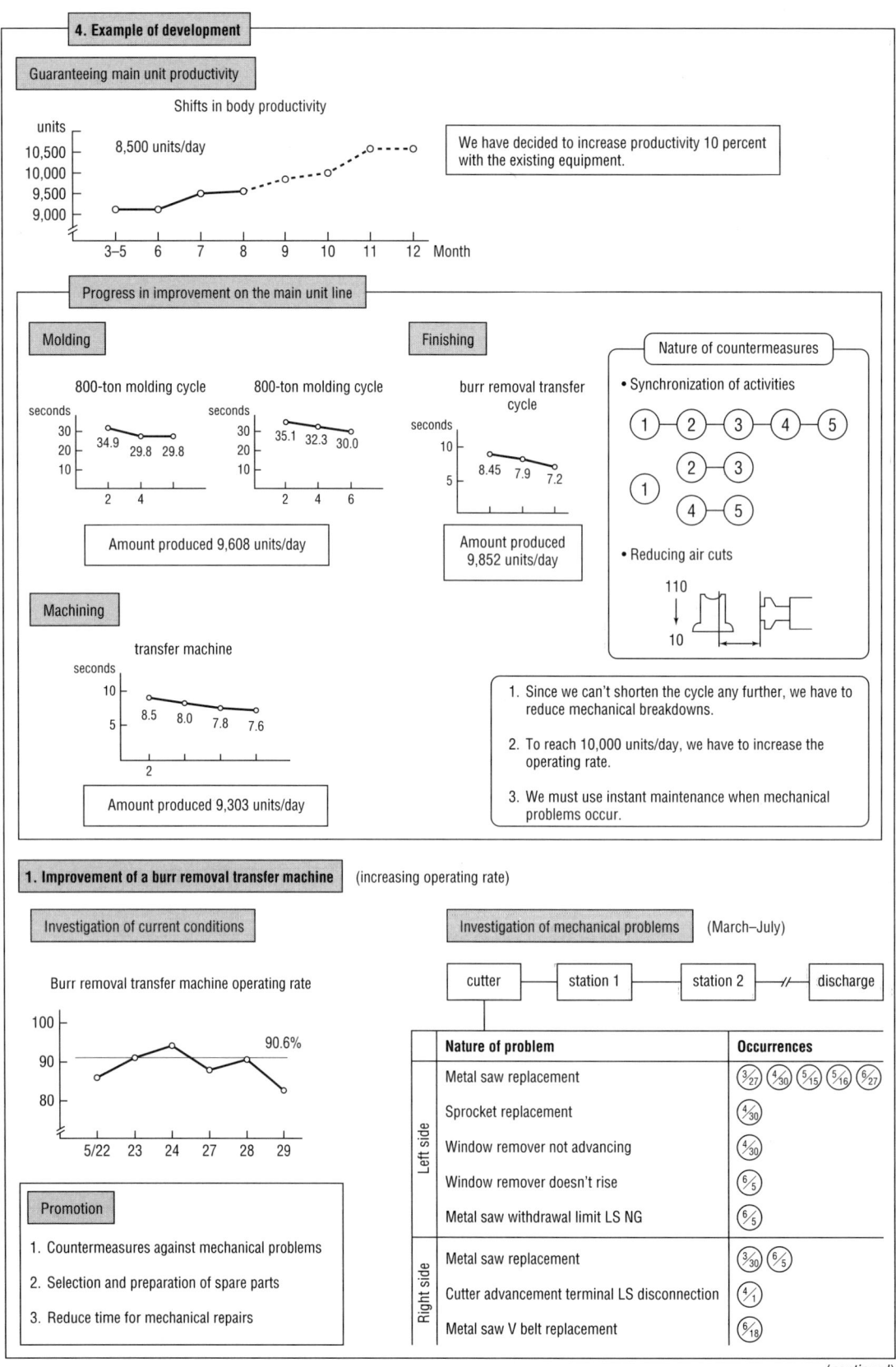

Figure 9-14. Collected Improvement Examples, continued

(continued)

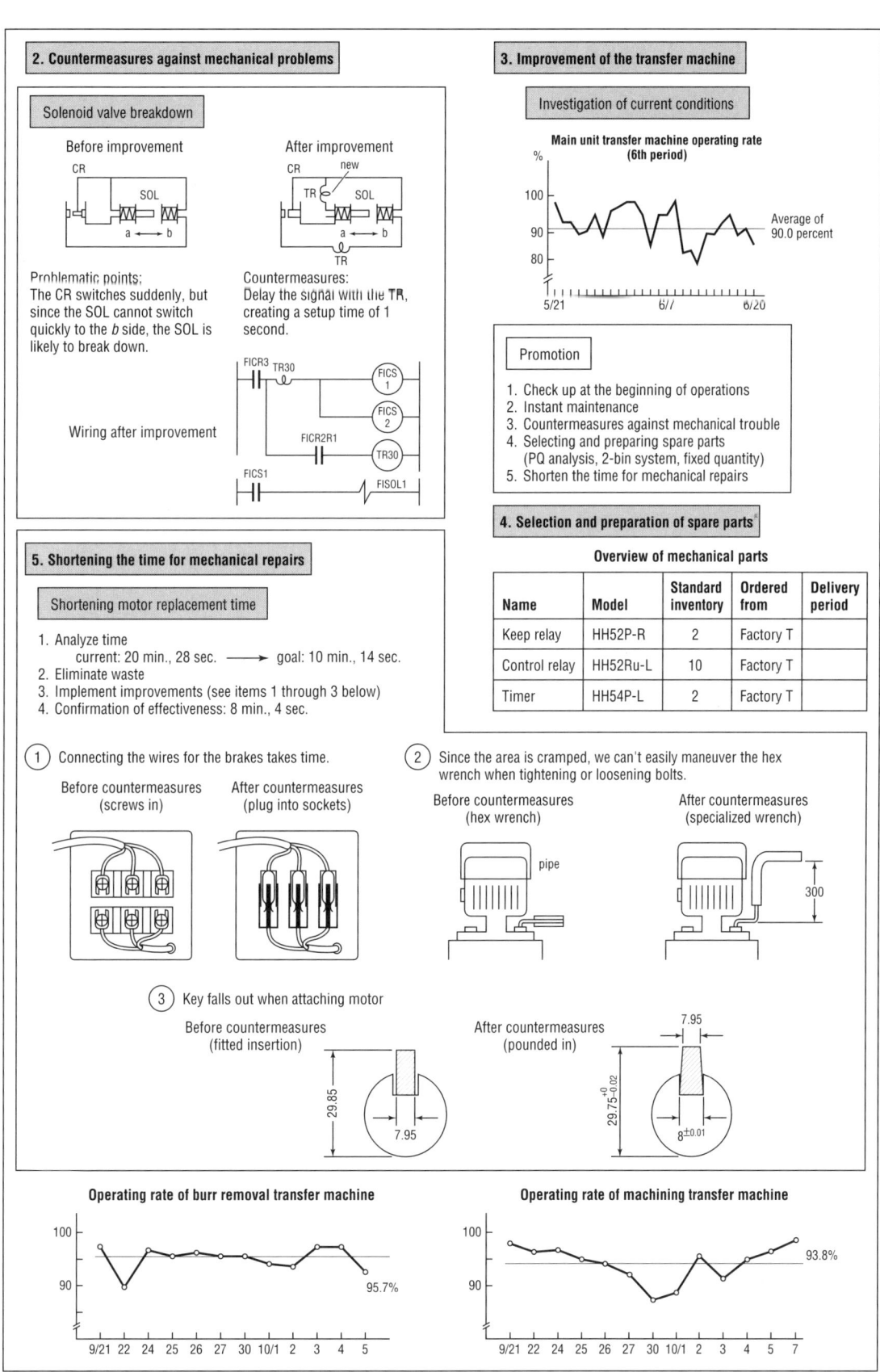

2. Countermeasures against mechanical problems

Solenoid valve breakdown

Before improvement   After improvement

Problematic points:
The CR switches suddenly, but since the SOL cannot switch quickly to the *b* side, the SOL is likely to break down.

Countermeasures:
Delay the signal with the TR, creating a setup time of 1 second.

Wiring after improvement

5. Shortening the time for mechanical repairs

Shortening motor replacement time

1. Analyze time
   current: 20 min., 28 sec. → goal: 10 min., 14 sec.
2. Eliminate waste
3. Implement improvements (see items 1 through 3 below)
4. Confirmation of effectiveness: 8 min., 4 sec.

① Connecting the wires for the brakes takes time.

Before countermeasures (screws in)   After countermeasures (plug into sockets)

② Since the area is cramped, we can't easily maneuver the hex wrench when tightening or loosening bolts.

Before countermeasures (hex wrench)   After countermeasures (specialized wrench)

③ Key falls out when attaching motor

Before countermeasures (fitted insertion)   After countermeasures (pounded in)

3. Improvement of the transfer machine

Investigation of current conditions

Main unit transfer machine operating rate (6th period)

Average of 90.0 percent

Promotion

1. Check up at the beginning of operations
2. Instant maintenance
3. Countermeasures against mechanical trouble
4. Selecting and preparing spare parts
   (PQ analysis, 2-bin system, fixed quantity)
5. Shorten the time for mechanical repairs

4. Selection and preparation of spare parts

Overview of mechanical parts

Name	Model	Standard inventory	Ordered from	Delivery period
Keep relay	HH52P-R	2	Factory T	
Control relay	HH52Ru-L	10	Factory T	
Timer	HH54P-L	2	Factory T	

Operating rate of burr removal transfer machine

95.7%

Operating rate of machining transfer machine

93.8%

Figure 9-14. Collected Improvement Examples, continued

165

Table 9-12. Inspection Standards for Electric Motor with Eddy Current Joint

Inspection Standards for Electric Motor with Eddy Current Joint

Equipment location   Plant A, line B
Place manufactured   K Electronics Company

Part/Inspected area	Inspection items	Inspection method	Judgment standards	Treatment method	Inspection cycle Ops.	Inspection cycle Maint.
Electric motor main unit						
Cooling and ventilation window (cooling water pump in water-cooled systems)	Intake and discharge (water)	Touch intake and discharge window. (Grip outflow pipe to check temperature. Check water cutoff detector.)	Make sure intake and discharge is normal (passing water through at 25°C or less).	Contact maint. dept., disassemble, check	Startup	(M)
	Main unit	Abnormal noise, vibration, overheating	Test bearings and frame by touch.	Make sure there are no sporadic or metallic noises; less than 25 μ; room temperature + 30°C or less.	Contact maint. dept., disassemble, check	During operations	(M)
High speed detection generator						
Drive V pulley	Wear	Operate by hand to make sure there's no vibration	No rattling or loosening	Adjustment, repair		M
	Operating status	Variation or unsteadiness	Variation or unsteadiness	Adjustment, repair		(M)
Drive V belt	Stretching	Watch for slipping	Operating smoothly	Replacement and repair		(M)
	Damage	Look for wear, cracking, and other damage	No damage	Replacement		M
Induction wire	Outer appearance	Visually check for damage or deterioration	No damage or deterioration	Replacement		(M)
Output terminal	Slackening	Tighten with a screw driver				M

(continued)

Table 9-12. Inspection Standards for Electric Motor with Eddy Current Joint, continued

Inspection Standards for Electric Motor with Eddy Current Joint

Equipment location: Plant A, line B
Place manufactured: K Electronics Company

Part/Inspected area	Inspection items	Inspection method	Judgment standards	Treatment method	Inspection cycle	
					Ops.	Maint.
Speed controller						
Interior of control panel	Temperature	Confirm the color of the thermotape.	No change in color	Replacement and repair		Ⓜ
Tachometer	Operation status	Turn the speed setting volume and observe the tachometer.	Indicator needle rises and falls smoothly.	Adjustment, repair		Ⓜ
	Zero point	Confirm the zero point of the indicator needle.	Make sure the zero points are uniform.	Adjustment		M

Notes

1. The inspection cycle in the operations departments is once per shift.
2. Maintenance cycles: M = once a month ◯ = during operation ☐ = while turned off or after turning off

Person responsible	Date revised	Revisions and reasons	Person responsible
JW			

Filled out by	TW	Date instituted	10/15	Issued by	Mgr., Maintenance Dept. 1

167

Date generated: _____   Name: _____

Items checked by operators

Label no.	Part inspected	Standard
△1	Filter inspection	No debris adhering
△2	Pressure plate	No debris on rubber plate
△3	Fallen screws	No screws around machine

Items to be checked by chief engineer

G1	Replace O-ring in bit	Degree of screw adsorption
G2	Replace hose	No nicks or dirt inside hose
G3	Torque adjustment	Check with torque meter

Devices for chief engineer to inspect and measure

Indicator lamp circuit

R1 — Filter check
R2 — Pressure plate check
R3 — Screw sensor
R4 — Emergency stoppage detection
PL — Inspection completed lamp

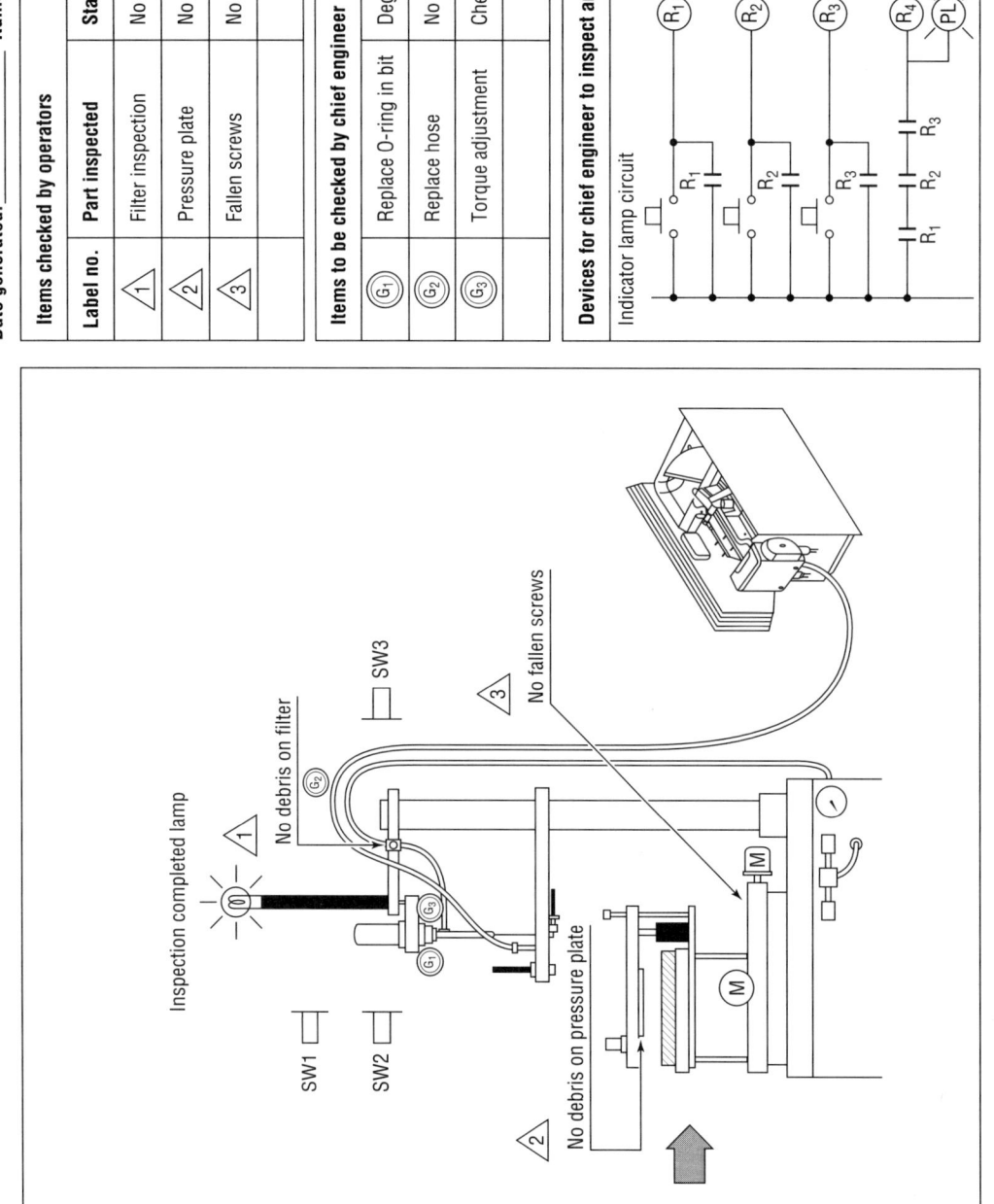

Figure 9-15. Daily Maintenance Sheet

Table 9-13. Pre-Shift Check Sheet

Before improvement

Safety department → Safety office → Supervisor
(maintenance)

Pre-shift check sheet

Month _____ Machine number _____

		Section manager	Center manager	Team leader	Supervisor

Daily inspection items

Inspection date

Inspection item	Summary	1	2	3	4	5	6	7	8	9	10	11	12	13	14	15	16	17	18	19	20	21	22	23	24	25	26	27	28	29	30	31
1. Filter inspection		✓	✓	✓	✓	✓	✓	✓	✓																								
2. Pressure plate		✓	✓	✓	✓	✓	✓	✓	✓																								
3. Fallen screws		✓	✓	✓	✓	✓	✓	✓	✓																								
4. Original air pressure, 6 kg/cm²		✓	✓	✓	✓	✓	✓	✓	✓																								
5. Presence of screws in feeder		✓	✓	✓	✓	✓	✓	✓	✓																								
6. Screw sensor lamp lit		✓	✓	✓	✓	✓	✓	✓	✓																								

Defective areas	Special notes about defect locations
	Processing conditions
1	Screws blocked

Summary of notations

X = Needs repair
⊗ = Repair adjustment completed
✓ = No abnormality
(Report any abnormalities discovered to the team leader.)

How to use this check sheet

1. The supervisor will record the daily inspections.
2. Team leaders will approve the check sheets once a week.
3. The safety office will keep this record for one year.

Daily Inspection Chart for Molds

Mold   13651

June   Molding machine   3

	Rank			Inspection time			Key to results	Revision	Nature of and reason for revisions	Date	Approval	Approval	Approval
A	Every 3 hours	○		= At startup		○	= Passed	1					
B	Weekly	△		= During repairs		X	= Failed	2					
C	Quarterly					⊕	= Passed after treatment	3					
						|	= Operations halted	4					

Responsible	Line	Inspection items		Areas (drawing)	Time	6 ①	②	3	4	5	6	7	⑧	⑨	10	11	12	13	14	⑮	16	17	18	19	20	21	㉒	㉓	24	25	26	27	28	㉙	㉚
Op.	A	Stripper	Dirty PL surface	1	○	○	○	○																												
	A		Scoring on PL surface	1	○	○	○	○																												
Super.	B		PL surface blow check	1	○	○	○	○																												
Op.	A		Dirty taper lock	1	○	○	○	○																												
	A	Cavity	Dirty PL surface	2	○	○	○	○																												
	A		Scoring on PL surface	2	○	○	○	○																												
Super.	B		PL surface blow check	2	○	○	○	○																												
Op.	A		Dirty taper lock	2	○	○	○	○																												
Super.	C	Guide pin	Scoring check	3	△					○																										
	C		Bending check	3	△					○																										
	C	Cooling water pipes	Blockage check	4	△					X																										
	C	Ejector lot	Scoring check	5	△					○																										
	C		Bending check	5	△					○																										
	C		Check for wear on tip	5	△					○																										
	C	Spoor bush	Scoring check	6	△					○																										
	C		Sagging check	6	△					○																										
	C		Deformation check	6	△					○																										
	C		Resin leakage check	6	△					○																										
	C	Junction box	Teminal loosening check	7	△					○																										
	C		Lead check	7	△					○																										
	C		Wire deterioration check	7	△					○																										
Op.	B	Wear ring	Dirty taper lock	8	○	○	○	○																												
Super.	B		Taper lock sagging	8	○	○	○	○																												
		Inspection	Operator, supervisor	1/mo.				S		S		S																							
		Confirmation	Supervisor, person in charge of molding	2/mo.				U		U		U																							

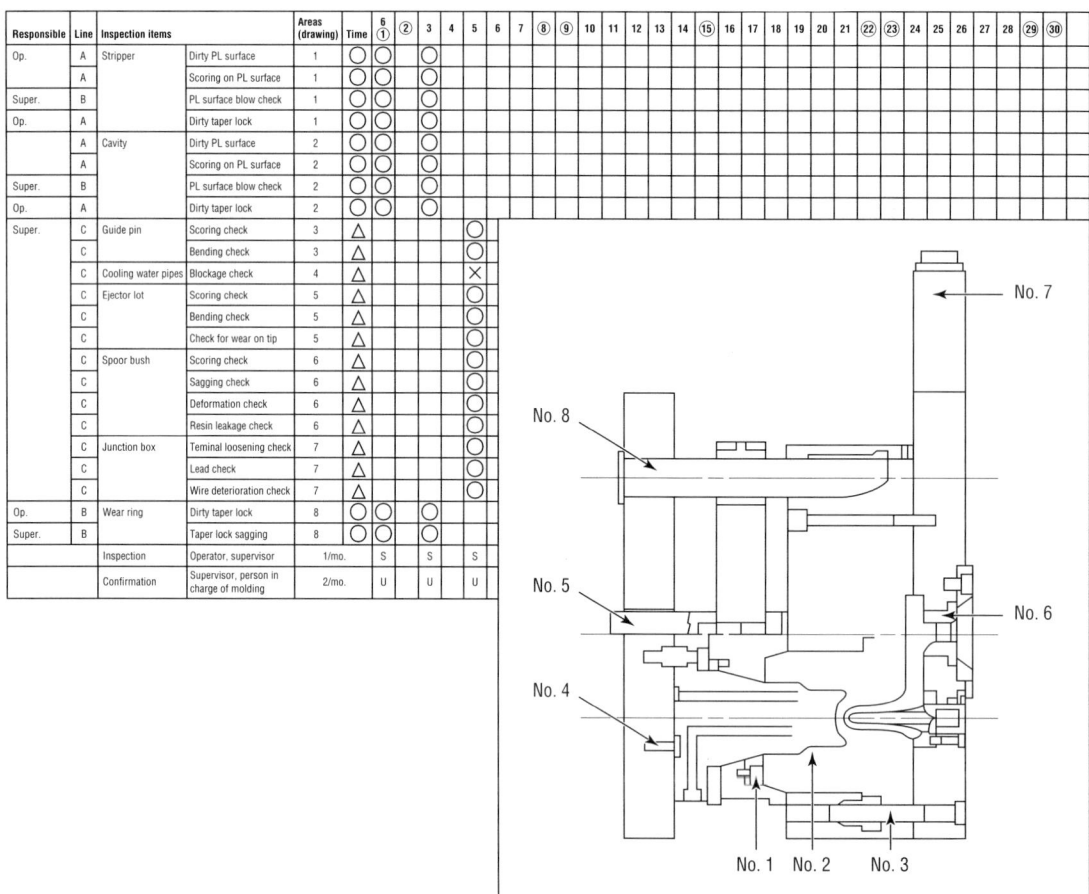

Inspection Standards

No.	Areas to be inpected	Inspection items	Inspection methods	Measurement tools	Inspection standards	Treatment	Cycle	Inspector
1	Stripper	Dirty PL surface	Visual	Sight	No plastic gas	Wipe off	1/3 hr.	Op.
		PL surface scoring	Visual	Sight	Not out of grease, no scoring	Increase grease	1/3 hr.	Op.
		PL surface blow check	Look at striking surface during blow check	Blowing	Not striking air vent	Shim panel adjustment	1/wk.	Super.
		Dirty taper lock	Visual	Sight	No plastic gas	Increase grease	1/3 hr.	Op.
2	Cavity	Dirty PL surface	Visual	Sight	No plastic gas	Shim panel adjustment	1/3 hr.	Op.
		PL surface scoring	Visual	Sight	Not out of grease, no scoring	Increase grease	1/3 hr.	Op.
		PL surface blow check	Look at striking surface during blow check	Blowing	Not striking air vent	Shim panel adjustment	Once a week	Super.
		Dirty taper lock	Visual	Sight	No plastic gas	Wipe off	1/3 hr.	Oper.
3	Guide pin	Scoring check	Visual	Sight	Not out of grease, no scoring	Treat surface with sandpaper, increase grease	1/3 mo.	Super.
		Bending check	Measure the gap gauge	Gap gauge	Bent less than 0.5 mm	Replacement	1/3 mo.	Super.
4	Cooling water pipe	Blockage check	Run water through	Flow meter	Over 3.2 kg/cm^3	Disassemble and check	1/3 mo.	Super.
5	Ejector lot	Scoring check	Visual	Sight	No scoring	Check nuts and bolts	1/3 mo.	Super.
		Bending check	Visual	Sight	No bending	Replacement	1/3 mo.	Super.
		Check for tip wear	Visual	Sight	No wear	Replacement	1/3 mo.	Super.

Figure 9-16. Daily Inspection Chart

Mold Maintenance Check Chart

Mold: CN-111   Customer: Company S   Date: 7/8

The machine operators perform daily maintenance and inspections at afternoon shift change. When finished, they date and initial the form. At the end of production, the person who replaces the molds confirms the inspection items. Samples are taken as required.

Division	Inspector	No.	Maintenance items	7/1	/3	/5	/	/	/	/	/	/	/	/	/	/	/	/	/	/	/	/	/
Q	Operator	1	PL surface gas removal	○	○	○																	
P	"	2	Knock pin lubricant coating	○	○	○																	
P	"	3	Grease coating on guide and angular pin	○	○	○																	
			Inspection item																				
P	"	1	No abnormal noise when knock is activated	○	○	○																	
Q	"	2	No dirt on knock pin	○	○	○																	
P	"	3	Slide action normal	○	○	○																	
P	"	4	Knock panel returns smoothly	○	○	○																	
P	"	5	Confirm looseness of stopper pin bolts	○	○	○																	
			Inspector's initials	MK	MK	MK																	

Division   Q = Quality   P = Production

Items to be checked at the end of production
1. Nicks in the mold (yes—no)
2. Grain off (yes—no)
3. Cooling water leakage (yes—no)
4. Scoring in mold (yes—no)
5. Nozzle radius needs to be rebent (yes—no)
6. Mold release needs polishing (yes—no)
7. Other necessary adjustments and corrections

Confirmation

Confirmation

Confirmation at end of production   Chief clerk →   Maintenance →   Section Mgr. →   Section Mgr. →   File

Figure 9-17. Mold Maintenance Check Chart

171

Solder tank maintenance inspection items

Item / Frequency	Day 0.5 / Time 4 HR.	1 / 8	2 / 16	6 / 50	25 / 200	100 / 800	150 / 1200	300 / 2400	Comments
1. Solder tank									
a. Eliminate oxidized film	○								
b. Clean solder cast box	○								
c. Adjust skinner chain							○		
d. Oil supply to guide shaft and solder dip cylinder				○					
e. Grease and oil skinner chain						○			
f. Solder replacement							○		
g. Cleaning solder heater					○				
2. Fluxer management equipment									
a. Clean bubbling tube					○				
b. Flux replacement					○				
c. Specific gravity management	○								
d. Clean sensor attachment system mgmt. equip.					○				
e. Clean mesh plate under pipe flux/alcohol tanks					○				
3. Pre-heater									
a. Replace aluminum foil on reflective plate					○				
4. Air									
a. Drain	○								
b. Regulator air pressure mgmt.		○							
c. Replicator oil mgmt.					○				
d. Clean silencer					○				
5. Rail conveyor									
a. Cleaning and oiling of carrier rail		○							
b. Clean conveyor belt		○							
6. Chain									
a. Grease and oil carrier feed chain						○			
b. Adjust carrier feed chain							○		
c. Grease and oil elevator chain						○			
7. Cylinder									
a. Oil each cylinder guide shaft				○					
8. Stop-up									
a. Oil stop-up bearing			○						
b. Clean stop-up carrier cradle			○						
9. Carrier									
a. Clean carrier			○						
b. Clean carrier bracket		○							
c. Clean canopy		○							
10. Conveyor									
a. Clean all conveyor belts		○							
b. Adjust all conveyors							○		
11. Left elevator									
a. Oil slide axle		○							
b. Elevator support oil supply		○							
c. Oil supply for elevator conveyor drive shaft		○							
d. Clean all conveyors		○							

Figure 9-18. Solder Tank Maintenance Inspection Items

Upper surface

D = daily
M = every Monday

Date _____

	Maint.	Inspector
		HY

No.	Equip.	When	Elements	Maintenance/inspection items	Check	Remarks
1	CP-3	D	Parts receptacle box	Remove parts that are in error	○	
2		M	X-Y table	Oil the LM guide	○	
3	CM60	D	Moving head	Remove debris embedded in sliding parts	○	
4				Remove cream solder adhering to nozzle	○	
5		M	Moving head	Remove debris embedded in nozzle guide of nozzle holder	○	
6	PA10	D	Ball feeder	Clean interior of ball feeder	○	
7				Is the pin inside the ball feeder?	○	
8			Underside solder recovery	Clean underside solder collector	○	
9			Underside pin holder	Clean underside pin holder nozzle	○	
10		M	Shutter	Oil shutter slide	○	
11	MR89 -UVC	M	Profile	Temperature profile measurement	○	

Note: If there's a malfunction during inspection, write an X in the appropriate column, and circle it once the problem has been remedied.
Do not input any products until corrections have been completed.

Underside

D = daily
M = every Monday
2M = every 2nd Monday

Date _____

	Maint.	Inspector
		HY

No.	Equip.	When	Elements	Maintenance/inspection items	Check	Remarks
1	CP-3	D	Leftover tape receptacle	Remove leftover tape	○	
2			Prism box	Clean prism surface	○	
3		M	Device table	Oil LM guide	○	
4				Oil toggle clamp	○	
5	CM60	D	Air device	Remove water inside MR unit filter	○	
6		M	Conveyor	Belt wear?	○	
7	PA10	D	Air device	Remove water inside air filter	○	
8		M	Air device	Clear convam filter	○	
9	Compr.	2M	Oil gauge	Amount of baby condenser foil	○	

Notes:
1. If there's a malfunction during inspection, write an X in the appropriate column, and circle it once the problem has been remedied.
Do not input any products until corrections have been completed.
2. For M and 2M inspections, if Monday is a holiday, conduct the inspection on the next working day.
3. For item 9, report to the supervisor if the oil gauge indicates less than half full.

Keep for one year.

Figure 9-19. Check Sheet for Beginning of Shift

TPM Equipment Inspection Chart

Equipment name: CNC Lathe Team: Western Person responsible: Nakai

Step	No.	Areas	Cycle	1	②	③	4	5	6	7	8	9	⑩	11	12	13	14	15	⑯	⑰	18	19	20	21	22	23	㉔	25	26	27	28	29	30	Time (min./month)
Cleaning	1	Top of main unit	2/week															Mn			Mn	Mn			Mn	Mn		Mn	Mn	Mn		Mn		
	1-1	Limit switch	2/week															OK			OK	OK			OK	OK		OK	OK	OK		OK		
Related to minor stoppages	2	Upper part of head surface	3/week	Mn			Mn		Mn		Mn			Mn		Mn		Mn			Mn	Mn	Mn		Mn			Mn		Mn		Mn		
	2-1	X axis, centering, sliding surface	3/week	OK			OK		OK		OK			OK		OK		OK			OK	OK	OK		OK			OK	OK	OK		OK		
	3	Bottom of main unit	1/week					Md							Md							Md							Md					
	3-1	Leveling bolts (7 places)	1/week					OK							OK							OK							OK					
	3-2	Rear oil pressure unit gauge						OK							OK						OK	OK							OK					
	3-3	Rear oil pressure pump						OK							OK						OK	OK							OK					
Lubrication	4	Chuck and grease nipple (3 places)	1/week/Th	S							S						S								S							S		
Related to 3-1	5	Sliding surface oil gauge	2/week	③				④			③				③			③				④			③				③				③		
	6	Machining liquid tank gauge	1/week/W						③														③											
Inspection	7		1/week/ 4th Tues.																															
		Inspection																																

Mn = Minor   Md = Medium   S = Supply

Figure 9-20. NC Lathe Inspection Chart (Before and After Revision)

After revision

TPM Equipment Inspection Chart																																			Equipment name: CNC Lathe / Team: Western / Person responsible: Nakai

Step	No.	Areas	Cycle	1	2	3	4	5	6	7	8	9	10	11	12	13	14	15	16	17	18	19	20	21	22	23	24	25	26	27	28	29	30	Time (min./month)
Cleaning	1	Top of main unit	1/week							Md							Md							Md							Md			
	1-1	Limit switch	1/week/Tues.							OK							OK							OK							OK			
	2	Upper part of head surface	2/week		Md					Md		Md					Md		Md					Md		Md					Md		Md	
	2-1	X axis, centering, sliding surface	2/week/T-Th		OK					OK		OK					OK		OK					OK		OK					OK		OK	
	3	Bottom of main unit	1/week/Th		Md							Md							Md							Md							Md	
	3-1	Leveling bolts (7 places)			5																													
	3-2	Rear oil pressure unit gauge																																
	3-3	Rear oil pressure pump			OK							OK							OK							OK							OK	
Lubrication	4	Chuck and grease nipple (3 places)	1/week/Th		S							S							S							S							S	
	5	Sliding surface oil gauge	1/week/F			2							2							1							1							
	6	Cutting fluid tank gauge	1/month/beginning		2																													
Inspection	7	Oil pressure gauge (check)	1/month/4th Tues.																					20										
	8	Oil pressure gauge (tool)	1/month/4th Tues.																					30										
	9	Oil pressure gauge (centering)	1/month/4th Tues.																					4										
		Inspection	Every day		8	3			7	3	8	1					7	1	11	1			3	10	1	80	1			5	7	1	8	
		Team leader checks	Once a day																															
		Site manager checks	Once a week																															94 min.

Cleaning – Major-medium-minor • **Lubrication** – Numbers (amounts of oil) • **Inspection** – Numbers (pressure temperature, etc.) OK, X

Figure 9-20. NC Lathe Inspection Chart (Before and After Revision)

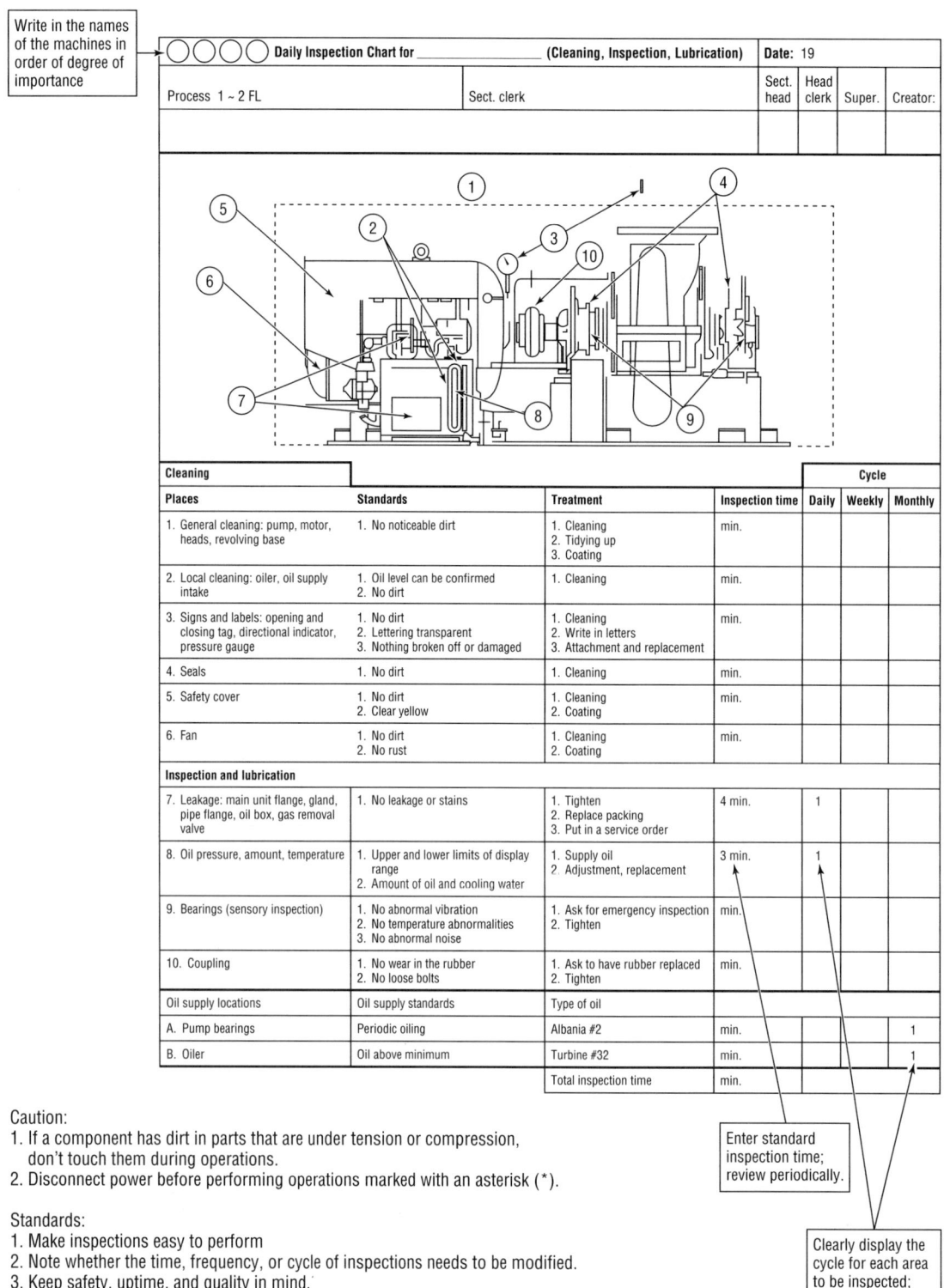

Write in the names of the machines in order of degree of importance

Daily Inspection Chart for _____ (Cleaning, Inspection, Lubrication)			Date: 19			
Process 1 ~ 2 FL		Sect. clerk	Sect. head	Head clerk	Super.	Creator:

Cleaning | | | | | **Cycle** | | |

Places	Standards	Treatment	Inspection time	Daily	Weekly	Monthly
1. General cleaning: pump, motor, heads, revolving base	1. No noticeable dirt	1. Cleaning 2. Tidying up 3. Coating	min.			
2. Local cleaning: oiler, oil supply intake	1. Oil level can be confirmed 2. No dirt	1. Cleaning	min.			
3. Signs and labels: opening and closing tag, directional indicator, pressure gauge	1. No dirt 2. Lettering transparent 3. Nothing broken off or damaged	1. Cleaning 2. Write in letters 3. Attachment and replacement	min.			
4. Seals	1. No dirt	1. Cleaning	min.			
5. Safety cover	1. No dirt 2. Clear yellow	1. Cleaning 2. Coating	min.			
6. Fan	1. No dirt 2. No rust	1. Cleaning 2. Coating	min.			

Inspection and lubrication

7. Leakage: main unit flange, gland, pipe flange, oil box, gas removal valve	1. No leakage or stains	1. Tighten 2. Replace packing 3. Put in a service order	4 min.	1		
8. Oil pressure, amount, temperature	1. Upper and lower limits of display range 2. Amount of oil and cooling water	1. Supply oil 2. Adjustment, replacement	3 min.	1		
9. Bearings (sensory inspection)	1. No abnormal vibration 2. No temperature abnormalities 3. No abnormal noise	1. Ask for emergency inspection 2. Tighten	min.			
10. Coupling	1. No wear in the rubber 2. No loose bolts	1. Ask to have rubber replaced 2. Tighten	min.			
Oil supply locations	Oil supply standards	Type of oil				
A. Pump bearings	Periodic oiling	Albania #2	min.			1
B. Oiler	Oil above minimum	Turbine #32	min.			1
		Total inspection time	min.			

Enter standard inspection time; review periodically.

Clearly display the cycle for each area to be inspected; review periodically.

Caution:
1. If a component has dirt in parts that are under tension or compression, don't touch them during operations.
2. Disconnect power before performing operations marked with an asterisk (*).

Standards:
1. Make inspections easy to perform
2. Note whether the time, frequency, or cycle of inspections needs to be modified.
3. Keep safety, uptime, and quality in mind.

Source: Seiichi Fukunichi, "Revising Inspection Charts," *TPM Age*, 2 (3).

Figure 9-21. Daily Inspection Chart (Cleaning, Inspection, Lubrication)

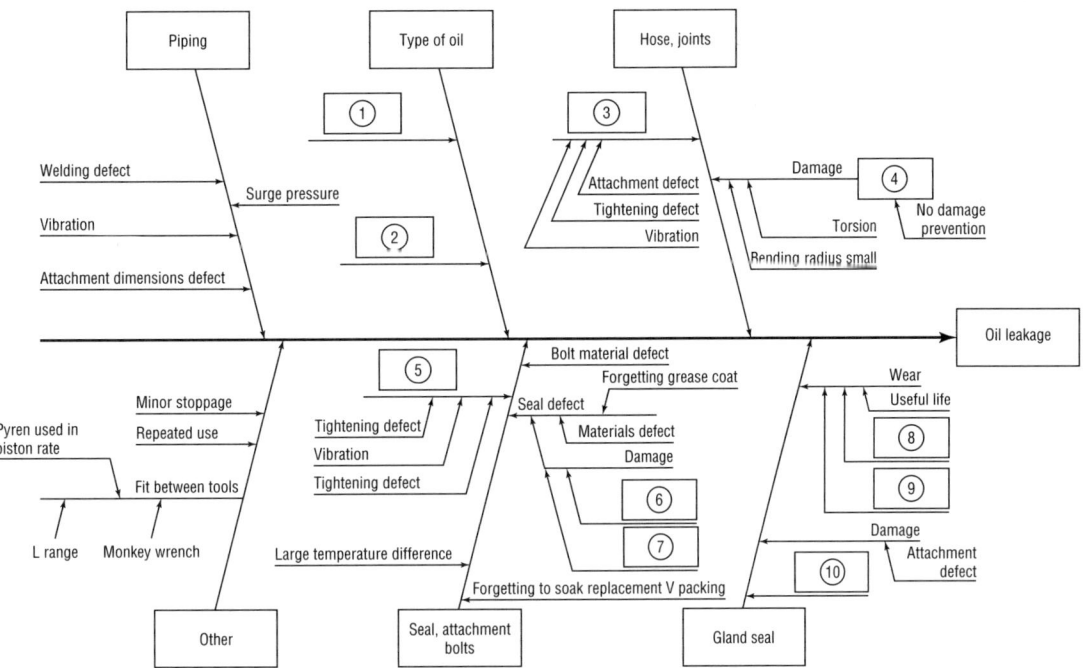

Source: Kunihiko Nawada, "Daily Inspections of High-Pressure Gas Production Installations," *TPM Age*, 1 (6).

Figure 9-22. Cause and Effect Diagram for Oil Leaks

Test Yourself: A Lean TPM Quiz 10

Company B, a plastics molding company, had a problem with large mold presses that would just stop working. When we asked what the machines' useful life was, company officials replied that it was 10 years. That's short compared to the human life span.

We were surprised when we first opened up the oil filters on the machine. They weren't just clogged; they were completely stopped up. When we asked the team leader when these filters had last been cleaned, he couldn't remember.

He also told us, "Recently the use cycles of these machines have decreased, and we don't use them very much. Since they're in poor condition and their yield isn't very good, we use them only when absolutely necessary to make some product. Still, it's a mess if they break down ...."

The cause of all this trouble was that the machines were running out of oil, and a machine lives or dies depending on the proper circulation of oil (see Figure 10-1). The operators didn't clean their machines, which means that they didn't inspect them, either. Even when they experienced the occasional minor stoppage they didn't think much of it. On the rare occasions when they oiled their machines, they must have noticed that the oil door was blocked, that the oil did not go all the way in but splashed back and spilled, and that the pipes were breaking from corrosion. The machines' "sudden death" was only the way the problem manifested itself on the surface. The real problem was the lack of daily inspections and maintenance.

These machines would still be useful if the operators had inspected them daily. The realization was a painful lesson for the employees in the merits of quick-and-dirty TPM and a motivation for reforming the way they did things.

To keep you thinking about the subject, try a little mental exercise in the form of the following TPM quiz. Try not to peek at our answers until you've tried to answer for yourself.

We'll warn you in advance that our answers may deviate from conventional wisdom and reflect our own biases and idiosyncrasies. Please view each answer as another way of looking at the situation.

Figure 10-1. Lubrication Problems

Question 1: Discipline as the Focus of the 5Ss

The central focus of the 5Ss is discipline. True or false?

Answer 1:

What we call "discipline" in TPM consists of carrying out predetermined tasks in a predetermined way. If we can't do our work with the proper discipline, then the other four Ss, which are the preconditions for discipline, become something that we do just for show. For example, we may misinterpret "organization" to mean "throw things away," and end up with the Jive Ss.

The Japanese word for "discipline" is related to the word for the basting stage in sewing traditional clothing. The master tailor lays down a basting thread as a guide, and the apprentice follows that standard in finishing the garment. In the same way, manufacturing managers must set the example and spirit for everyone else in the plant. That's why we answer "true."

We practice TPM because we're busy.

There's an old saying, "If you want to get something done, give it to a busy person." If we just take five minutes a day, we can reduce breakdowns and defects, and that will liberate us from our busy-ness.

What kind of TPM is possible in five minutes? Well, once you've gotten into the habit, you can take care of basic TPM in that short period of time, such as cleaning, spot checks, and minor countermeasures. The most necessary activity in the mix may be a TPM pre-setup routine.

Question 2: The Operator's Role in Machine Life Span

The operator plays a huge role in determining how long a machine will last. True or false?

Answer 2:

The operator is the one who has the closest contact with the machine from day to day. If that operator ignores dirt and doesn't follow the standards for changeover, operations, and daily inspections, the machine will begin experiencing minor stoppages and will eventually deteriorate and break down. When we see an artificial distinction between the people who run the machines and the people who fix them, we also find that the machines tend to have a shortened useful life. That's why we answer "true."

Knowledge and skill at the operator level can help prevent breakdowns. Necessary activities and abilities include:

1. Correct operation and adjustment of the machines
2. Spot checks and cleaning on the occasion of such setup and changeover activities as changing molds, jigs, or materials, or replacing cutting tools
3. Informing the maintenance department of the results of daily inspections
4. Performance of minor repairs (simple parts replacement) and handling emergency procedures
5. Use of instant maintenance techniques
6. Use of the senses to discover abnormalities
7. Mastery of basic technical skills for repairs. Examples include:
   a. skills in precision measurement
   b. basic knowledge of safety procedures for replacing simple parts
   c. simple diagnostic skills such as how to use a vibration meter or a thermolabel
8. Recording breakdowns and other process disruptions and helping to prevent their recurrence

We're sure you can add others on your own.

Question 3: What's Wrong with This Diagram?

Something is not quite right with the diagram in Figure 10-2. What is odd about it, and why?

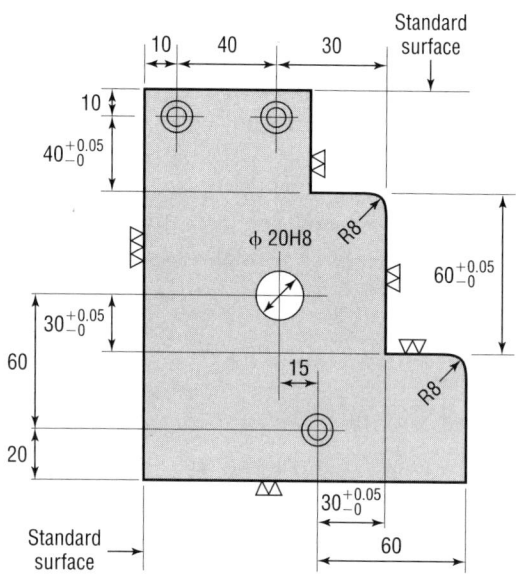

There's no special mention of the tolerance; the dimensions are assumed to be +0.1, −0

Figure 10-2. A Machining Diagram

Answer 3:

There is no standard written on this diagram for the position of the ɸ 20 hole. This means that the operator has to calculate it, which can be a lot of trouble. Here's how to figure the distance from the standard surface to the central position in the x (horizontal) and y (vertical) directions:

$$x = 10 + 40 + 30 - 30 - 15 = 35$$
$$y = 10 + 40 + 60 - 30 = 80$$

However, the tolerances have been left out of these dimensions. How can they be calculated? You can try calculating them yourself, but you may not be able to come up with anything useful.

Question 4: Standards for Parts Fitted on a Shaft

The standards for parts fitted on a shaft are usually based on the size of the hole. True or false?

Answer 4:

True.

1. In general, the tolerance for the measurements on the h axis is zero. If you choose an appropriate hole according to this standard, you'll get the required fit.

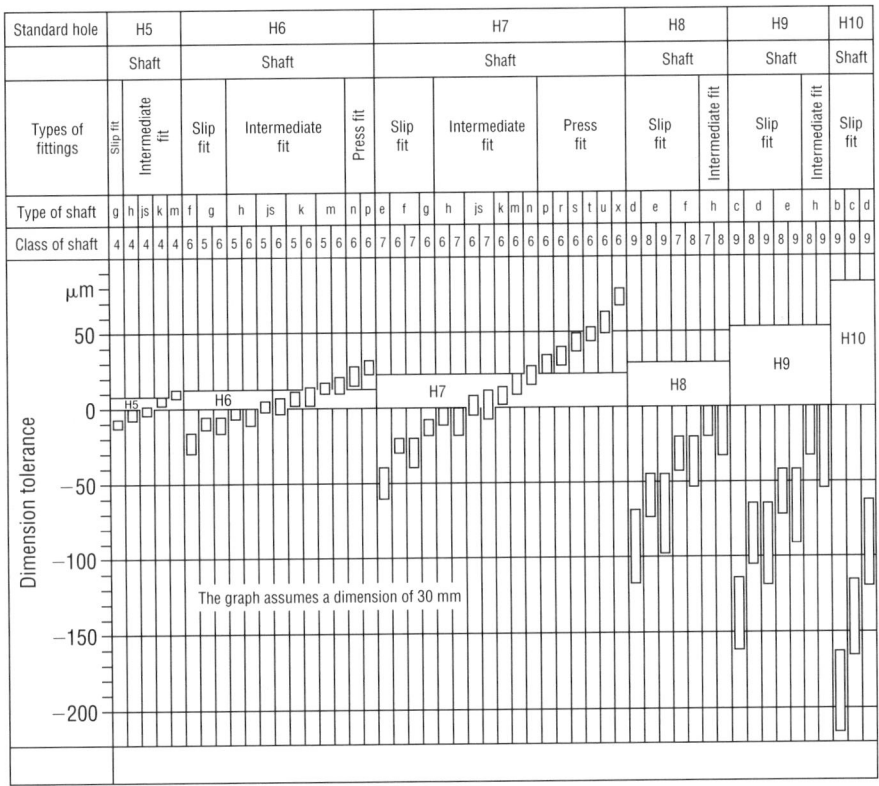

Figure 10-3. Fit Chart for Commonly Used Hole Standards

However, if you have a gap fit, a closed fit, or an intermediate fit mixed in there, it's better to use the axis as the standard. Otherwise you will need to put the shaft through additional machining.

2. From the point of view of operations, it's easier to machine (cut or finish) and measure the dimensions of the shaft.

3. When you adjust the hole according to the standard of the shaft, you need a limit gauge and a lot of reamers for the hole.

4. Polishing rods that have an intersection of about 8 *h* at the time of purchase are available, and you can use them with no further adjustment.

Question 5: Which Senses Detect What Situation?

Every machine gradually deteriorates from day to day. That process turns minor stoppages into medium stoppages, and waste due to speed loss and adjustments also occurs. It's important to find these phenomena early and carry out appropriate maintenance.

The operator's senses play an important role in discovering deterioration. A vague feeling that something is strange can lead to quick repairs and preventive maintenance. Therefore, it is vital for operators to go through the daily inspection routine asking themselves, "Is there anything odd here?"

The following is an example from company N. Next to each observation, we've listed one or two of the senses that were used to detect the situation. Are these the appropriate ones?

1. The sleeve surface of a vertical fraise is worn; the surface roughness in the scraper is also gone.	Sight Hearing
2. The press cylinder is sluggish; it's using up the lubricant too quickly.	Sight Hearing
3. A burning odor is coming from the machine. The motor seems scorched.	Smell Touch
4. The machine is giving off some steam; nuts and bolts are loose.	Sight Hearing
5 Air is leaking from somewhere, making a faint hiss.	Hearing
6. The oil gauge is hard to see. The oil may be dirty and may have deteriorated.	Sight
7. Something seems strange about the thermocouple in the heat panel sensor. It may be broken.	Sight Touch

Answer 5:

They're all correct. They're simple examples, so we don't have to analyze them. There's no better sensor than a human being. If something seems abnormal to an operator, it probably is abnormal. Even if the operator is wrong, he or she needs to notify the leader immediately and take appropriate measures; following through is especially important. To make this happen, you need to prepare a manual that takes operators from the point where they suspect trouble to restoration of the problem. They also need training in proper procedures and carrying out minor repairs by themselves. Keep in mind the following points:

1. Setup for maintenance

   Make sure parts, tools, measuring instruments, and other items needed for maintenance (for example, protective clothing, face masks, or adhesives) are close at hand.

2. Leak repair knowledge

   When repairing leaks (from the main unit, nozzle fittings, gaskets, pipes, or valves), operators need to know

   • how to tighten bolts

   • how to replace gaskets

   • how to apply emergency measures (band construction methods, application of repair compound, etc.)

   To perform such repairs effectively, they need to know the temperature, pressure, and composition of the substances they are handling.

3. Correct bolt attachment methods

4. Correct methods for disassembling, rechecking, and reassembling the machine. This includes such points as

   • shaft and bearing maintenance methods

   • electric motor maintenance methods

   • key-fitting methods

   • maintenance methods for other kinds of parts

Question 6: Avoiding Replacement of Guide Rails During Setup

Figure 10-4 shows guide rails that direct a plate into a mold. The company manufactures press materials with plate thickness of between 1 mm and 0.05 mm. Press operators change the guide rails depending on the plate thickness, particularly when dealing with the thinner plates. Is there a way to avoid replacing the guide rails?

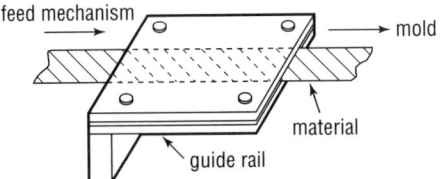

Source: "From a Collection of Ideas for Rationalizing and Improving Press Operations," *Puresu Gijitsu* (Press Technology) 28 (3).

Figure 10-4. Guide Rail Before Improvement

Answer 6:

Yes. Figure 10-5 shows an actual example of how company H improved the guide rail by making it adjustable. As a result, they no longer needed to replace the guide rail, thus reducing the setup time to a mere five seconds.

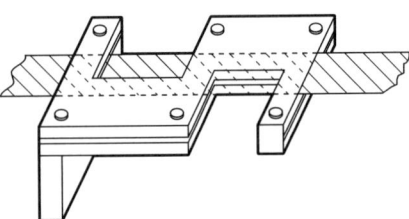

Figure 10-5. Guide Rail After Improvement

Question 7: The Most Practical Type of Chucking Jaws for an NC Lathe

Figure 10-6 shows several ways of configuring chucking jaws for an NC lathe. Choose the one you think is most practical.

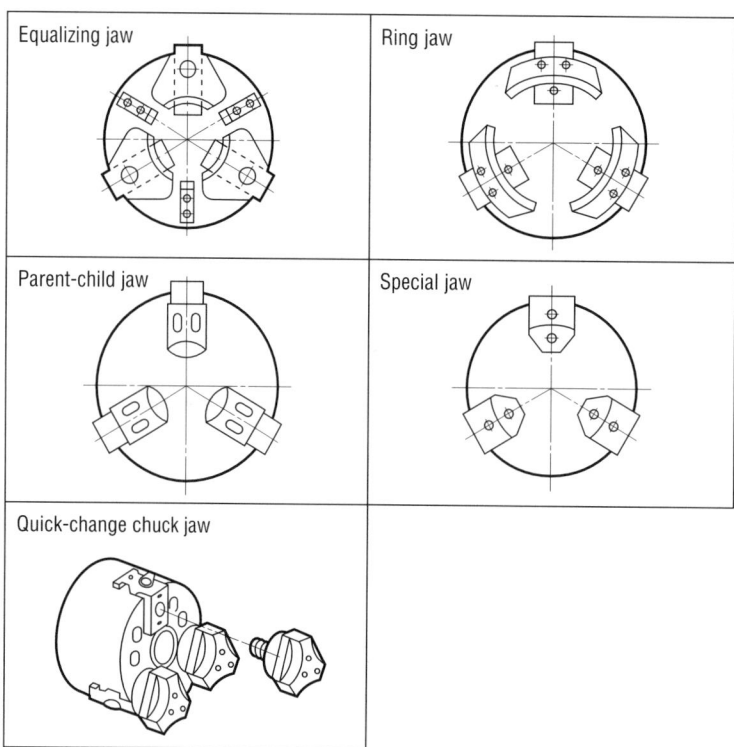

Source: Esaki and Kanemori, "High-Efficiency Machining and Setup Improvement in an NC Lathe," *Tool Engineer* 28 (8).

Figure 10-6. Chucking Jaws

Answer 7:

The chucks pictured in Figure 10-6 have different effects, as Table 10-1 describes. We tend to prefer the parent-child chuck. However, Figure 10-6 shows the small jaws attached with two bolts, which we'd reduce to one. The ideal is the quick-change chuck jaw, in which the small jaws are not removed at all.

Table 10-1. Effects of Various Chucking Jaws on Precision

Jaw	Major uses	Effects	Remarks
Equalizing jaw	• Improving the degree of roundness of low-relief shapes • No machining in front of the chucking part	• The jaw hits the work in six spots, and even if the degree of roundness in the chuck is poor, there is little chucking strain.	• Not suitable for heavy cutting. • A metal surface other than the jaw is needed.
Ring-shaped jaw	• Improving the degree of roundness of low-relief shapes • Pre-machining already completed on the chucking part	• The jaw strikes a large area, so there is little local chucking strain. With low-relief items, the degree of roundness is equal to or better than that of the chucking part.	• Appropriate when the chucking part has a good degree of roundness.
Parent-child jaw	• Simplified setup and changeover • Limited variety of similar articles	• Since only the small jaws are replaced, replacement is easy. • There's no need to shape the jaws at the time of replacement. • Producing the small jaws is relatively easy.	• Depending on range of grip, larger jaw is hard to use for a variety of purposes and is completely inappropriate in some cases. • Jaws are the soft jaw, the hard jaw, and the spike jaw.
Special jaw	• Large-diameter, heavy objects where chucking hardness is demanded.	• Due to its strong gripping force, it allows stable heavy cutting.	• Due to its specialized nature, it requires extra time to make and install.
Quick-change jaw chuck	• Reduces setup and changeover time for small-lot or single-item articles.	• One-touch system allows quick jaw replacement and adjustment of the gripping diameter. • One jaw adaptable to many kinds of products.	• Repeated gripping precision of 0.02 mm.

Question 8: Studying Current Conditions Before Making an Inspection Chart

When creating a daily inspection chart, the first step in implementing it is studying the current conditions. True or false?

Answer 8:

True.

The inspection techniques begin with getting a clear picture of the actual current situation. In essence, this means reviewing the history of all the equipment problems that have occurred.

1. Examine the records of past minor, medium, and major stoppages and identify areas where problems frequently occur. (You can look at the current situation with a cumulative record of equipment breakdowns.)

2. Look at rotating parts, sliding parts, vibrating parts, and other parts that could cause danger if they malfunction.

3. Look at anything else that may affect safety.

4. Review customer complaints, defect reports, and notes about last-minute adjustments; include product quality checkpoints on your inspection form.

Question 9: Which Sentence Does Not Describe a Type of Waste?

Most of the following sentences describe the types of waste found in plants where the equipment and machines don't work very well, but at least one sentence describes a good situation. Which sentence reflects positive conditions in a lean factory?

1. The speed is faster than the cycle time, so the work is accumulated in one place after machining and then washed in a large, high-performance washing machine.
2. The number of types of machining operations that can be performed on the equipment is limited.
3. Since the operators are having trouble achieving the desired precision, they've slowed the machines down.
4. Since the force is either too strong or too weak, the work gets nicked during the course of machining, and this defect is carried all the way down the line.
5. Operators are careful not to run their machines at greater than the stipulated speed.
6. The factory managers could reduce the number of operators needed to run the line if they only added some functions to the machines, but they don't do it.
7. Operators go about setup and blade replacement with a leisurely attitude, treating the time almost like an unofficial break.
8. Since products get stuck or get out of position or the machine malfunctions, operators can never leave their machines unattended.
9. The plant sometimes experiences minor stoppages, but people can fix them in about three minutes, so they don't look into the causes. If they took the time to investigate, the minor stoppage would turn into a medium stoppage.
10. The blade has too little play when it advances or withdraws.
11. Machining is completed, but the equipment on the line is still idling.
12. There seems to be too much motor capacity for the items being machined.

Answer 9:

We will leave you to think further about this set of sentences. As a hint, consider that lean manufacturing does not mean running the equipment as fast as possible to build up a stockpile of in-process inventory.

Most of the other wastes are listed in Figure 10-7, a chart concerning overall equipment effectiveness (OEE), equipment-related losses, and focused improvement technology.

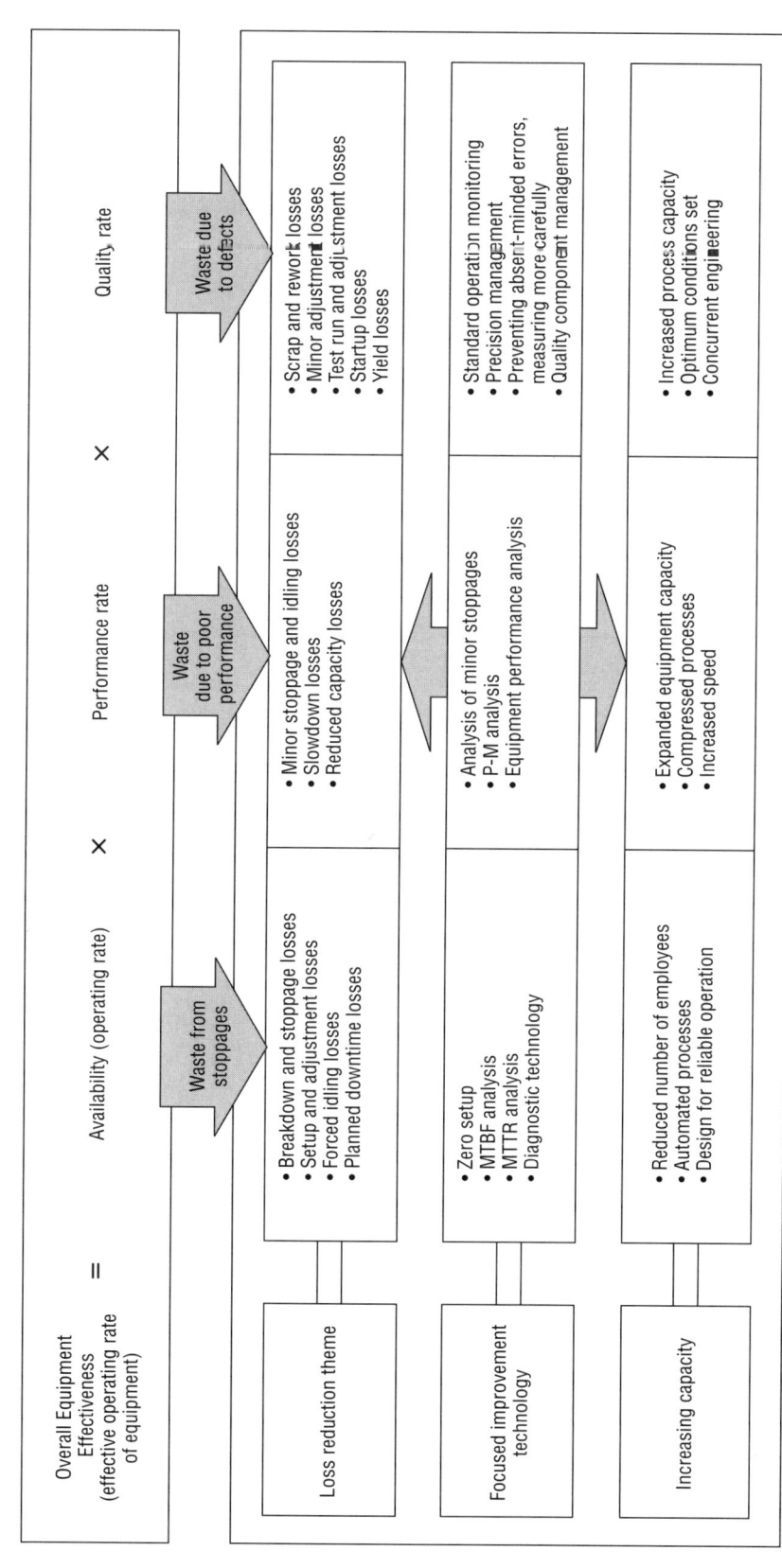

Source: Sekine, Arai, and Yamazaki, *Toyota Implementation Formulas: the Tiger Volume* (Nikkan Kogyo Service Center).

Figure 10-7. Overall Equipment Effectiveness, Losses, and Focused Improvement Technology

Question 10: How's Your 5S Score?

Table 10-2 is part of Mynac's 5S evaluation chart, which was presented in Chapter 3. The entire chart has 30 items with 5 points possible for each item, or 150 points for a perfect score.

1. Could your plant score 100 points or more?
2. One of the items under "discipline" is how workers greet visitors. At Mynac, employees are asked not to greet visitors. Can you explain the reasoning behind this policy?

Table 10-2. Mynac 5S Evaluation Chart

Worksite	
Section	
Line	

Date	
Members	

Checkpoints	Evaluation Level				
	1	2	3	4	5

Organization

	1	2	3	4	5
Warehouse (inventory management)	The storage area is too cluttered to walk around freely.	Items are placed irrationally.	Items have designated storage places, but they're often ignored.	Items are stored in proper locations, but no standard criteria indicate when to reorder.	Items are managed for a just-in-time supply and tracked on inventory boards.
Aisles	Work-in-process and other items stand in a roped-off area in the aisles.	Items are set on the sides of the aisles so employees can pass, but carts and dollies cannot pass.	Items protrude into the aisles.	Items protrude into the aisles but have warning labels.	There is no work-in-process, so the aisles are completely clear.
Work areas	Items lie scattered around for months, in no particular order.	Items lie around for months, but they don't get in the way.	Unneeded items have been red-tagged and a disposal date	Only items to be used within the week are kept	Only items needed the same day are kept around.

Answer 10:

1. If you're evaluating your own plant, you're going to be generous, so just about any company can score 100 points. Note, however, that when you bring in a more objective third party to do the evaluations, it's not as easy to get a high score.

2. The management at Mynac believes that concentrating on one's work is the best attitude to show to visitors. However, if an employee happens to make eye contact with a visitor, or encounters a visitor while walking down the hall, the customary response is a smile and greeting.

"Whether the moon looks large or small to you depends on the size of your heart."

It's often difficult for people to see something that's better than they are.

Most of the time, people don't recognize that they are looking at other people and things through the lens of their own measures of value. That's why they can't see anything that's better than they are.

To become more perceptive, we need to enlarge our points of view through our efforts and actions.

Question 11: The Definition of "Design TPM"

The definition of design TPM is which one of the following?

A. Design TPM refers to incorporating Q (quality), C (cost), and D (delivery) into the design.

B. Design TPM refers to setting up quality improvement teams within the design department and instituting improvement activities and campaigns that require everyone's participation and benefit everyone.

C. Design TPM refers to designing equipment and machines that won't break down.

Answer 11:

Definition C is appropriate. Definition A seems plausible, but if you put quality first, you may ignore cost and delivery and delay the product's entry into the market. Similarly, if you put cost first, you may end up ignoring the other factors. Design is the technique of combining current technical knowledge, so the most straightforward method is the one that puts delivery first.

Definition B simply will not lead to success. Every individual in the design department is highly intelligent, and the simple improvement activities applied on the factory floor don't work well here. On the other hand, when we design equipment and machines that don't break down (definition C), we don't experience minor, medium, or major stoppages. Of course, we also don't get product defects or legal actions from customers. Design TPM is therefore the highest form of preventive maintenance.

Question 12: The Definition of Value Analysis

Design VA (value analysis) refers to cost reduction activities during mass production in which you try to find substitutes for purchased parts that have no value from the point of view of product function, improve the tooling to reduce the machining allowance, and institute other design changes. What is the biggest problem with regard to involving the operations personnel in design VA?

There's no point in having a theory unless you implement it.

Even the most magnificent theory is ineffective unless it's accompanied by actions. Endlessly discussing, revising, and arguing an untested theory is just a form of amusement.

As practical people, we should construct our theories based on the opinions and observations of experienced workers in the field, consider the techniques, and then implement them. To do that, you need to be decisive and energetic.

Remember, administrators may understand a theory, but unless they put their understanding into action, they're just fooling around.

Answer 12:

VA causes problems when it's applied to design because of the time required from people working on the shop floor who must review the materials. For example, at some plants, the supervisor is supposed to have studied the VA materials and given a response within a week after receiving them. This makes the supervisor feel overwhelmed, and when that happens, slowdowns start to occur around the plant, and the lead time for design ends up growing longer. Therefore we suggest that design VA be used with caution.

Question 13: Is There Such a Thing as an Economic Lot Size?

There's no such thing as an economic lot size. True or false?

Figure 10-8 represents the following formula, which was developed in the early decades of the twentieth century. The formula assumes that the production cost (L) of a part is the sum of its setup costs and its inventory maintenance costs.

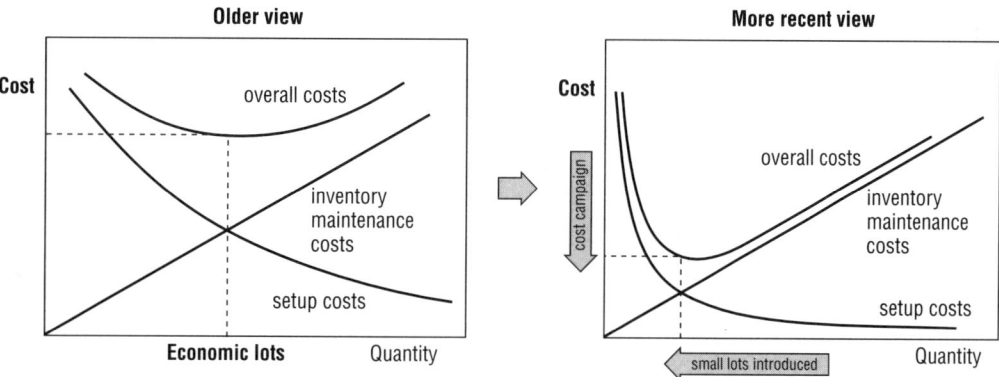

Figure 10-8. A Change in Thinking About Economic Lots

$$L = A \times \frac{x}{Q} + \frac{Q}{2} \times C \times i$$

where A = Setup costs = setup charge/minute × setup minutes/day

x = Sales record for each product (units per year)

Q = Economic lot size (excluding loss due to stockouts)

C = Production costs (more correctly, with setup costs removed)

i = Rate of inventory loss, generally 20 to 30 percent of production costs per year

Thus if we think of the loss from setup and the loss due to inventory as being equal, we get

$$\frac{Ax}{Q} = \frac{QCi}{2}$$

If we solve for Q we get

$$Q = \frac{QCi}{2Ax}$$

$$\therefore Q = \sqrt{\frac{2Ax}{Ci}}$$

This becomes the famous operations research formula for inventory management. Put into words, it comes out as

$$\frac{\text{Economic}}{\text{lot}} = \sqrt{\frac{2 \times \text{Setup costs/Time} \times \text{Quantity/Year}}{\text{Production costs} \times \text{Inventory loss}}}$$

Two things are evident from these formulas and graphs.

1. Larger production lots mean lower setup costs, but inventory loss increases in a linear fashion.
2. Smaller production lots mean greater setup losses.

If we consider the meaning of this formula more deeply, it's evident that if the setup time is made infinitely small, we can have a production lot size of one unit. This results in one-piece flow through the process, so the concept brings both unfinished goods and inventory down to zero.

Thus, the notion that there is no such thing as an economic lot is the concept behind the nonstock production formula. Is this true or false?

Answer 13:

False. There is such a thing as an economic lot. The weak point in the formula for obtaining the economic lot size (Q) is that loss due to parts shortages is not taken into consideration. In the past, we assume, people must have made up for missing items at the inventory level, but essentially, we need to calculate Q using the sum of idleness loss due to parts shortages as well as inventory loss.

The following example deals with loss due to changeover, but usually we appropriate only the loss due to changing molds. For example, in a plant that subcontracts for an auto manufacturer, the number of changeovers per day averages ten. If we analyze the setup and changeover for a 2,000-ton press at company A, we find:

Mold changeover time: within 9 minutes, single setup

Pre-setup time: 40 minutes

Cleanup and maintenance time: 30 minutes

Based on this, setup time = 9 + 40 + 30 = 79 minutes. If machining time averages 30 minutes, the ratio between machining and setup is 30 : 79, yielding a 109-minute production cycle. Furthermore, since company A is following a formula of just-in-time delivery at regular intervals in irregular amounts, its transportation costs, with 12 shipments per day, make up 5 to 10 percent of its sales. If we display the situation graphically, we end up with something like Figure 10-9.

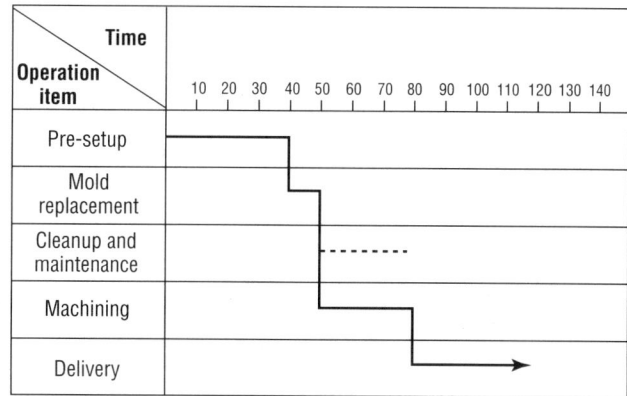

Figure 10-9. Cumulative Time for Changeover, Operation, and Delivery

In this graph, we count only mold changing time as changeover. During actual operations, however, changeover includes finishing the pre-setup for the next process or cleaning up (mold maintenance, etc.) to prepare for the next pre-setup.

Furthermore, we also have to take transportation costs into account. The load efficiency of JIT delivery (percent of cargo space used in each trip) according to a periodic, flexible-quantity schedule averages around 30 percent; with a fixed-quantity, irregular schedule, it can average 60 percent (see Figure 10-10). That 30 percent difference can make a difference in your costs for fuel, drivers, and so on.

The reasoning in this example suggests that an economic lot size does exist that takes into consideration pre-setup and transportation costs.

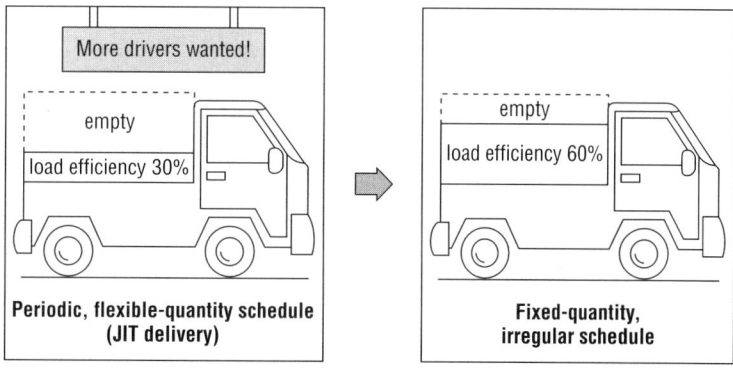

Figure 10-10. Delivery Schedules and Load Efficiency

Golf Is a Four-Unit Lot, One-Piece Flow Game

No doubt many of you enjoy a game of golf, so ask yourselves this: When you play golf, what size group do you play in?

Some people play in groups of three, four, or five people; others, because they hate to wait, prefer to play in pairs. Still others find even this too annoying and go out early in the morning to play a round alone.

To investigate the concept of idle time, we measured the time taken by groups of one to four people and plotted the results in Figure 10-11. We found that the time required was proportional to a multiple of the number of players. We had predicted that idle time would rise at a fixed rate, but in fact, as the figure shows, it turned out to be a rising second-degree curve. This illustrates the Second Law of Thermodynamics, or entropy, applied to people: The more people there are, the more disorderly they become, and they don't return to their previous state. However, as the group gradually moves around the course, the fewer people there are, so the less entropy there is. Thus if we take the playing time of someone golfing alone as the average, anything above that line is idle time.

We worked out Figure 10-11 with the cooperation of others whose golfing skills were about equal to ours. If even one person in the group is noticeably less skilled than the others, extra time is wasted on setup (ball replacement), hesitation, bad form, and aimless swings, and entropy increases. To achieve consistency and reduce waiting time, golfers must practice to make good technique their standard.

When we apply this improvement measures for our plants, we appropriately apply the principle of unchanging machining standards, which may be called a facet of changeover technology. By training the operators to follow invariable standards, we achieve certainty, and both time and entropy decrease. (See Sekine and Arai, *Kaizen for Quick Changeover* [Productivity Press, 1992], p. 48, for more about changeover standards.)

To determine the ideal number of people for a golf group, you might consider many factors—OB and putts, bogies, and such things as the players' physical condition. We might think of this as the optimal group size (similar to a lot) based on experience.

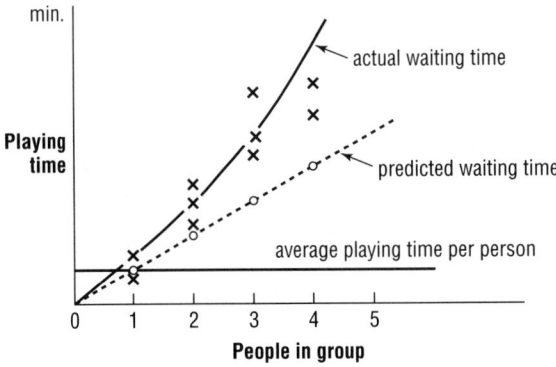

Figure 10-11. Waiting Time in Golf

1. When playing a round alone, the player doesn't have to stand around idly. On the other hand, the player is just playing against his or her own score. Even if a player manages to improve the old score, there's no one there to see, so the person must be content with knowing he or she did a good job.

2. In rounds with two people, one may be a beginner who is being coached by the more experienced player. This isn't much fun for the experienced player, who has to wait.

3. In rounds with three people, someone might be irritated at having to wait, but for players in poor physical condition this may be the minimum "lot size."

4. Rounds with four people may be the most interesting, although not everyone agrees.

Thus, whether we're talking about golf or manufacturing, there is such a thing as an optimal lot. Economic production lots do exist. Therefore, there can't be any truly non-stock production formulas. The idea of zero inventory is also bizarre.

In the Toyota production system formula, the idea is to have the inventory the company needs. However, Toyota defines "necessary inventory" much more narrowly than most other companies do. That concept is illustrated in Figure 10-12.

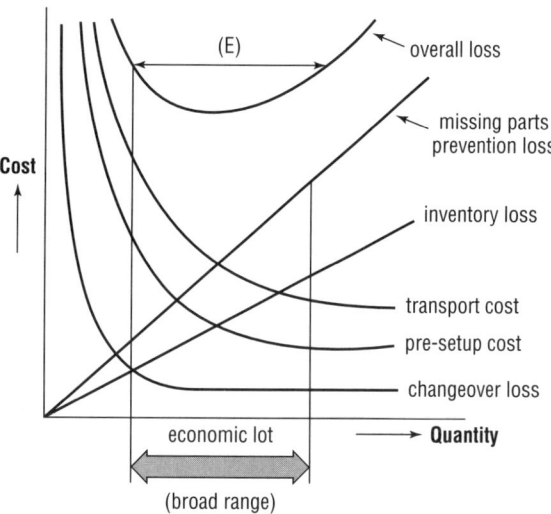

Figure 10-12. An Optimum Economic Lot Size Does Exist

Question 14: Machining Centers and Mass Production

Manufacturers of machining centers sometimes claim that these centers are useful for large-lot mass production. The president of a certain company believed these words and bought two machining centers after seeing them in action in another plant.

However, when his plant personnel actually tried using the machining centers and conducted an operation analysis, they found that air cuts (movement to return to home position) took one-third of the working time and tool changes took another third; thus the machine actually worked on the product only one-third of the time. They reported back to the executives that the machining centers weren't very useful for mass production. But the top management didn't believe them. "Check your figures again," they ordered.

Is it true that machining centers are not suited for large-lot mass production? See if you can develop a persuasive answer for the top managers.

Answer 14:

True.

Machining centers are inappropriate for large-lot and medium-lot production (the A and B group products in a PQ chart). Let's look at an example from company O for material to persuade the company president.

Which is best: a machining center, simple U-line production, or the step-by-step formula?

The goods to be machined at company O were cases for electronic products. The director of technology believed that machining centers were good because of the precision of the impressed surface in the center. With two machining centers lined up next to each other to be handled by one operator, it should have been clear to anyone that waste due to idleness would be a problem. The handling time was mostly taken up by attaching, detaching, and inspecting the work, which take about 20 seconds. If we assume 30 seconds as the machine center cycle time, we're looking at 300 seconds, which requires 10 machines.

With each machine costing ¥30 million (approximately $240,000), they couldn't buy ten on a moment's notice. Furthermore, they wanted to assign two shifts, but when they were shorthanded, it would be hard to persuade workers to take overtime. An added concern was that many workers lacked flexibility because of obligations at home.

Then manager M got the idea of conducting tool action analysis and came up with the bit action analysis displayed in Table 10-3. It covered 20 processes with 106 elemental operations (abbreviated here to save space). If we plot the results as in Figure 10-13, we get a graph classified by elemental operations.

Next, M summarized the results of the analysis and created a machining center waste elimination chart like the one in Table 10-4. The results were just as he had estimated:

Tool replacement	32.5 percent
Cutting time	33.5 percent
Air cut	34.5 percent
Total	100.0 percent

The plan for improvement consisted of creating a standard work combination sheet, and then determining how to shorten the long horizontal-line operations, such as the 21 seconds of machining time in operations 8 and 9 (process 2). Try plotting these operations on the standard work combination sheet in Figure 10-14.

Table 10-3. Machining Center Bit Action Analysis

Process order	Tool	No.	Action	Name of tool	Simplified drawing of tool	Speed (rpm)	Cycle (cm/min)	Feed (mm/sec)	Cutting (mm)	Cum. time	Actual time	Notes
1		(1)	Shutter closing							1sec		
		(2)	Chuck closing							2	1 sec	
		(3)	B shaft revolution							5	3	
		(4)	Tool change							10	5	
2	□90	(5)	Movement revolution	Mealing chuck BT40-C32-90 Strong twisting end mill Sumitomo MES-235φ35	φ29, 90, 90, GL180	5000		fast feed		11	1	
		(6)	Z 30.0			↓		↓		12	1	
		(7)	Z 0			↓		↓		12	0.5	
		(8)	Approach out			↓	455.5	1000		11	2	
		(9)	□90 end surface			↓	↓	↓	0.5	33	19	
		(10)	Z direction withdrawal			↓		fast feed		33	0.5	
3	φ83	(11)	Movement			↓		↓		34	0.5	
		(12)	Z-9.75			↓		↓		34	0.5	
		(13)	Approach cutting			↓	455.5	800		35	0.5	
		(14)	Arc cutting			↓	↓	↓	φ0.6	47	12	
		(15)	Tool withdrawal			↓		fast feed		48	1	
		(16)	Return to home					↓		54	6	
4	φ16 rough	(17)	Movement/ revolution	Boring bar with BT40 shank MA3H-560-1020 (Microcut unit with M1L-2.E)	φ15.7, 80, 30, φ30, GL110	6000		fast feed		55	1	
		(18)	Z 30.0			↓		↓		56	1	
		(19)	Z-52.0			↓		↓		56	0.5	
		(20)	cutting			↓	259.9	480	φ0.5	58	1.5	
		(21)	Z-62.4			↓	↓	60		59	1	
		(22)	Return to home			↓		fast feed		1 min 6 sec	7	
5	φ17 rough	(23)	Movement/ revolution	Boring bar with BT40 shank MA3H-560-1030 (Microcut unit with M1L-2.E)	φ16.7, 80, 30, φ30, GL110	5700		↓		1 06	0.5	
		(24)	Z 30.0			↓		↓		1 08	1.5	
		(25)	Z-52.0			↓		↓		1 09	1	
		(26)	Z-59.0			↓	299	456	φ0.5	1 10	1	
		(27)	Z-59.45			↓	↓	60		1 10	0.3	
		(28)	Z			↓		fast feed		1 10	0.3	
		(29)	cutting			↓		↓		1 11	0.1	
		(30)	Z-52.0			↓		↓		1 11	0.5	
		(31)	Z-59.0			↓	299	456	φ0.5	1 12	1	
		(32)	Z-59.45			↓	↓	60		1 13	1	
		(33)	Return to home					fast feed		1 20	7	
6	φ22 rough	(34)	Movement/ revolution	Boring bar with BT40 shank MA3H-560-1100 (Microcut unit with M1L-2.E)	φ21.7, 90, 30, GL120	4400		↓		1 21	1	
		(35)	Z 30.0			↓		↓		1 22	0.5	
		(36)	Z-52.0			↓		↓		1 23	0.5	
		(37)	Z-59.0			↓	300	352	φ0.5	1 24	1	
		(38)	Z-59.45			↓	↓	60		1 25	1	
		(39)	Return to home					fast feed		1 31	6	
7	φ28 35 rough	(40)	Movement/ revolution	Boring bar with BT40 shank and instrument MA3H-560-1010 (BR-406) (Microcut unit with M1-style-2.i.1 and extra-hard chip with style 2.i)	φ34.7, φ27.7, 90, 30, φ40, 7.7, GL130	2750		↓		1 32	1	
		(41)	Z30.0			↓		↓		1 33	1	
		(42)	Z-55.0			↓		↓		1 34	1	
		(43)	Z-69.5			↓	239	220	φ0.5	1 38	4	
		(44)	Z-70.0			↓	↓	50		1 39	1	
		(45)	Return to home					fast feed		1 45	6	

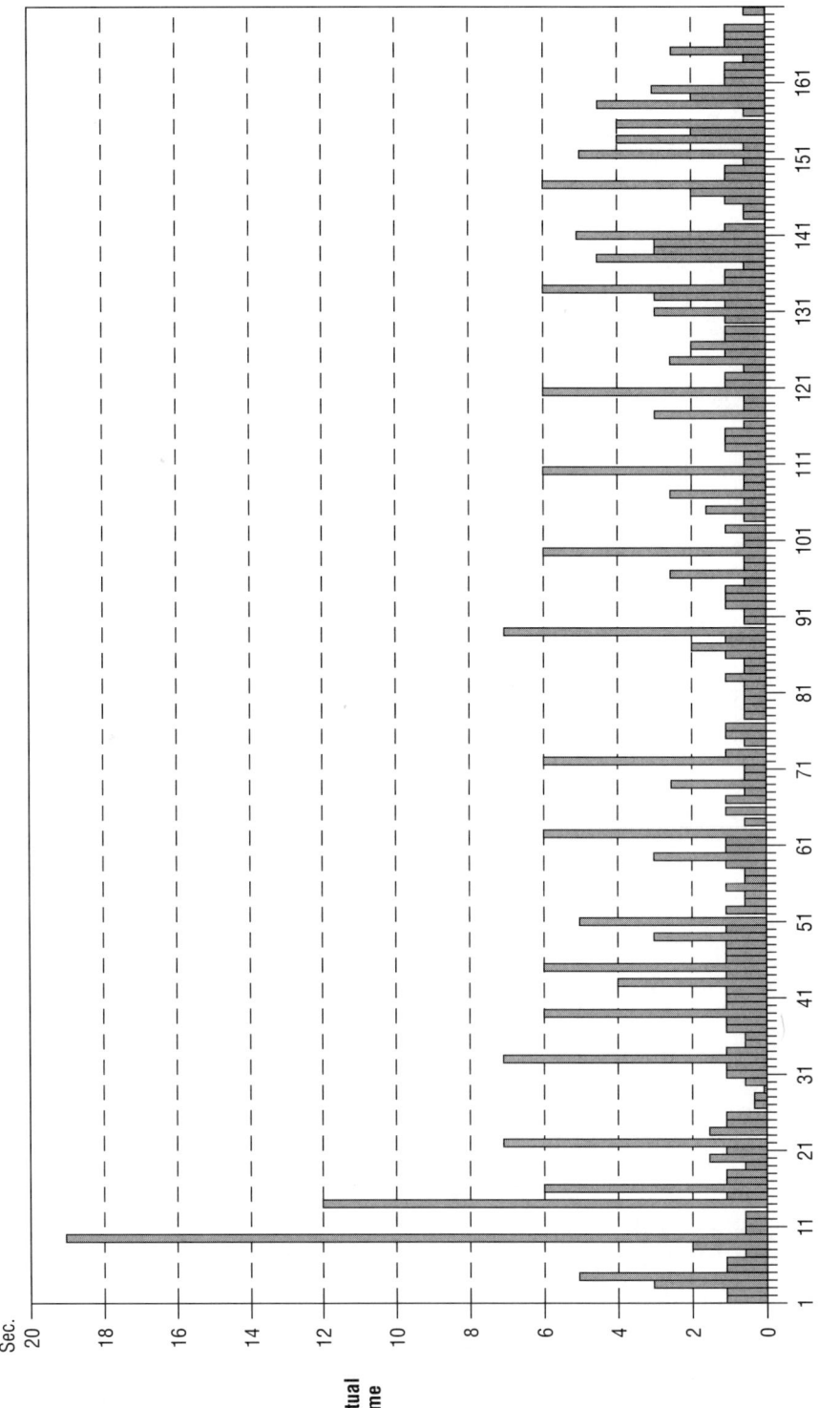

Figure 10-13. M/C Bit Action Analysis Graph

Table 10-4. M/C Waste Elimination Chart

Category		Time	Plan for waste reduction
Tool replacement		97.0 seconds (32.5%)	Zero replacement formula (common attachment and detachment U-shaped line)
Cutting time	□ 90	21 seconds	Minimize end surface machining R
	φ 83 rough	12.5 seconds	Make machining allowance smaller
	φ 16 rough	2.5 seconds	Give up rough machining
	φ 17 rough	3.3 seconds	
	φ 22 rough	2 seconds	
	φ 28 rough φ 35 rough	5 seconds	
	R19.7	4 seconds	
	φ 83 finished	4 seconds	
	φ 16 finished	1.5 seconds	
	φ 17 finished	3 seconds	
	φ 22 finished	1.5 seconds	
	φ 35 finished	2 seconds	
	φ 28 finished	1.5 seconds	
	Center drill	11.5 seconds	1. Boring formula
	φ 3.3 drill	4.5 seconds	
	φ 3.5 reamer	2 seconds	2. Set up a line where 2 shafts can be used.
	φ 3.6 drill	17.5 seconds	
	Total	99.3 seconds (33.0%)	
Movement, etc. (air cut)		104.7 seconds (34.5%)	1. Air cut prevention program 2. Set up smoothly flowing lines
Total		301.1 seconds (100%)	

As a side note, changing the programming took about 10 minutes, because part of it needed to be modified. The key is to program so that no modifications are necessary, and to manage that, all you have to do is implement invariable machining standards.

To persuade top management, M prepared the machining center waste elimination chart shown in Table 10-4. Then he proposed two possible model lines.

Plan A: Instead of machining centers costing ¥30 million each, we can create an entire line of general-purpose machines with one operator at ¥30 million per line. After making the economic calculations, we can compare to see which one is better. (Conditions assumed include a six-year lease with a 1.3 lease coefficient, and 20 eight-hour working days a month.)

Plan B: Create two lines run by a robot (¥27 million, or about $216,000), along with a U-shaped line run by one human operator. If not, use a robot suspended from the ceiling (see Figure 10-15).

Product name/no.	Gear case machining	**Standard Work Combination Sheet**	Created on		Quota per day		Units	—— Manual ········· Automatic feed
Process	M/C		Affiliation		Number needed		min sec	∿∿ Walking

Order of operation	Name of operation	Time Other	Time MT	Operation time
1	TC start	9	0	
2	☐ 90 Movement revolution	3	21	
3	φ 83 rough	8	12	
4	φ 16 rough	2	4	
5	φ 17 rough	10	4	
6	φ 22 rough	8	2	
7	φ 28 φ 35 rough	9	5	
8	R19.7	8	4	
9	φ 83 finished	11	4	
10	φ 16 finished	13	2	
11	φ 17 finished	16	4	
12	φ 22 finished	13	2	
13	φ 35 finished	12	2	
14	φ 28 finished	12	2	
15	Center drill	16	9	
16	φ 3.3 drill	15	4	
17	φ 3.5 drill	9	3	
18	φ 3.6 drill	4	6	
19	φ 3.6 drill	14	9	
20	Air blowing, all shafts to original position	10	0	
	Total	190	99	

Operation time scale: 18″   90″   3′   4′30″   6′

Operation time — Units of 3 seconds

Figure 10-14. Standard Work Combination Sheet

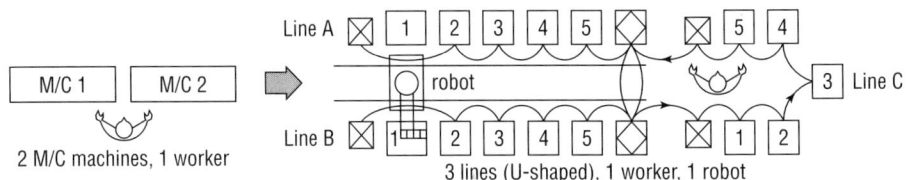

M/C 1 M/C 2
2 M/C machines, 1 worker

Line A ⊠ 1 2 3 4 5 ◈ ⊠ 5 4
robot → 3 Line C
Line B ⊠ 1 2 3 4 5 ◈ ⊠ 1 2
3 lines (U-shaped), 1 worker, 1 robot

This kind of line is better than a machining center for medium-size lots and larger.

Figure 10-15. M/C Line vs. Robot-Equipped Line

Implementation Overview for TPM Innovation

Part 1: Announcing the Top Management Decision

The company president (or the plant manager) announces the plan to introduce revised TPM, based on the New 5Ss and instant maintenance.

Part 2: Orienting All Employees for Implementation of TPM Innovation

Deliver a short-term orientation on TPM Innovation to all employees, including management.

- *TPM Introduction and Education Plan, Worksheet 1*

    Executives: Procedures for TPM Innovation and the role of upper-level management (two-day retreat)

    Managers: Procedures for TPM Innovation, specific means of promoting it, and the role of managers (two retreats of two days each)

    Promotion team leaders (chosen from among section managers and administrators): Practical training in on-site leadership (two practice sessions, two days each)

    On-site leaders (implementation team members): Practical training in equipment-related waste and the New 5Ss (three practice sessions, two of them with the promotion team leaders)

Part 3: Forming a TPM Review Team

1. Form a TPM Review Team. This group will deal with all sorts of matters within the entire company (or plant). Team members include the following people:

    Implementation manager: Represents upper-level management in promoting TPM and leads meetings of the Productivity Audit and Diagnosis Committee.

    Promotion Team: Organized by workplace, consists of four to five people, including the leader and a subleader chosen from the members.

    In plants where setup is a bottleneck, a Setup Study Group should also be organized.

    - *TPM Team Organization Chart, Worksheet 2*

2. Hold a kickoff meeting for the entire TPM Review Team to determine the procedures and schedules.

    - *TPM Team Meeting Evaluation, Worksheet 3*

3. Delegate individual duties to the members of the Productivity Audit and Diagnosis Committee.

   • *TPM Top Management Diagnosis—Self-Diagnosis Sheet, Worksheet 4*

Part 4: Setting Basic Policies and Goals for TPM Innovation

1. Summarize the background material for the introduction of TPM Innovation in terms of the external and internal environments. For example:

   External   International competition, 30 percent cost reduction, cutting delivery periods in half

   Internal   a. Handling process defects, equipment breakdowns, and increases in minor stoppages
               b. The fact that improvement is hindered by decreases in the working ratio due to increased setups

2. Administer *TPM Surveys (Worksheet 5)*

3. Set goals. For example:

   1. Zero accidents
   2. All breakdowns restored within three minutes
   3. Process capability of 1.33 or higher
   4. Zero customer complaints (quality or delivery)
   5. Effective operating rate of 80 percent or higher
   6. Work-in-process and inventory reduced to smallest possible amount
   7. 100 percent on-time delivery

4. Referring to the goals, have each TPM Team determine the issues for the Improvement Study Group.

   • *Summary of Improvement Study Group Issues, Worksheet 6*

5. Have the TPM Innovation Team set up a TPM Management Board

   • *TPM Management Board, Worksheet 7*

6. Confirm the basics of TPM.

   1. Daily maintenance (workers take care of their machines and production lines themselves)
   2. Periodic maintenance (annual or semiannual scheduled work; like a complete physical exam for machines)
   3. Instant maintenance (fixing small problems within three minutes)

Part 5: Studying Current Conditions (Documents)

The activities of the TPM Innovation Team start with their own version of "presetup": getting a grasp of the current situation through the documentary records.

1. Collect production records and materials related to TPM.

   • *Hourly Production Record, Worksheet 8*
   • *Cumulative Availability, Performance Rate (Yield), and Quality Rate Chart, Worksheet 9*
   • *Monthly Summary of Non-Working Time, Worksheet 10*
   • *Manufacturing Process Planning Sheet, Worksheet 11*
   • *Process Quality Management Chart, Worksheet 12*

2. Use industrial engineering methods to find types of waste that are not immediately visible.

- *Operation Observation Sheet (Work Sampling), Worksheet 13*
- *Line Balance Measurement Sheet, Worksheet 14*
- *Main Line Check Sheet, Worksheet 15*

Part 6: Studying Current Conditions (Observing Waste in the Workplace)

Once you thoroughly understand the preliminary documented data, go to the workplace and observe the actual site and the things there.

- *Malfunction Cause and Treatment Chart, Worksheet 16*

Find the seven types of waste in TPM, seek out their causes, and develop counter-measures.

Seven TPM Wastes

1. Minor stoppages, medium stoppages, major stoppages

2. Lengthy setup times

3. On-the-spot readjustments, defects, faulty products, and low yields

4. Planned stoppages

5. Incomplete 2S application

6. Overproduction by heavy equipment

7. Equipment problems at production startup

Part 7: Improving 5S Conditions through Organization and Orderliness (Chapter 2)

1. Discipline (the fifth S) is the center of the 5S system. The first two Ss—Organization and Orderliness—ensure success in the other Ss.

2. It is easiest to introduce Organization and Orderliness through the pre-setup process. Create U-shaped pre-setup lines.

3. Make the introduction of the 2Ss enjoyable and lighthearted. Determine the procedures and use an approach like taking the plant "into the sauna to sweat off grime."

4. You don't have to work particularly on Cleanliness and Standardized Cleanup if you have mastered Organization and Orderliness, keeping everything in a standard position according to standard labels.

Part 8: Introduce the New 5Ss (Chapter 3)

1. Create the will to implement the 5Ss.

2. Be sure you understand how to promote the 5Ss and know the basic foundations (such as Mynac's 10 foundations in Chapter 3).

3. Evaluate the level of 5S application at each work site. Collect the results and improve the places and items that don't score well.

- *5S Evaluation Chart, Worksheet 17*

4. Maintain the 5Ss as a system.

Part 9: Learn the Basics of Instant Maintenance (Chapter 4)

Instant maintenance is a maintenance technology in which breakdowns are restored to normal within three minutes.

STEP 1: STUDY CURRENT CONDITIONS

Use industrial engineering techniques to find workplace waste that isn't immediately visible.
- *PQ analysis chart (P stands for "product line," and Q for "quantity")—use Pareto Chart, Worksheet 18 as format*
- *Layout flow diagram (records the flow of the operation), Figure 5-1*
- *Monthly Line Efficiency Report (documents maximum and standard efficiency of equipment), Worksheet 19*
- *Equipment Operation Analysis Chart, Worksheet 20*
- *Minor Stoppage Cause Sheet, Worksheet 21*

STEP 2: SUMMARIZE THE PROBLEMS WITH EACH LINE AND MACHINE

Study the current situation for each line and piece of equipment.
- *Check sheet for equipment problems, by machine, Table 4-3*
- *Summary of problems by line and by machine, Table 4-4*
- *Comparison of differences between equipment, Table 4-5*
- *Summary of Defects by Line, Table 4-6, Worksheet 22*

STEP 3: ANALYZE MINOR STOPPAGE MECHANISMS AND THEIR CAUSES

1. Observe the workplace and analyze the mechanisms of minor stoppages.
   - *Mechanism analysis chart, Figure 4-1*
   - *Minor Stoppage Analysis Sheet, Worksheet 23*
2. Create hypotheses about the causal mechanisms.

STEP 4: FORM A CLEAR PICTURE OF THE PHENOMENON, MECHANISMS, AND CAUSES

1. Learn the true causes of minor stoppages by using Why-Why Analysis to pursue the phenomenon to its causes and mechanisms, and then to ways of eliminating them. Using the Five Ws and One H is sometimes helpful for identifying basic facts.
   - *Why-Why Analysis, Figure 4-2, Worksheet 24*
2. Immediately come up with theories about the causes.

STEP 5: SET UP AN INSTANT MAINTENANCE SYSTEM

1. Describe situations requiring instant maintenance on a small tool improvement sheet.
   - *Small Tool Improvement Sheet, Worksheet 25*
2. Train the implementation team members.
3. Choose an example and give a public demonstration of instant maintenance.
4. Make implementation plans for long-term maintenance and problem-solving (large and medium improvements).
   - *Plans for Resolving Equipment Problems, Table 4-7, Worksheet 26*

STEP 6: PROVIDE SUPPORT FOR AUTONOMOUS MAINTENANCE ACTIVITIES

1. Create daily maintenance sheets for autonomous maintenance.
   - *Daily Maintenance Sheet, Figure 4-3, Worksheet 27*

2. Place autonomous maintenance inspection labels on every part of the equipment that requires daily inspections. After the inspection, turn the label over.
   - *Instant Maintenance Labels, Figure 4-4, Worksheet 28*

3. For areas without autonomous inspection labels attached, use pre-shift check sheets.
   - *Pre-Shift Check Sheet, Table 4-8, Worksheet 29*

4. Classify the inspection items. (Delegate only minor inspections to operators so they're not overburdened.)

STEP 7: CREATE INSTANT MAINTENANCE MANUALS AND DEMONSTRATE TECHNIQUES

1. When equipment trouble occurs, make a record of equipment problems, noting the phenomenon, a guess about its cause, and a report of the countermeasures taken.
   - *Record of Equipment Problems, Table 4-9, Worksheet 30*

2. After repairs, document what was done on a repair record, Figure 4-5.

3. Compile the best on-site repair methods into an instant maintenance sheet.
   - *Instant Maintenance Sheet, Table 4-10, Worksheets 31 and 32*

4. Perform a public demonstration of instant maintenance.

5. Summarize the results of instant maintenance in a report.
   - *Equipment Maintenance Card, Worksheet 36*
   - *Sporadic/Planned Equipment Repair Report, Worksheet 37*

Part 10: Implementing Instant Maintenance–An Example (Chapter 5)

You can't implement instant maintenance from behind a desk. The section reviews an example of instant maintenance as applied to an electronic component insertion machine. Use this example and the previous section as guidelines for instant maintenance procedures for your own company's lines and equipment.

STEP 1: STUDY CURRENT CONDITIONS
   - *PQ analysis chart—use Pareto Chart, Worksheet 18, as format*
   - *Layout flow diagram, Figure 5-1*
   - *Monthly Line Efficiency Report, Worksheet 19*
   - *Equipment Operation Analysis Chart, Worksheet 20*
   - *Minor Stoppage Cause Sheet, Worksheet 21*

STEP 2: SUMMARIZE THE PROBLEMS WITH EACH LINE AND MACHINE

Make a record of problems by line (by equipment)
   - *Minor Stoppage Analysis Chart, Table 5-2, Worksheet 23*
   - *Graph of monthly occurrences, by machine, Figure 5-3*
   - *Machining Center Minor Stoppage Cause Sheet, Worksheet 33*

STEP 3: ANALYZE MINOR STOPPAGE MECHANISMS AND HYPOTHESIZE CAUSES

1. Observation in the workplace

   Look at the situation with regard to stoppages, presence of foreign matter, circumstances of defects and failures, standards and deviations from them.

2. Mechanism analysis, Table 5-3
   - *PM Notes, Worksheet 34*
   - *P-M Analysis, Worksheet 35*

STEP 4: FORM A CLEAR PICTURE OF THE PHENOMENON, MECHANISMS, AND CAUSES

Apply Why-Why Analysis to clarify the outlines of minor stoppage phenomena, the mechanisms of their causes, and ways of eliminating them. Use the Four (or Five) Ws and One H to help identify basic facts.

- *Why-Why Analysis, Figures 5-5, 5-6, and 5-7, Worksheet 24*

STEP 5: SET UP AN INSTANT MAINTENANCE SYSTEM

1. Determine the parts subject to instant maintenance through use of a cause and effect diagram.
   - *Cause and Effect Diagram, Figures 5-8 and 5-9, Worksheet 36*

2. Parts for instant maintenance should be stored using a two-bin system with kanban.

3. Create instant maintenance sheets.
   - *Instant Maintenance Sheets, Figures 5-11, 5-12, 5-13, 5-14, and 5-15, Worksheets 31 and 32*

4. The members and leaders of the Promotion Team should train themselves in these techniques.

STEP 6: DEMONSTRATE THE MAINTENANCE TECHNIQUES TO EMPLOYEES

1. Remember that everyone is a teacher and a student.

2. Conduct the demonstration in an informal way, lightheartedly and with humor, with lots of opportunity for friendly interaction.

STEP 7: SUMMARIZE THE RESULTS

Summarize the results of instant maintenance in a report.
   - *Equipment Maintenance Card, Worksheet 37*
   - *Sporadic/Planned Equipment Repair Report, Worksheet 38*

Part 11: Improving Setup Operations (Chapter 6)

1. A well-done setup can demonstrate the effectiveness of TPM.

2. Figure out exactly how much time is devoted to setup, Table 6-1.

3. Learn what keeps setup from going smoothly.

Seven Causes of Aimless and Disorganized Setup

1. Pre-setup without standards (Figure 6-2)
2. Jig and mold replacement without standards
3. Work machining diagrams without standards
4. Blade replacement without standards
5. Programming without standards
6. Machines without positioning standards
7. Cleanup without standards

STEP 1: STUDY CURRENT SETUP LOSSES

1. Investigate the actual time spent in setup and changeover.
2. Apply PQ analysis to understand reasons behind setup loss.
   • *Figure 6-9. Pareto Chart, Worksheet 18*

STEP 2: FORM A SETUP IMPROVEMENT PROMOTION TEAM

Form a setup improvement promotion team. Choose a leader and members for the team.

STEP 3: PERFORM ON-SITE OBSERVATION AND OPERATION ANALYSIS

Conduct the analysis using industrial engineering methods. Consider videotaping to aid your setup operation analysis.

• *Setup Operation Analysis, Worksheet 39*

STEP 4: APPLY WASTE ELIMINATION CONCEPTS

1. Divide the various kinds of waste into waste during preparation, waste during replacement, waste during adjustment, and miscellaneous waste, and then eliminate them.
2. Divide the waste reduction plan into small improvements, medium improvements, and major improvements.

STEP 5: DEPLOY IMPROVEMENT PLANS

Create plans with clear goals for implementing improvements.

• *Improvement Deployment Chart, Worksheet 40*

   • Minor improvements: those that can be carried out immediately
   • Medium improvements: those that require a bit of time and money
   • Major improvements: those that require improvements to equipment or study of technology

STEP 6: IMPLEMENT IMPROVEMENTS

1. Implement them immediately.
2. Summarize on a standard work sheet for changeover.
   • *Standard Work Combination Sheet, Worksheet 41*

STEP 7: EVALUATE AND SPREAD HORIZONTALLY

Once you've reaped some successes, expand the improvements to other equipment (machines) and lines.

Part 12: Eliminating the Waste of Planned Downtime (Chapter 7)

1. Learn to recognize the problem of excessive planned downtime (machines not used for processing).
   - *Monthly Line Efficiency Report, Table 7-1, Worksheet 19*
   - *Efficiency management graph, Figure 7-1*
   - *Classification of employee time management losses, Table 7-2*

2. Unless you raze the process, you won't get different results.
   - *Due to changing customer needs, companies must produce more types of products in smaller quantities (group C products in PQ analysis), as opposed to mass production of a few types of products (group A products). This requires a new look at product families and flexible line layouts.*
   - *Did you choose a two-line or a three-line layout in your analysis of company A's process paths? See Figure 7-4.*

STEP 1: STUDY CURRENT CONDITIONS

1. Learn the techniques of process design.

2. Use industrial engineering techniques to get a picture of the actual situation.
   - *PQ analysis, Table 7-3 and Figure 7-6*
   - *Pareto Chart, Worksheet 18, for format*
   - *Process path analysis, Figure 7-4*

STEP 2: ELIMINATE VISIBLE WASTE

Rethink the entire situation from the ground up as you eliminate waste, Table 7-4.

STEP 3: PURSUE THE GOALS OF PROCESS DESIGN

Find the causes for allowing planned stoppage to happen.
   - *Four (or Five) Ws and one H, Table 7-5*
   - *Why-Why Analysis, Figure 7-7, Worksheet 24*

STEP 4: DISCOVER THE TROUBLE SPOTS

1. First cutting edge: Planned downtime
   Second cutting edge: Setup
   Third cutting edge: Minor stoppages

2. $$\text{takt time} = \frac{\text{daily work time}}{\text{daily required quantity}}$$

$$\frac{\text{total human work time per piece}}{\text{takt time}} = n \text{ employees}$$

If you end up with a fraction for number of employees, round the number down by using simple automation and other waste reduction methods to reduce the operator time required.

STEP 5: DESIGN A LAYOUT FOR THE NEW LINE

The new line may be L-shaped, U-shaped, straight, or parallel, whichever suits the situation.

STEP 6: CREATE THE LINE AND TRY RUNNING IT

1. Create the line and practice operating it.
   - *Process Capacity Table, Table 7-6, Worksheet 42*
   - *Standard Work Combination Sheet, Table 7-6, Worksheet 43*

2. Demonstrate to the operators how the line works.

Part 13: Eliminating Abnormalities within the Process (Chapter 8)

Most minor stoppages happen as a result of abnormalities in the process. Chapter 8 describes soldering defects and the steps by which their causes were traced and removed.

STEP 1: STUDY CURRENT CONDITIONS

Create a summary of defects by product type and machine. This chart will help you understand the actual situation for each defect.
- *Defect Reduction Activity Chart, Table 8-1, Worksheet 44*

STEP 2: CREATE FAMILIES OF CIRCUIT BOARDS

Classify the boards into families, based on a combination of their size and the density of components, Table 8-2.
   a. Use PQ analysis to rank the families of boards in the order of the quantity manufactured of each (groups A, B, and C in Table 8-2).
   b. Circle the items in each family that have had the fewest defects.

STEP 3: INVESTIGATE THE PRIMARY FACTORS FOR SOLDER DIP DEFECTS

1. Find the factors for defects using a cause and effect diagram.
   - *Cause and Effect Diagram, Table 8-3, Worksheet 36*

2. Summarize the factors you find on a control factors chart, Table 8-4.

STEP 4: LOOK FOR OPTIMAL CONDITIONS

1. Determine optimal conditions during daily production activities.
2. Use one of the following quality engineering techniques:
   - *ordered factor elimination method, Table 8-5*
   - *orthogonal array, Table 8-6*

A. Perform Experiments for Finding Optimal Conditions
   1. Learn the lot size through PQ analysis, Figure 8-2
   2. Create experimental data that determines the level of the factors, Table 8-7

B. Summarize the Data

Use the main points of Table 8-7 to determine which factors have an effect.

C. Put the Findings into Graphic Form

Create a graph of each factor.

D. Identify the Effective Factors and Levels

Note the factors that have an influence and their levels.

STEP 5: ESTIMATE THE DEFECT RATE UNDER OPTIMAL CONDITIONS

Estimate a value based on the computation formula.

STEP 6: TEST WITH A CONFIRMATION EXPERIMENT

1. Run another experiment to confirm the experimental results.

2. Experimental results that are not close to the designated values suggest the existence of another factor or level other than the ones tested. Plan a second test.

Part 14: Performing Daily Equipment Inspections (Chapter 9)

From introduction to implementation.

STEP 1: SUMMARIZE THE PROBLEMS AT EACH STATION

1. Review the problems and accidents occurring in the last six months for each station.
   - *Record all breakdowns of five minutes or more.*
   - *Compile a record with breakdowns listed in order of frequency.*

2. Summarize the data compiled in a chart for the station. (Figure 9-1)

STEP 2: ANALYZE THE BREAKDOWN MECHANISMS

1. Construct a Pareto analysis of the mechanisms (in order of decreasing frequency).

2. While conducting the analysis, investigate all the broken parts (components) and make on-site observations of breakdowns.

3. Use Why-Why Analysis to seek out the mechanisms of the breakdowns.
   - *Why-Why Analysis, Worksheet 24*

STEP 3: IMPLEMENT MEASURES TO ELIMINATE THE CAUSES

Implement the countermeasures that will eliminate breakdowns, beginning with the true factors discovered through mechanism analysis.

STEP 4: ATTACH DAILY INSPECTION LABELS TO EQUIPMENT

1. Attach inspection labels to the equipment in the order that the inspections are to be carried out.
   - *Instant Maintenance Labels, Worksheet 28*

2. If the daily inspections uncover abnormalities, perform instant maintenance.

STEP 5: DIVIDE PARTS AND COMPONENTS INTO THREE GROUPS

Create a Pareto chart showing the frequency with which parts and components break down; divide the scores into 3 groups:

    A: parts to always have on hand

    B: parts contracted for immediate delivery

    C: parts for general purchase

STEP 6: MANAGE THE SPARE PARTS INVENTORY

1. Use a two-bin system with kanban to manage the spare parts inventory.

2. Manage all other "A" parts by calculating specific order points:

    Breakdowns per day $\times$ delivery time $\times$ safety margin (buffer)

STEP 7: IMPROVE PARTS REPLACEMENT PROCEDURES

1. Eliminate waste from parts replacement operations through on-site observations.

2. Promote improvement with immediate implementation and plans for promotion of the countermeasures decided upon.

STEP 8: INVESTIGATE BROKEN PARTS

Return to step 1 for daily inspections and continue investigations.

STEP 9: THOROUGHLY TRAIN EMPLOYEES IN DAILY MAINTENANCE

1. The leader should conduct daily visual inspections.

2. Operators learn to use their senses (sight, hearing, smell, touch, taste).

3. Daily inspection charts should be filled out daily at a predetermined time.

    • *Daily Inspection Chart, Worksheet 45*

Part 15: Eliminating Minor Stoppages (Chapter 9)

1. Automatic shutdowns and idling are the two most common types of minor stoppages.

2. Minor stoppages happen because of the work or actions of designers, production managers, or operators.

3. Production problems due to minor stoppages include the following:

    1. The equipment's performance efficiency drops, reducing productivity and overall effectiveness.

    2. In case of stoppages, operators must watch the equipment, so each machine requires its own operator.

    3. Idling racks up cycles on the counter but doesn't increase productivity.

    4. Due to a decrease in equipment efficiency, depreciation cannot be done.

    5. Product defects tend to increase.

    6. Decreased product quality requires more effort in inspections.

    7. Effective deployment of employees is not possible.

4. How minor stoppages differ from breakdowns

    1. Minor stoppages arise from variability in human work, materials, machines, and methods.

    2. Breakdowns mean loss of functionality and a reduction in useful life. Restoring them requires extra labor, time, parts, and money.

STEP 1: STUDY CURRENT CONDITIONS

Conduct on-site investigations of stoppages in the following order:
1. Find the daily direct operating time (DT) for each machine (piece of equipment).
2. Find the total daily time on the job (TT).
3. Determine the operating rate = DT/TT.

When the operating rate is low, investigate how much downtime the machines and equipment have each day.
- *Monthly Line Efficiency Report, Worksheet 19*
- *Equipment Operation Analysis Chart, Worksheet 20*
- *Minor Stoppage Analysis Sheet, Worksheet 23*
- *Daily Inspection Chart, Worksheet 45*
- *Minor Stoppage Investigation Sheet (Automated Equipment), Worksheet 46*

STEP 2: RETHINK THE PROCESS FROM THE GROUND UP

1. Thoroughly observe the workplace.

2. Create a correct hierarchy of the phenomena.

3. Look into the frequency of and time between breakdowns.

4. Analyze the causes and figure out how the system ought to be.

   - *TPM Top Management Diagnosis—Self-Diagnosis Sheet, Worksheet 47*

STEP 3: PURSUE THE GOAL

To bring minor stoppages to zero:
1. Avoid running out of materials or variation in parts.
2. Avoid mixing of defective materials or parts with good ones.
3. Position parts likely to get dirty or worn so that they can be replaced easily.
4. Protect the machinery and materials from contamination by debris or foreign matter.
5. Synchronize low, medium, and high-speed changeovers among the units.
6. Create a mistake-proofing system.
7. Make standard setting formulas unchangeable so no adjustments are needed.

STEP 4: ROOT OUT THE CAUSES

Use Why-Why Analysis to root out the true causes behind the stoppages.
- *Why-Why Analysis, Worksheet 24*

STEP 5: PLAN COUNTERMEASURES

Common countermeasures include
1. Getting rid of debris
2. Eliminating variation
3. Standardizing settings

- *Line Improvement Follow-up Sheet, Worksheet 48*

STEP 6: CREATE A PLAN FOR IMPROVEMENT

Sort out small, mid-scale, and large improvements according to the time and effort they will require.

STEP 7: IMPLEMENT THE PLAN IMMEDIATELY

Start right in with the measures you can implement today.

Part 16: Using Sensory Inspection to Detect Machine and Equipment Abnormalities (Chapter 9)

Make use of vision, hearing, touch, and smell. Here are some of the principal ways you can apply these senses for equipment inspection:

Vision

1. Oil leaks or changes in color
2. Abnormal meter readings
3. Wear, corrosion, play, thermal action, breakage
4. Product abnormalities
5. Equipment malfunctions, brake slippage
6. Leaking or sagging pipes
7. Smoke

Hearing

1. Vibration noise, chattering
2. Abnormal noises
3. Other noise

Touch

1. Overheating
2. Vibration
3. Deterioration of oil
4. Handles and levers that don't move easily
5. Play

Smell

1. Burning odors from coil short circuits or rubber
2. Chemical leakage

Inspection Cycle

The inspection cycle for a given machine is determined by its frequency of breakdowns: daily, weekly, monthly, or quarterly.

Part 17: The TPM Team Presentation

Team presentations are a good way to share learning about TPM and equipment inspections.

Team-Based Inspection Contest

1. Each team on the line participates in a contest. The teams report on their equipment inspection activities, competing for the best presentation of their inspection methods.

2. Management reviews the team presentations and publicly recognizes the team that performs the best inspections.
   - *TPM Team Presentation Evaluation, Worksheet 49*
   - *TPM Team Line Contest Evaluation, Worksheet 50*

Worksheets for TPM Innovation

The worksheets in this chapter are actual check sheets, forms, and charts collected at Japanese plants that have implemented TPM. You can browse them page by page to find something to serve your situation, or refer to the "Implementation Overview for TPM Innovation" (pages 203–216), which keys the worksheets to specific topics and steps from the chapters of this book. Blank versions of most of these worksheets are included for reference. If a worksheet does not quite fit your situation, you can use it as a model for a form that suits your company's needs.

List of Worksheets

24. Why-Why Analysis

25. Equipment Problem-Solving Plan

26. Small Tool Improvement Sheet

27. Daily Maintenance Sheet

28. Instant Maintenance Labels

29. Pre-Shift Check Sheet

30. Record of Equipment Problems

31. Instant Maintenance Sheet (Type 1)

32. Instant Maintenance Sheet (Type 2)

33. Machining Center Stoppage Cause Sheet

34. PM Notes

35. P-M Analysis Chart

36. Cause and Effect Diagram

37. Equipment Maintenance Card

38. Sporadic/Planned Equipment Repair Report

39. Setup Operation Analysis

40. Improvement Deployment Chart

41. Standard Work Combination Sheet—Changeover

42. Process Capacity Table

43. Standard Work Combination Sheet

44. Defect Reduction Activity Chart

45. Daily Inspection Chart

46. Minor Stoppage Investigation Sheet (Automated Equipment)

47. TPM Top Management Diagnosis—Self-Diagnosis Sheet

48. Line Improvement Follow-up Sheet

49. TPM Team Presentation Evaluation

50. TPM Team Line Contest Evaluation

Worksheet 1   *TPM Introduction and Education Plan*

Use this sheet to document the various responsibilities assigned for the early stages of the program (introduction and education prior to actual implementation).

Worksheet 2   *TPM Team Organization Chart*

This sheet documents the team leader and membership assignments for TPM improvements.

Worksheet 3   *TPM Team Meeting Evaluation*

This sheet helps TPM teams record notes from their meetings as well as evaluate the effectiveness of their time together.

Worksheet 4   *Productivity Diagnosis Deployment Plan*

This sheet is used to record the productivity audit committee's deployment plans for carrying out productivity diagnosis from the top down. This committee usually includes managers from the president to the section managers.

Worksheet 5   *TPM Survey*

Use this survey to gather information about shopfloor problems and to collect improvement ideas.

Worksheet 6   *Summary of Improvement Study Group Issues*

This is a public display of the improvement study group's concerns; it serves as a sign of the group's determination to see the process through. Be as specific as possible.

Worksheet 7   *TPM Management Board*

TPM Management Boards technically are not worksheets, but are large-size displays of TPM and productivity statistics for the workplace. It may be useful to make management boards for several levels within the organization, so that each work area knows how its TPM scores contribute to the whole company's overall equipment effectiveness (OEE).

Worksheet 8   *Hourly Production Record*

The aim of this management chart is to compare planned and actual production in terms of units produced per hour. It can also be used for recording actual time worked and delegation of operations. (Example is from an auto parts company.)

Worksheet 9   *Cumulative Availability, Performance Rate (Yield), and*
Quality Rate Chart

This chart consolidates availability (operating rate) with the performance rate (efficiency or yield) and quality information for a particular machine that influence it, making a handy instrument for maintenance and improvements. (Example from company Y.)

Worksheet 10 *Monthly Summary of Non-Working Time*

Non-working time is hard to analyze because it has many different elements. A summary chart can help you see how much time is spent in non-work activities, so you can start developing countermeasures. The items in the top row can be modified to suit your company's needs. (Example from company Y.)

Worksheet 11. *Manufacturing Process Planning Sheet (by Machine Model)*

This is a QC process chart that is helpful in documenting the process control system for ISO 9000 certification.

Worksheet 12 *Process Quality Management Chart (General Use)*

This chart combines a process flowchart with the items to be managed at each step to assure quality output. The chart specifies how and when each quality management item is to be confirmed.

Worksheet 13 *Operation Observation Sheet (Work Sampling)*

This handy work sampling method classifies operations in advance into net, semi-wasteful, and wasteful operations. The user simply identifies the type of operation and enters the data in the appropriate column.

Worksheet 14 *Line Balance Measurement Chart*

This sheet is used to compute the amount of waste due to idleness that occurs in line operations due to imbalance among the operators in terms of the time required for individual operations. The graph area is used to show the line organization's efficiency graphically and mathematically. The example is from conveyor operations.

Worksheet 15 *Main Line Check Sheet*

This is a check sheet used during assembly. It represents an operating standard for assembling precision parts.

Worksheet 16 *Malfunction Cause and Treatment Chart*

This form lets you summarize causes and countermeasures for malfunctions and difficult operations so you can track and eliminate waste by process.

Worksheet 17 *5S Evaluation Checklist*

This checklist was developed at Mynac for evaluating its workplace every month. This basic format can be modified to express the workplace 5S standards you want to sustain. (Note: There is no blank version for this example. See Table 3-3, page 44, for the full example.)

Worksheet 18 *Pareto Chart*

A Pareto chart is often used in problem solving to rank problems or other events according to number or frequency. It is also used in a specialized way in the PQ charts described in this book.

Worksheet 19   *Monthly Line Efficiency Report*

This chart allows you to compare line efficiency at a glance to help you immediately make adjustments (such as process redesign) to fix it. It is compiled from daily operation reports (by line or by individual machine or equipment).

Worksheet 20   *Equipment Operation Analysis Chart*

It's important to lay out the current operating conditions for each machine and piece of equipment so they can be understood at a glance. This chart provides a monthly summary of the working conditions of machines and equipment.

Worksheet 21   *Minor Stoppage Cause Sheet*

This sheet focuses on the causes of minor stoppages and serves as an instrument for classifying them accordingly. You may compile them by the month or by the week.

Worksheet 22   *Summary of Defects by Line*

Equipment problems cause an unexpectedly large number of defects. This will become clear to you by using this worksheet together with Worksheet 30.

Worksheet 23   *Minor Stoppage Analysis Sheet (Hand-Fed Machines)*

When observing hand-fed machines, the thing you notice most is the number of minor stoppages. Since these minor stoppages may stem from a variety of causes, you need to create a summary sheet to classify them by cause.

Worksheet 24   *Why-Why Analysis*

This sheet helps you clarify the phenomena involved in minor stoppages and defects, the mechanisms of their causes, and methods for eliminating them. When minor stoppages or defects occur, ask "why?" several times, writing the questions and responses in the order of the arrows.

Worksheet 25   *Plans for Resolving Equipment Problems*

Summarize permanent maintenance items as shown on this form. Note any improvement items that you reached after using Why-Why Analysis (Worksheet 24).

Worksheet 26   *Small Tool Improvement Sheet*

This form helps shopfloor employees implement and report on improvements they have worked out individually or in teams.

Worksheet 27   *Daily Maintenance Sheet*

This clarifies the nature of the daily maintenance to be performed by the operators and team leaders. Concentrate on creating these standard sheets for the problems most likely to occur in daily operations.

Worksheet 28   *Instant Maintenance Labels*

This is a small tool for codifying instant maintenance and daily maintenance in an easily understandable form. Revise this example to suit the nature of your own equipment.

Worksheet 29   *Pre-Shift Check Sheet*

Use this check sheet to ensure that no daily inspection items are overlooked. Before each shift, follow the sequence of inspection labels and determine whether each item is in need of repair, under repair, adjusted, or normal. Make notations for each machine and piece of equipment. At the end of the month, submit the sheet for review.

Worksheet 30   *Record of Equipment Problems*

Use this sheet to record the causes of and countermeasures for the equipment problems you observe.

Worksheet 31   *Instant Maintenance Sheet (Type 1)*

This sheet is used for instructing operators in instant maintenance procedures so they can carry them out in three minutes or less. Display these forms as close to the work site as possible, such as by the pre-setup line. You may want to post instant maintenance manuals directly on machines and equipment that need this treatment frequently.

Worksheet 32   *Instant Maintenance Sheet (Type 2)*

This sheet, based on observation of minor stoppages, gives operators a clear picture of causes and treatment procedures.

Worksheet 33   *Machining Center Stoppage Cause Sheet*

This sheet compiles machining center stoppages by cause to help you understand which causes are the worst. The example is for a hand-fed machine.

Worksheet 34   *PM Notes*

This sheet is used for keeping a chronological record of minor stoppages and their maintenance. Use this along with the Why-Why Analysis (Worksheet 24).

Worksheet 35   *P-M Analysis Chart*

This sheet organizes the P-M analysis process and can be used in equipment improvement to prevent further problems. Drawings help the reader understand the situation more clearly.

Worksheet 36   *Cause and Effect Diagram*

The familiar fishbone diagram is used to help sort out the possible causes of problems. For a systematic review of causal factors, the diagram often uses for its main "ribs" the four manufacturing elements: machines, methods, materials, and human activity.

Worksheet 37   *Equipment Maintenance Card*

This card records the equipment's maintenance status; it may be used for preventing problem recurrence or for instant maintenance if problems do recur.

Worksheet 38   *Sporadic/Planned Equipment Repair Report*

Reports of breakdowns can be hard to understand after the fact. For example, if the cause is not listed, the problem is often hard to investigate afterwards. The first thing to do is to create a blank form so the information is easy to record. (Example is from company Y.)

Worksheet 39   *Setup Operation Analysis*

Measure setup time continuously with a stopwatch. The goal is a grasp of the actual working situation. Often used as a step in improving setup, this analysis is a "secret weapon" of the on-site IE expert.

Worksheet 40   *Improvement Deployment Chart*

This chart is used to deploy improvements by breaking the overall goal into the elements that are expected to contribute to it. The example is for improvement in automatic equipment changeover time. The graph at the top left tracks the total time required for the changeover (down from 17 minutes in April to 9 minutes in June through August). The graphs below break the changeover time into three main elements and show how each element has changed over this period. The graphs across the top indicate minutes shaved from changeover time as a result of improvements at three levels. In the center grid, the team describes the improvement activities for the three changeover elements.

Worksheet 41   *Standard Work Combination Sheet—Changeover*

This standard work form is used to chart the time taken by machines and people during setup and changeover. Like all standard work forms, it reflects a living standard that should be seen as the springboard for improvements.

Worksheet 42   *Process Capacity Table*

To use this chart:

①, ②, ③: Enter the machining processes in their sequence, along with the machine number.

④: Enter the manual operation time (excluding walking time), the automatic feed (operation) time, and the completion time (manual + automatic times).

⑤: Enter the employee time for tool replacement.

⑥: Enter the minimum machining capacity in parentheses. In the column below, enter the processing capacity (= working hours ÷ completion time per workpiece + tool replacement time per workpiece).

Worksheet 43   *Standard Work Combination Sheet*

The formula

$$\frac{\text{human time/piece} + \text{machine time/piece}}{\text{takt time}} = n \text{ people}$$

is a simplified tool for setting the optimum machining conditions.

①, ②: Enter the machining operations in their sequence, along with the machine number.

③: Enter the manual operation and automatic feed time from the process capacity table.

④: Operating time:

Draw a solid line to show manual operation time.

Draw a dashed line to show automatic feed time.

Draw a wavy line to show walking time.

⑤: Enter the takt time. Indicate it on the graph by drawing a vertical red line.

⑥: When the last operation is completed, indicate the time spent walking back to the first operation.

⑦: If the length of the cycle coincides with the takt time, it's optimum; if it's longer, the process needs improvement. When it is shorter than the takt, it indicates idleness.

⑧: If any part of the time exceeds the line drawn for the takt time, continue it from the left axis as shown.

Worksheet 44   *Defect Reduction Activity Chart*

Zero defects is an eternal preoccupation in manufacturing plants, and a main theme in improvement campaigns. This charts helps people understand defects in the course of daily manufacturing activities and figure out their causes by observing the phenomena.

Worksheet 45   *Daily Inspection Chart*

This is a daily inspection record for machines, molds, dies, jigs, and other equipment and tools (the example is for a mold). This form will be easier to use if you blow it up to wall chart size.

Worksheet 46   *Minor Stoppage Investigation Sheet (Automated Equipment)*

This has the same purpose as Worksheet 23, with more items. Use past minor stoppages at your plant as a guide in deciding which items to include.

Worksheet 47   *TPM Top Management Diagnosis—Self-Diagnosis Sheet*

This sheet is used by managers for recording and submitting their diagnoses of problems. The evaluation uses a five-point system. There is also space for favorable comments and suggestions for improving points that are only slightly unsatisfactory.

Worksheet 48   *Line Improvement Follow-Up Sheet*

Line improvement is not only the manager's responsibility; the operators' cooperation is also necessary. You need to determine the improvement items, time period, and responsibility, and then lay them out in a simple way so everyone can cooperate in carrying out the improvements. Make notations about improvements that need to be made today and in the near future, then post them in the workplace.

Worksheet 49   *TPM Team Presentation Evaluation*

This sheet is used for scoring presentations by TPM teams to report on results of their activities. The evaluation helps teams develop skills for managing workplace problems and documenting situations and improvement results.

Worksheet 50   *TPM Team Line Contest Evaluation*

This sheet is used for scoring work areas in competition with each other on improvements in the eight items listed in the left column. The values for each item are recorded at registration for this contest, then compared with values at a later inspection. The amount of change is translated into a rank from 0 to 5, using the key shown on the right side.

TPM Introduction and Education Planning Sheet

Date created: _____ / _____ / _____

Person in charge Instructor	Executives Department head or higher	Managers Section manager or clerk	Team Members	Other company employees	Subject of education or training (Summary)	Days	Place
Beth Clark	George Yamada	Jeff Goldstein	Juanita Ruiz Tim Perry	Pat Ryan Maria Sanchez	Introduction to 5S	2	Bldg. 3 conf. room

Worksheet 1

TPM Introduction and Education Planning Sheet

Date created: ____ / ____ / ____

| Person in charge

Instructor | Executives

Department head or higher | Managers

Section manager or clerk | Team Members | Other company employees | Subject of education or training

(Summary) | Days | Place |
|---|---|---|---|---|---|---|---|
| | | | | | | | |
| | | | | | | | |
| | | | | | | | |
| | | | | | | | |
| | | | | | | | |
| | | | | | | | |
| | | | | | | | |
| | | | | | | | |
| | | | | | | | |
| | | | | | | | |

Worksheet 1a

TPM Team Organization Chart

Date created: _____ / _____ / _____

```
                        ┌─────────────┐
                        │             │
                        └──────┬──────┘
                               │
  ┌ ─ ─ ─ ─ ─ ─ ─ ─ ─ ┐       │        ┌──────────────────┐
    Specialist Division│       ├────────│ Head of          │
  │                            │        │ Promotion Office │
    ─ ─ ─ ─ ─ ─ ─ ─ ─ ─│      │        ├──────────────────┤
  │                            │        │ Productivity Audit│
    ─ ─ ─ ─ ─ ─ ─ ─ ─ ─│      │        │ Committee        │
  │                            │        │                  │
    ─ ─ ─ ─ ─ ─ ─ ─ ─ ─│      │        └──────────────────┘
  │                            │
  └ ─ ─ ─ ─ ─ ─ ─ ─ ─ ┘       │
                               │
  ┌────┬────────┬──────────┬───┴────┬──────────┐
```

Team	Team	Team	Team	Team
Amazing 5				
Leader	**Leader**	**Leader**	**Leader**	**Leader**
Juanita Ruiz				
Subleader	**Subleader**	**Subleader**	**Subleader**	**Subleader**
Tim Perry				
Ed Kwan				
Marcia Doherty				
Steve Twerski				

Worksheet 2

TPM Team Organization Chart

Date created: _____ / _____ / _____

```
                    ┌──────────────┐
                    │              │
                    └──────┬───────┘
                           │
  ┌ ─ ─ ─ ─ ─ ─ ─ ─ ─ ┐    │      ┌──────────────────┐
  │ Specialist Division│   ├──────│     Head of       │
  │                   │   │      │ Promotion Office   │
  ├ ─ ─ ─ ─ ─ ─ ─ ─ ─ ┤   │      ├──────────────────┤
  │                   │   │      │                   │
  ├ ─ ─ ─ ─ ─ ─ ─ ─ ─ ┤   │      │ Productivity Audit│
  │                   │   │      │    Committee       │
  ├ ─ ─ ─ ─ ─ ─ ─ ─ ─ ┤   │      │                   │
  │                   │   │      └──────────────────┘
  └ ─ ─ ─ ─ ─ ─ ─ ─ ─ ┘   │
```

Team	Team	Team	Team	Team
Leader	Leader	Leader	Leader	Leader
Subleader	Subleader	Subleader	Subleader	Subleader

Worksheet 2a

TPM Team Meeting Evaluation

Created by	Leader	Facilitator	Manager

Team name: A		Concerns: Countermeasures against debris		Report	No.

Time/date	2/5/97	(AM) PM	7–8	Place: **Mita Plant**	Topic:	

Cumulative hours met	During work		Total		Cumulative frequency of meetings	In attendance		Absent
	Outside work		Total					

Chair	B	Attendance rate $= \dfrac{Attendees}{Members} \times 100 = \dfrac{8}{10} = 80\%$	A, B, C, D, E, F, G, H	I, J
Scribe	C	Speaking rate $= \dfrac{Speakers}{Members} \times 100 = \dfrac{10}{10} = 100\%$		

Activity	1. Discuss where to go from here (Select themes)	2. What's the current situation? (Investigate and analyze	3. What should be done and when? (Set goals)	4. Why do these things happen? (Go after the causes)	5. What various means should be used? (Plan counter-measures)	6. Try implementing them	7. Did implementing go well? (Confirm the results)	8. What we need in order to do better in the future. (Prevent and check recurrences)	9. Reflections	10. Future plans
Check	✓	✓	✓	✓	✗	✗	✗	✗		

	Activity notes:
	1. Selection of themes went well.
	2. Analysis of the current situation also went well.
	3. We set about half our goals—OK.
	4. Our efforts to find the causes and analyze the mechanisms went poorly.
	5. The first idea we got was improvement of the equipment guards. We can't think of anything else.

Plans for next time	**Nature of discussion items to be confirmed:** We'll ask individual members to share their ideas.	**Time and date** month ____ / ____ day AM PM Place

Requests for the manager	Advice from the manager concerned
Please give us some hints	

Requests for the manager		Manager
Facilitator: Please note we have too many concerns to deal with	**Promoter** Initials	Initials

Worksheet 3

TPM Team Meeting Evaluation

Created by	Leader	Facilitator	Manager

Team name:		Concerns:		Report ¦ No.	

Time/date	AM PM		Place:		Topic:	

Cumulative hours met	During work	Total	Cumulative frequency of meetings		In attendance	Absent
	Outside work	Total				

Chair		Attendance rate $= \dfrac{\text{Attendees}}{\text{Members}} \times 100 =$	
Scribe		Speaking rate $= \dfrac{\text{Speakers}}{\text{Members}} \times 100 =$	

Activity	1. Discuss where to go from here (Select themes)	2. What's the current situation? (Investigate and analyze	3. What should be done and when? (Set goals)	4. Why do these things happen? (Go after the causes)	5. What various means should be used? (Plan counter-measures)	6. Try implementing them	7. Did implementing go well? (Confirm the results)	8. What we need in order to do better in the future. (Prevent and check recurrences)	9. Reflections	10. Future plans
Check										

Activity notes:

Plans for next time	Nature of discussion items to be confirmed:	Time and date
		month _____ / _____ day
		AM
		PM
		Place

Requests for the manager	Advice from the manager concerned

Requests for the manager	Promoter		Manager
	Initials		Initials

Worksheet 3a

Productivity Diagnosis Deployment Plan

Date planned: 10/11   Productivity Sub-Group

Frequency of diagnosis	1–50 Times	51 Times	52 Times	53 Times	54 Times	55 Times	56 Times	57 Times	58 Times	59 Times
Diagnosis scheduled (Day/time)	—	10/9 13:00–15:30	10/9 14:00–17:00	10/16 13:00–16:30	10/16 14:00–17:30	10/22 14:00–16:00	10/22 14:00–16:00	10/31 13:00–17:00	10/31 14:00–18:00	11/7 13:00–17:00
Plant I	—	Main office	Plant H	Main office	Plant O	Plant T	Plant K	Main office	Plant O	Main office Plant H
Number of lines diagnosed	—	4 line	5 line	6 line	6 line	3 line	3 line	7 line	7 line	10 line
Department / Audit committee	—	Ops. 1	Ops. 1 Parts 2	Parts 2	Parts 2	Parts 1	Mecha-tronics	Ops. 1 Ops. 2 Parts 1	Parts 2 Casting	Ops. 1 Ops. 2 Mecha-tronics
Director A	9 times									
Dept. Mgr. B	14	●								
Asst. Dept. Mgr. C	11		●							
Asst. Dept. Mgr. D	5									
Asst. Dept. Mgr. E	3									
Asst. Dept. Mgr. F	10		●							
Asst. Dept. Mgr. G	8								○	○
Asst. Dept. Mgr. H	5									
Section Mgr. I	10		●			○		○		
Section Mgr. J	10	●		○				○		
Section Mgr. K	7									
Section Mgr. L	8									
Section Mgr. M	7	●	●	○	○	○	○	○	○	○
Section Mgr. N	9									
Section Mgr. O	7									
Section Mgr. P	11	●								
Section Mgr. Q	3									
Section Mgr. R	36	●	●	○	○	○	○	○		○

☐ Numbers inside squares are diagnostic frequency
● Black dots indicate people who will implement diagnosis
○ White circles indicate people scheduled to make diagnoses

Worksheet 4

Productivity Diagnosis Deployment Plan

Date planned:_____ / _____ / _____   Productivity Sub-Group

Frequency of diagnosis	Times	Times	Times	Times	Times	Times	Times	Times	Times	Times
Diagnosis scheduled (Day/time)										
Plant I										
Number of lines diagnosed										
Department / **Audit committee**										

☐ Numbers inside squares are diagnostic frequency
● Black dots indicate people who will implement diagnosis
○ White circles indicate people scheduled to make diagnoses

Worksheet 4a

TPM Survey

Process: _____

Responding person: _____

Name of process	Main steps	Particularly troublesome phenomena	Good ideas that can be implemented quickly and easily	What should be done?	
				Change procedures	Improve equipment
Machining	Cutting	Overshooting	I'd like the bit covered.	Reduce the machining allowance.	M/C style
Machining	Oil on floor	Operators slip and fall.	I'd like to see rubber skids on the floor.	Change to water-soluble lubricant.	Attach an oil pan.
				Interview as much as you can and elicit complaints.	

Worksheet 5

TPM Survey

Process: _____

Responding person: _____

Name of process	Main steps	Particularly troublesome phenomena	Good ideas that can be implemented quickly and easily	What should be done?	
				Change procedures	Improve equipment

Worksheet 5a

Summary of Improvement Study Group Issues

Date: 9/18

Issue:	Eliminate complaints due to running out of thread
Process:	Line 5

<table>
<tr><td>Process/layout personnel required</td><td>

sewing machine

main unit

round tip machining

wrapper

Workers: 1
Quota per day: 35 cases
C/T: 1457″

</td></tr>
</table>

Current problems	1. Sometimes the machine doesn't stop, even when out of thread.
	2. Sometimes items "sewn" without thread keep flowing as good quality items.

Improvement goals	Goal items	Current values	Goal values	Results (enter in red ink)
	A. 1. The machine should stop right away when out of thread.	B. Sometimes it doesn't shut down.	Have it stop immediately.	It does stop immediately.
	2. Create guidelines for operations when the machine is out of thread.	None	Create guidelines	Guidelines created
	3. Reduce frequency of running out of thread.	3.6 times/3 shifts	Once/3 shifts	None in 2.5 hours

Members	Leader	A	Subleader	B
	C D E F G H			

Note: Write clearly. Use black ink for entries before improvement, red ink for entries after improvement.

Worksheet 6

Summary of Improvement Study Group Issues

Date: _____ / _____ / _____

Issue:	
Process:	

Process/layout personnel required	
	Workers: Quota per day: C/T:
Current problems	

Improvement goals	Goal items	Current values	Goal values	Results (enter in red ink)
Members	Leader		Subleader	

Note: Write clearly. Use black ink for entries before improvement, red ink for entries after improvement.

Worksheet 6a

TPM Management Board

Operating Rate

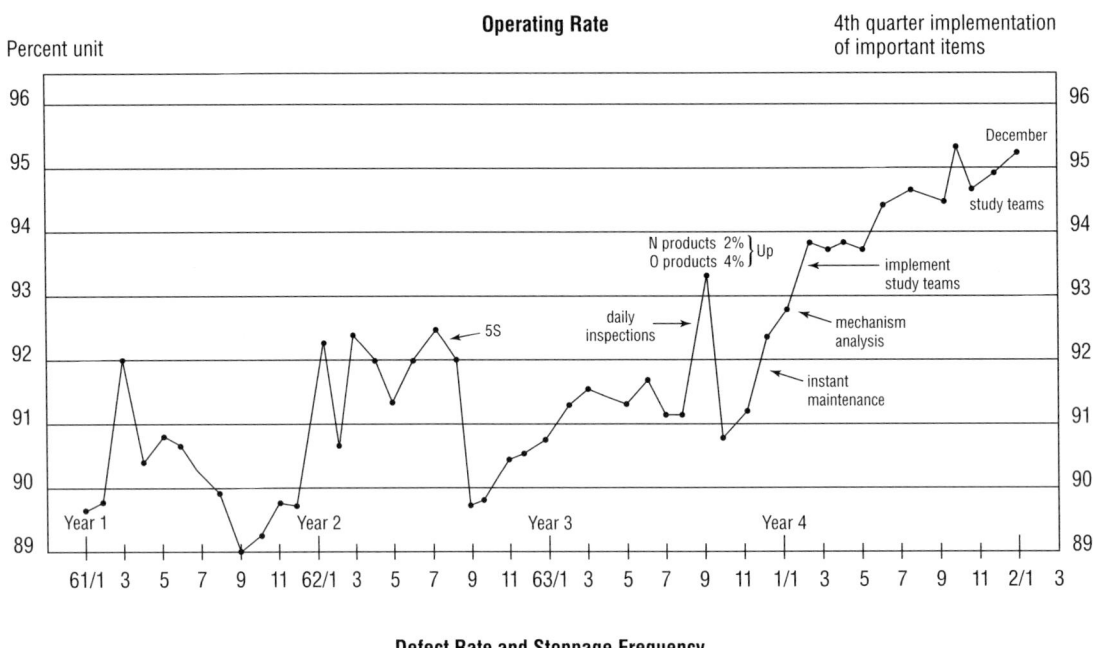

Defect Rate and Stoppage Frequency

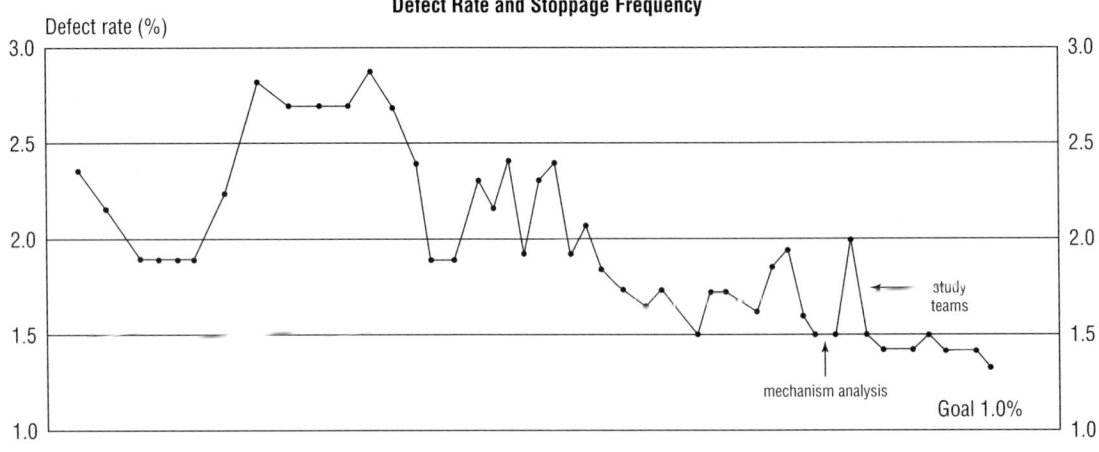

Frequency of Stoppages

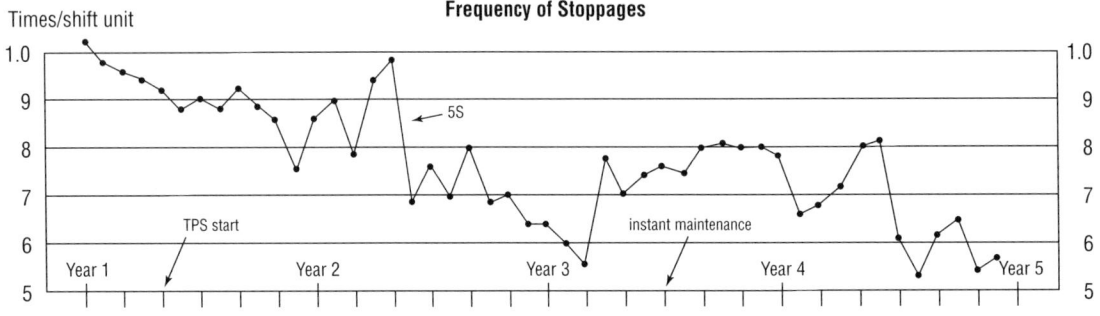

Worksheet 7

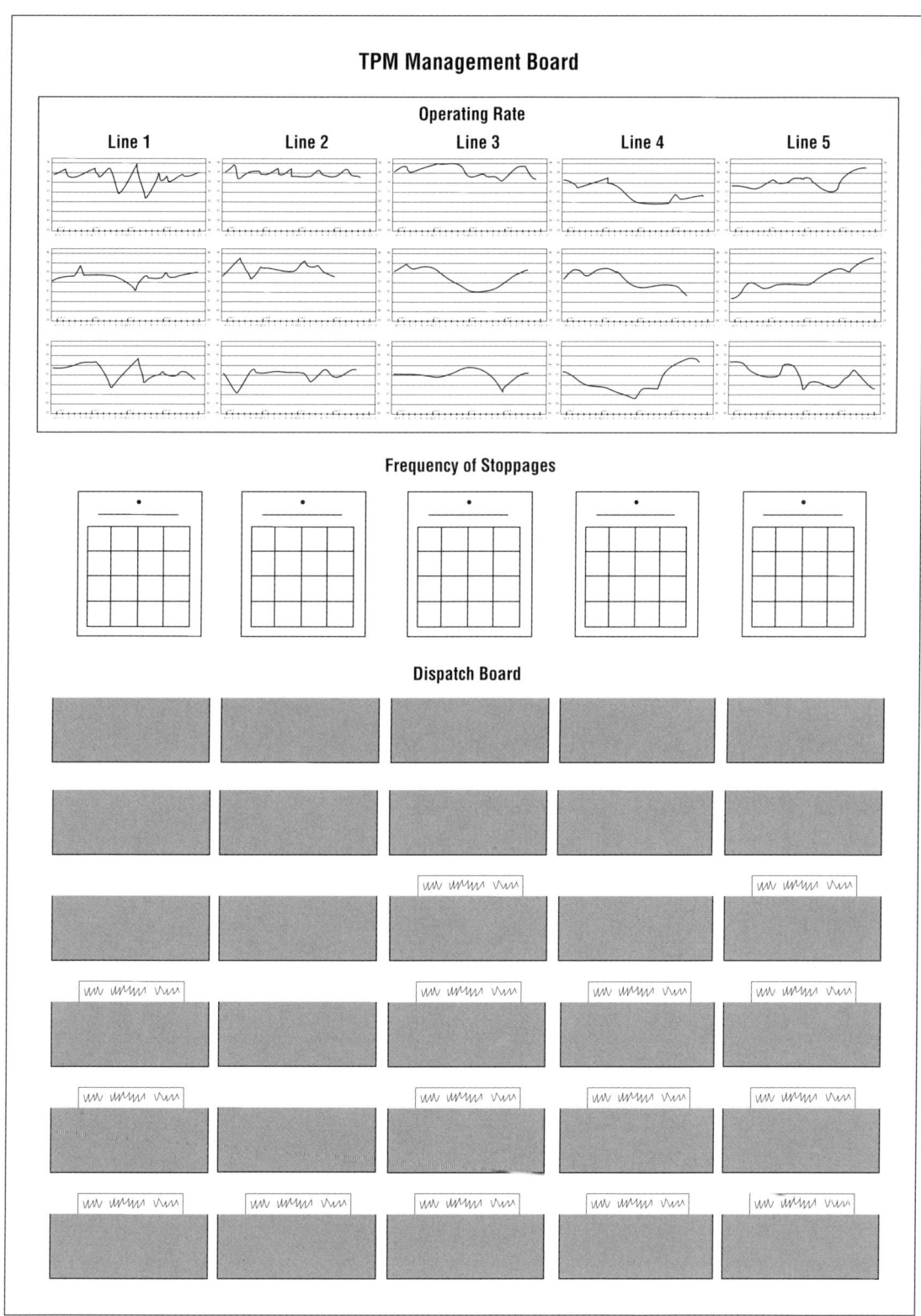

Worksheet 7a

October 5, 1991

Hourly Production Record

Line: 91E ASSY

Units/day	1,000

Approval	Staff	Super.
NK	GA	KN

Month/day	10/3 Day 4 / Night 3 people		Total 7 people	10/4 Day 3 / Night 3 people		Total 6 people	10/5 Day 3 / Night 3 people		Total 6 people
Time	Planned	Actual	Remarks	Planned	Actual	Remarks	Planned	Actual	Remarks
9:15	30	20	A12502	30	20		30	25	
10:15	70	80		70	50		70	70	
11:15	110	136		110	60	Prototypes came back	110	115	
12:15	150	210		150	95	Minor adjustments	150	170	
13:00	180	235		180	120		180	191	Replace cutting tool
14:00	220	280		220	149		220	230	
15:00	260	335		260	153	Minor adjustments	260	270	
16:00	300	385		300	30	A12301 Changeover	300	320	
17:00	340	430		340	90		340	350	Replace cutting tool
18:00	420	520		380	160		380	400	
19:00	40	55	A12503K Changeover	420	230		400	453	

Worksheet 8

240

Date: ____ / ____ / ____

Hourly Production Record

Line: _____

Units/day [ ]

	Approval	Staff	Super.

Month/day	Day Night	people	Total people	Day Night	people	Total people	Day Night	people	Total people
Time	Planned	Actual	Remarks	Planned	Actual	Remarks	Planned	Actual	Remarks
9:									
10:									
11:									
12:									
13:									
14:									
15:									
16:									
17:									
18:									
19:									
20:									
21:									
22:									
23:									
0:									
1:									
2:									
3:									
4:									
5:									
6:									
7:									
8:									
Total									

Worksheet 8a

Cumulative Availability, Efficiency Rate, and Quality Rate Chart

Year: _____   Machine: _____   Plant: _____   Date: ____ / ____ / ____

Month	September			October			November		
Working days	27 days			25 days			26 days		
Total possible operating time	12,420 min			23,370 min			24,540 min		
Actual operating time	8,389 min			17,295 min			19,508 min		
Set rpm	150			150			150		
Actual rpm									
Production p/min	91 prm			100 prm			122 prm		
Production pieces	1,135,956/shift			2,339,392/shift			3,017,504/shift		
Machine counter	1,240,909			2,678,959			3,099,615		
3-ply counter	1,184,074			2,564,394			3,012,270		
Difference	56,835/shift			114,565/shift			87,345/shift		
B products									
Availability rate	67.5%			74.0%			79.4%		
Efficiency rate	91.5%			87.3%			97.3%		
Quality rate	60.9%			66.7%			81.9%		
Total performance loss rate	8.4%			12.6%			2.6%		
Packaging counter									
Packaging finished									
Packaging quality									
Complaints	Cause	9	%	Cause	7	%	Cause	7	%
Worst No. 1	Back sheet torn	3	33.3	No tape	2	28.5	Hot melt hardening	3	42.8
No. 2	No tape	2	22.2	Fastening defect	2	28.5	Tape not sticky	2	28.5
No. 3	Tape not sticky	2	22.2	Tape not sticky	1	14.2	No tape	1	14.2
No. 4	Hot melt hardening	1	11.1	Insects	1	14.2	Leakage	1	14.2
No. 5	R cutter defect	2	22.2	Gathering defect	1	14.2			
Strength ratio	Downtime	2,845		Downtime	4,327	18.5%	Downtime	4,672	19.0%
Worst No. 1	Other	1,210	42.5	Rotor EXT valve	692	15.9	Coater EXT valve	729	15.6
No. 2	Hot melt device	449	37.1	Human error	678	97.9	Human error	515	11.0
No. 3	3-ply bending	356	79.2	Tape cutter	501	73.8	R cutter	458	9.8
No. 4	Tape machine	144	40.4	3-ply bending	424	84.6	Wing	361	7.7
No. 5	Crimping machine	128	88.8	R cutter	328	77.3	3-ply bending	324	6.9
Frequency	Stoppages	383	%	Stoppages	727	%	Stoppages	822	%
Worst No. 1	UF tape	98	25.5	UF tape out	128	17.6	Coater EXT valve	179	21.7
No. 2	Other	76	19.8	Coater EXT valve	128	17.6	UF tape out	175	21.2
No. 3	3-ply bending	37	9.6	Tape cutter	86	11.8	Tape cutter	69	8.3
No. 4	Tape machine	35	9.1	R cutter	62	8.5	Human error, etc.	62	7.5
No. 5	Hot melt device	26	6.7	3-ply bending	54	7.4	3-ply bending	61	7.4
Remarks									

Worksheet 9

Cumulative Availability, Efficiency Rate, and Quality Rate Chart

Year: _____   Machine: _____   Plant: _____   Date: ____ / ____ / ____

Month			
Working days			
Total possible operating time			
Actual operating time			
Set rpm			
Actual rpm			
Production p/min			
Production pieces			
Machine counter			
3-ply counter			
Difference			
B products			
Availability rate			
Efficiency rate			
Quality rate			
Total performance loss rate			
Packaging counter			
Packaging finished			
Packaging quality			

Complaints	Cause		%	Cause		%	Cause		%
Worst No. 1									
No. 2									
No. 3									
No. 4									
No. 5									

Strength ratio	Downtime		#	Downtime		#	Downtime		#
Worst No. 1									
No. 2									
No. 3									
No. 4									
No. 5									

Frequency	Stoppages		%	Stoppages		%	Stoppages		%
Worst No. 1									
No. 2									
No. 3									
No. 4									
No. 5									

Remarks

$$\text{Performance (efficiency)} = \frac{\text{Pieces produced}}{\text{Machine counter}}$$

$$\text{Availability (operating rate)} = \frac{\text{Actual operating time}}{\text{Possible operating time}}$$

$$\text{Quality rate} = \frac{\text{Good pieces produced}}{\text{Possible working strokes/min}}$$

$$\text{Total performance loss rate} = \frac{\text{Pieces produced}}{\text{Machine counter}}$$

Worksheet 9a

Monthly Summary of Non-Working Time

Date: ___/___/___

Approval	Staff	Person in charge

Item / Line	(a) Total time on the job	(1) Setup	(2) Adjustment	(3) Breakdown	(4) Idled	(5) Waiting for materials	(6) Waiting for tools	(7) Other	(b) Sum of 1–7	(c) Operating time	(d) Operating rate	(e) Idle time	(f) Average rpm	(g) Actual production	(h) Non-working time 5,6,7	(i) Non-working time 1,2,3,4	(j) Actual working time without 5.6.7	(k) Estimated revolution rate	(l) Estimated operating rate
A	139,620	420	910	1,030	320	0	24,480	1,560	28,720	110,900	79.3	20.7	308	27,071	26,040	113,580	77.4	+19.0	96.4
B	118,140	2,010	900	3,450	3,740	0	4,900	2,180	17,180	100,960	76.2	23.8	191	14,690	7,320	110,820	69.4	+3.0	72.4
C	128,880	1,480	460	2,840	2,570	2,270	6,740	1,440	17,800	111,080	80.5	19.5	200	17,887	10,450	118,430	75.5	+5.0	80.5
D	128,880	570	240	1,100	2,090	0	7,020	1,440	12,460	116,420	78.1	21.9	200	18,187	8,019	120,861	75.2	+5.0	80.2
E	118,140	1,750	1,600	100	310	680	6,110	1,370	11,920	106,220	79.6	20.4	170	14,369	8,160	109,980	76.9	+6.3	83.2
F	97,460	2,270	1,020	390	470	0	19,150	1,180	24,480	72,980	84.5	15.5	170	10,484	20,330	77,130	80.0	+6.3	86.3
Sum of A–E	731,120	8,500	5,130	8,910	9,500	2,950	68,400	9,170	112,560	—	—	—	—	—	—	—	—	—	—
Average of A–E By item / Rate of occurrence	—	—	—	—	—	—	—	—	—	—	79.7	20.3	207	—	—	—	75.7	+7.4	83.1
S	107,400	1,910	3,930	930	1,360	0	9,240	1,320	18,690	88,710	71.8	28.2	92	58,622	10,560	98,840	65.8	+7.0	72.8
Sum of A–S	—	10,410	9,060	9,840	10,860	2,950	77,640	10,490	131,250	—	—	—	—	—	—	—	—	—	—
Average of A–S By item / Rate of occurrence	—	—	—	—	—	—	—	—	100	—	75.6	24.3	151	—	—	—	70.8	+7.2	78.0
Notes																			

Worksheet 10

Monthly Summary of Non-Working Time

Date: ___ / ___ / ___

Approval	Staff	Person in charge

Item / Line	ⓐ Total time on the job	① Setup	② Adjustment	③ Breakdown	④ Idled	⑤ Waiting for materials	⑥ Waiting for tools	⑦ Other	ⓑ Sum of 1–7	ⓒ Operating time	ⓓ Operating rate	ⓔ Idle time	ⓕ Average rpm	ⓖ Actual production	ⓗ Non-working time 5, 6, 7	ⓘ Non-working time 1, 2, 3, 4	① Actual working time without 5,6,7	ⓚ Estimated revolution rate	ⓛ Estimated operating rate
Total																			
Average																			
Total Sum																			
Notes																			

Worksheet 10a

245

Manufacturing Process Planning Sheet (by Machine Model)

Approval	Examination	Created

Process I			Management items					Quality characteristics			Rank	In charge of confirmation		Record	Treatment of abnormalities and persons responsible			Methods of handling abnormalities	Related standards
Process I chart	Process name	Machine used	Item	Indicated value	Method frequency	Name of characteristic	Standard value	Goal values	Measurement methods & tools	Frequency	Rank	Mfg.	Inspection	Type method	Operator	Supervisor	Super	Methods of handling abnormalities	Name
1	Materials receiving					Raw materials			Mill sheet confirmation	Each time	A		O				O	Return to manufacturer	Raw materials standards
2	Receiving inspection					Outer diameter	Raw materials standard		Micrometer	"	A		O				O	"	"
						Degree of hardness			Hardness meter	"	A		O				O	"	"
3	Storage					Outer appearance			Sight	"	A		O				O	"	"
4	Cold casting	Casting machine	Attachment conditions for jigs and tools	See separate chart							A	O		Record chart		O		Repairs	Operation standards
			Time period for attaching jigs and tools								A	O		"		O		"	"
			Number of revolutions		Revolution meter						A	O		"		O		"	"
						Length			N = 3 length gauge	Beginning/end	A	O		Operations chart		O			
						Hole diameter	According to attached chart		N = 3 stopper gauge	"	A	O		"		O			
						Outer diameter			N = 3 micrometer	"	A	O		"		O		Submit quality incident tags	Steps for handling non-conforming products
						Uneven thickness			N = 3 dial gauge	"	A	O		"		O			
						Outer appearance			N = 3 sight	"	A	O		"		O			

Manufacturing Process Planning Sheet (by Machine Model)

Approval	Examination	Created

Process I			Management items					Quality characteristics			Rank	In charge of confirmation		Record	Treatment of abnormalities and persons responsible			Methods of handling abnormalities	Related standards
Process I chart	Process name	Machine used	Item	Indicated value	Method frequency	Name of characteristic	Standard value	Goal values	Measurement methods & tools	Frequency		Mfg.	Inspec-tion	Type method	Operator	Supervisor	Super		Name

Worksheet 11a

247

Process Quality Management Chart (General Use)

Machine model	(for all machines)
Machine number	(for all machines)
Personnel lineup	5S people

Name of department creating	Approval	Confirmation	Created by	Confirming department	
Inserter	A	B	C	D	E

Process flow chart	No.	Process name	Manufacturing conditions/ Management quality characteristics		In charge of confirmation O = operator S = staff		Record
			Management items	Norm values	Confirmation method	Frequently	
		Chip mounting A1 surface process					
①	①	Furnace setting	Digital SW	Operations manual	O	Visual	Changeover	
②	②	Rail width modification	Rail width	Board width approx.+ 0.5 mm	O	〃	〃	
③	③	P plate, set plate making	Machine model Plate number	Production instructions	O	〃	〃	
④	④	Set up printer	Screen name Squeegee speed	Operations manual	O	〃	〃	
⑤	⑤	Test printing	No blurring, off center, or spreading	Operations manual	O	〃	〃	
⑥	⑥	Set up dispenser	Program backup pin	〃	O	〃	〃	
⑦	⑦	Test striking	Amount of bond	Limit sample	O	〃	〃	
⑧	⑧	Materials setup	Name of part Quantity	Materials setup chart	O	〃	〃	
⑨	⑨	Parts confirmation	Name of part	〃	O	〃	〃	
⑩	⑩	Set up mounter	Program name Backup pin	Production instructions	O	〃	〃	
⑪	⑪	Alignment check	Off-center, constant, polarity	Operations manual	O	〃	〃	
⑫	⑫	Furnace temperature profile	Furnace internal temperature	Template	O	Reflow checker	〃	

Worksheet 12

Process Quality Management Chart (General Use)

Machine model	
Machine number	
Personnel lineup	

Name of department creating	Approval	Confirmation	Created by	Confirming department	

Process flow chart	No.	Process name	Manufacturing conditions/ Management quality characteristics		In charge of confirmation O = operator C = staff		Record
			Management items	Norm values	Confirmation method	Frequently	

Worksheet 12a

Date: ____ / ____ / ____

Observed workplace	**Operation Observation Sheet (Work Sampling)**												Observer:		
Classification	**Nature of operation**	**Code**	**1**	**2**	**3**	**4**	**5**	**6**	**7**	**8**	**9**	**10**	**Sub-total**	**% of total**	**Notes**
Net operations (operations having earnings as their goal)	Attachment/standard joining	A1	1	1	15	15	15						47	7.3	
	Removal	A2						16	14				30	4.6	
	Packing/packaging	A3						21	14				35	5.4	
		A4													
		A5													
	Total												112		
Semi-wasteful (necessary, but don't add value)	Inspection/measurement	B1	3	18	17	3	6	3	22	12	10		94	14.5	②
	Selection	B2			8	8					25		41	6.3	
	Last-minute adjustment	B3													
	Adjustment	B4	3	1	4	1	4			10	5		28	4.3	
	Setup and changeover	B5	2	3	10	4	6			19			44	6.8	
	Hand wiping, oil stains	B6		2		2							4	0.6	
	Total												211		
Waste (unnecessary actions that don't add value)	Walking	C1	1	9	7	8	9	4	3	2	10	8	61	9.4	③
	Transport	C2	4			1		1	1	1	4	7	19	3.0	
	Transport by hand	C3	4	2	1	1	6	1	4		1	2	22	3.4	
	Storage/accumulation	C4	4					3	4	2	5		18	2.8	
	Waste from looking for things	C5				1	5	3	1	3			13	2.0	
	Idleness (missing parts)	C6	2									18	20	3.1	
	Slack time	C7	25	10	20	20	5	12	2	4	2		100	15.5	①
	Consultations/meetings	C8		3				2				4	9	1.4	
	Recording	C9		6									6	1.0	
	Cleaning and tidying	C10	3						1	9	2	20	35	5.4	
	Removing labels	C11	12	9									21	3.2	
	Total												324	100	

1 Current personnel = 64

2 Net operations = 112 (17.3%)   Semi-wasteful = 211   Waste = 324   Total incidents = 647

3 Personnel reduction goal = current personnel × (1 – net operating rate) = 53 fewer workers

Worksheet 13

Observed workplace	Operation Observation Sheet (Work Sampling)												Observer:		
Classification	**Nature of operation**	**Code**	1	2	3	4	5	6	7	8	9	10	**Sub-total**	**% of total**	**Notes**
Net operations (operations having earnings as their goal)	Attachment/standard joining	A1													
	Removal	A2													
	Packing/packaging	A3													
		A4													
		A5													
	Total														
Semi-wasteful (necessary, but don't add value)	Inspection/measurement	B1													
	Selection	B2													
	Last-minute adjustment	B3													
	Adjustment	B4													
	Setup and changeover	B5													
	Hand wiping, oil stains	B6													
	Total														
Waste (unnecessary actions that don't add value)	Walking	C1													
	Transport	C2													
	Transport by hand	C3													
	Storage/accumulation	C4													
	Waste from looking for things	C5													
	Idleness (missing parts)	C6													
	Slack time	C7													
	Consultations/meetings	C8													
	Recording	C9													
	Cleaning and tidying	C10													
	Removing labels	C11													
	Total														

1 Current personnel = ____

2 Net operations = ____ (____%)   Semi-wasteful = ____   Waste = ____   Total incidents = ____

3 Personnel reduction goal = current personnel × (1 – net operating rate) =____ fewer workers

Worksheet 13a

Line Balance Measurement Sheet

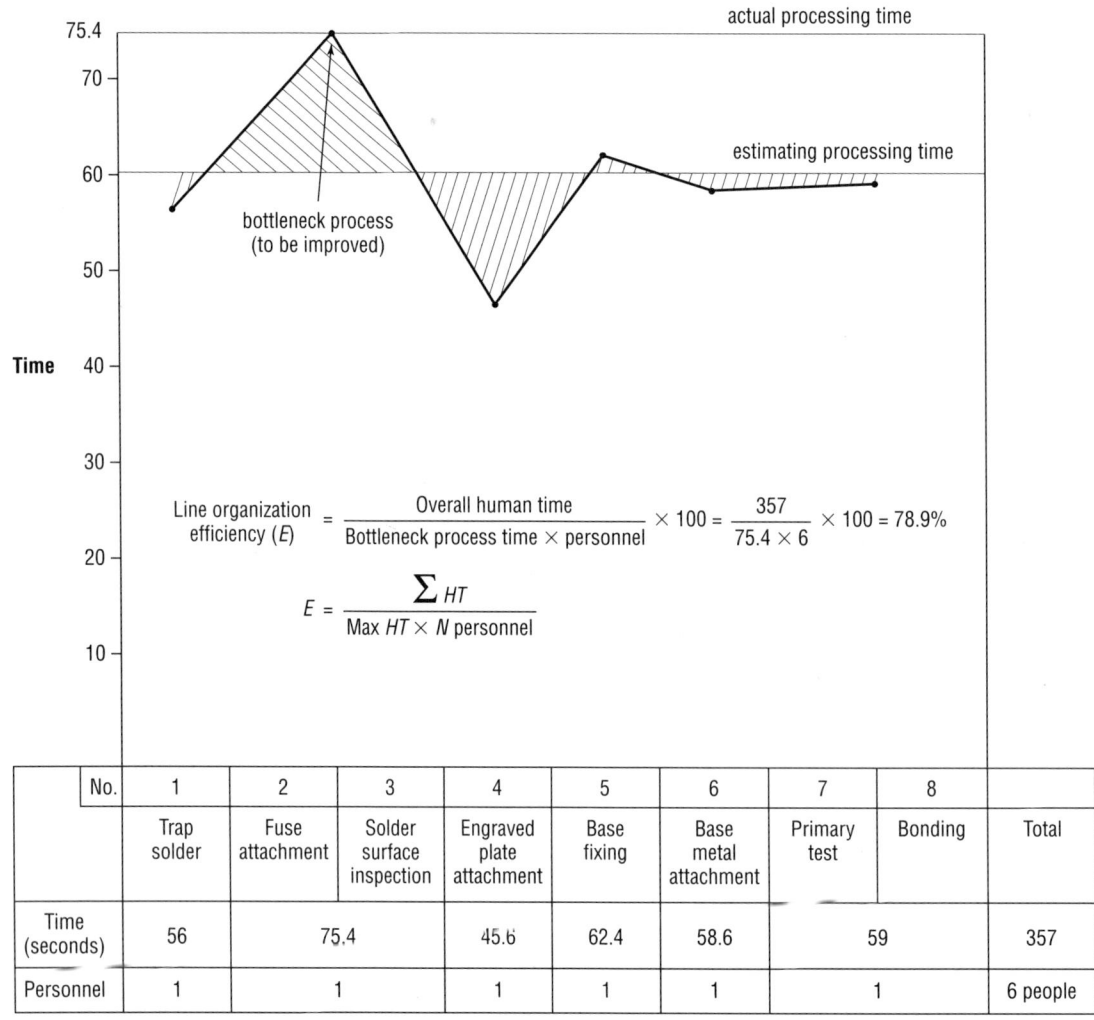

$$\text{Line organization efficiency } (E) = \frac{\text{Overall human time}}{\text{Bottleneck process time} \times \text{personnel}} \times 100 = \frac{357}{75.4 \times 6} \times 100 = 78.9\%$$

$$E = \frac{\sum HT}{\text{Max } HT \times N \text{ personnel}}$$

No.	1	2	3	4	5	6	7	8	
	Trap solder	Fuse attachment	Solder surface inspection	Engraved plate attachment	Base fixing	Base metal attachment	Primary test	Bonding	Total
Time (seconds)	56	75.4		45.6	62.4	58.6	59		357
Personnel	1	1		1	1	1	1		6 people

Worksheet 14

Line Balance Measurement Sheet

Plant		Takt: seconds/unit		Approval	Planner	Date created:
Line		Production: units/day				___ / ___ / ___

Units:													
Operator													
Sequence		1	2	3	4	5	6	7	8	9	10	11	12
Waste elimination time													
Before improvement	HT												
	MT												
After improvement	HT												
	MT												

[____] Waste elimination time   [____] Machine time (MT)   [____] Human time (HT)

Worksheet 14a

Main Line Check Sheet

Revision	Reason	Date		Line mounted	Date
⚠ × 1	Measures against customer complaints				
⚠ ×					Names of operators
⚠ ×			**Model: D40, A, P**		
			T/M machine no.		

Drawing		Check items			Standard	Check
	Sub	**Sub Assembly**				
	Sub	①	No nicks or debris in oil seal. Spread grease with finger.		Confirm with finger 40%–60%	
	Sub	②	Confirm and clean oil pores of original gears and gauge.		Watch out for spattering, air blow	
	Sub	③	Make sure O-ring doesn't bite in when attached.		• Use three-board • Check visually for gaps	
	Sub	④	Coupling all right? D50-15 mold common use, different from mold 16		Visual check	
	Sub	⑤	After original gear assembly, confirm Brg revolution		Shouldn't be heavy	
		Process 1				
	1	⑥	Make sure there's no casting sand or debris		Visual	
	1	⑦	Make sure that the O-ring doesn't bite in when the drain plus is attached		• Grease the case, too • Torque 13–18 kgm	
		Process 2				
	2	⑧	Gear direction of countershaft and intermediate shaft all right?		Visual	
	2	⑨	Add play to the collar of the countershaft and the intermediate shaft		0.3–0.6 mm	
		Process 3				
	3	⑩	Confirm snap ring of main axis and bearing		Is it definitely in?	
	3	⑪	Holder tightening torque, lock bending all right?		12 φ 5.5–12.5 kgm	
	3	⑫	Confirm backlash, particularly common bitmg gear		0.09–0.31 Others 0.17–0.4	
	3⚠	⑬	After greasing seal surface of the shaft, was the gauge attached?		No nicks or dents in seal surface	
		Process 3				
	4	⑭	No debris or foreign matter inside			
	4	⑮	Gasket, three-bond no. 4 spread		• Not out of position • No nicks	
	4	⑯	All speed changes smooth? Brg movement all right?			

Drawing annotations:

① Make sure there are no nicks or debris on the lip

Don't mistake it for D51-16

② ③ ④ oil pore, plug, collar Ⓐ

oil groove, oil pore

⑧ Countershaft
A = 8 1 2
= 5 6 1
= 1 3 1

Intermediate shaft
A = 5 6 2
= 3 6 2

Worksheet 15

254

Main Line Check Sheet

Revision	Reason	Date				Line mounted	Date		
			Model:			Names of operators			
			T/M machine no.						

Drawing		Check items	Standard	Check

Worksheet 15a

Malfunction Cause and Treatment Chart

Plant: __Our plant__   Family: _____   Team reporter: __HR__   Date: ___ / ___ / ___

Order	Process	Malfunction	Rank	Cause (Nature)	Improvement plan (countermeasures)
1	Storage tank	Attachment of short pipe to cast iron valve		The pipe is short and the bolt won't go in.	We improved it by switching to a stud bold, but in the future we'd like the pipe to be made longer than the bolt
2	Heat exchange	Nozzle flange strikes against leg		1. When we joined the part to the crossbeam of the stand, inspection of the air between the nozzle revealed leakage. 2. This was an adaptation from a drawing, and the actual dimensions were not shown.	If we generate a drawing through CAD, we'll soon discover a method.
3	Filter	Cannot replace filter mesh		The pipe goes straight up.	This is a simple design error, and we fixed it with 4 elbow joints. It doesn't look very good, so to avoid having to do this in the future, we'd like to put something about this in the checklist for design review.

Worksheet 16

Malfunction Cause and Treatment Chart

Plant: _____   Family: _____   Team reporter: _____   Date: ____ / ____ / ____

Order	Process	Malfunction	Rank	Cause (Nature)	Improvement plan (countermeasures)

Worksheet 16a

5S Evaluation Chart

Worksite	
Section	
Line	

Date	
Members	

Checkpoints	Evaluation Level				
	1	2	3	4	5

Organization

	1	2	3	4	5
Warehouse (inventory management)	The storage area is too cluttered to walk around freely.	Items are placed irrationally.	Items have designated storage places, but they're often ignored.	Items are stored in proper locations, but no standard criteria indicate when to reorder.	Items are managed for a just-in-time supply and tracked on inventory boards.
Aisles	Work-in-process and other items stand in a roped-off area in the aisles.	Items are set on the sides of the aisles so employees can pass, but carts and dollies cannot pass.	Items protrude into the aisles.	Items protrude into the aisles but have warning labels.	There is no work-in-process, so the aisles are completely clear.
Work areas	Items lie scattered around for months, in no particular order.	Items lie around for months, but they don't get in the way.	Unneeded items have been red-tagged and a disposal date has been set.	Only items to be used within the week are kept around.	Only items needed the same day are kept around.
Machine parts storage	Parts are jumbled together with paper scraps and rags.	Broken and unusable parts are being stored.	Frequently used parts are stored separately from those that will not be used soon.	All parts are stored in standard places with standard labels according to an easily understood system.	Nothing is found out of place.
Workbenches and tables	Tables are covered with unneeded materials.	Tables hold materials that are only used once every two weeks.	Tables hold extra pencils and other unneeded stationery items.	Items remain on the tables for as long as a week.	Only the minimum items needed are kept on the tables.
Equipment (mainly sewing machines)	Equipment is placed in no particular order, some of it rusted and unusable.	Usable and unusable equipment are kept together.	Unusable and unneeded equipment has been thrown out.	Equipment is managed according to its frequency of use and degree of importance.	The equipment is set up so anyone can find what they need to use at any time.
Line organization	The line is in disarray, and planned downtime occurs more than 40 percent of the time.	Planned downtime is as high as 40 percent, and the flow of the line is unclear.	Planned downtime is around 30 percent, and there is waste involved in transporting material.	Planned downtime is around 20 percent, and unfinished goods are kept on the line.	The line is well-organized and flows smoothly, with no more than 10 percent planned downtime.

Orderliness

	1	2	3	4	5
Reorder level (items arrive when needed, in the required quantities)	Parts are reordered when 90 percent have been used up, and parts shortages still occur.	Parts are reordered when 90 percent have been used up, standby occurs at assembly processes.	Parts are reordered when 95 percent have been used up. Problems sometimes occur during model changes.	Parts are reordered when 99 percent have been used up. New product orders generally arrive on time.	The work site practices just-in-time manufacturing, reordering only as inventory is used up.
Waste from searching during setups	Employees wander aimlessly searching for parts and materials.	Employees move in a zigzag pattern gathering up what they need.	Some effort is wasted in searching for things.	Setup carts are used.	No time is lost in searching for things.

Worksheet 17

(continued)

Checkpoints	Evaluation Level				
	1	2	3	4	5

Orderliness, *continued*

Checkpoints	1	2	3	4	5
Jig and tool storage	Jigs and tools are scattered all over the place.	Jigs and tools are stored in boxes.	Different kinds of jigs and tools are stored separately.	Jigs and tools are stored on shadow boards.	Jigs and tools are arranged by frequency and order of use.
Parts and materials	Defective and good parts are stored together.	Defective parts are kept on a separate shelf.	Only good parts are kept in storage; nicks, humidity damage, and other problems are avoided.	Storage shelves are clearly labeled and well organized.	Parts are delivered just-in-time, using kanban and tracking boards.
Drawings and charts	Current drawings are jumbled together with torn, outdated charts.	Drawings are organized and filed by category.	Drawings that are hard to read have been replaced with new ones.	Drawings are stored so they're easy to retrieve.	The system allows anyone to return drawings to their proper places.
Documents and other written materials	Documents are scattered randomly on tables and shelves; old documents lie forgotten in storage.	Documents have been straightened enough so that someone who looks long enough will find them eventually.	Documents of the same type are stored in the same place.	Documents and written materials are classified and color coded.	Visual storage is fully implemented. Anyone can easily retrieve documents and return them to their proper place.

Cleanliness

Checkpoints	1	2	3	4	5
Wires and pipes on the ceiling	Dusty wires and pipes dangle haphazardly from the ceiling.	Wires and pipes are laid out for each line, but they are hard to clean around.	Wiring for each line is bundled, making cleaning easier.	Few pipes or wires are evident, and there's no debris from overhead.	No pipes or wires hang from the ceiling.
Aisles	Aisles are littered with cigarette butts, thread, and metal shavings.	There are no large pieces of trash, but small paper scraps, debris, and dust are present.	The aisles are cleaned only in the morning.	Surface defects discovered during cleaning are quickly repaired.	Efforts are made to keep the aisles from getting dirty in the first place.
Machines and equipment	Machines and equipment are dirty and are used in that state.	Visible parts of the equipment appear to be cleaned occasionally.	Equipment is cleaned during setup and changeover.	Operators clean the equipment once a day.	Machines and equipment have inspection labels and are cleaned every morning.
Cleanliness of work areas	Pieces of thread, scraps of cloth, and cutting dust are scattered around.	There's no large debris, but smaller debris and dust are present.	The area has been cleaned.	The area is cleaned every day at the end of the shift.	Debris and dust are caught automatically to keep the area clean.
Work tables and desks	Surfaces are piled so high with documents, tools, and parts that they can't be cleaned.	Dust and debris have accumulated under the work tables.	Work tables and desks are cleaned once a day.	Even the legs of the tables and desks are clean, and all nicks and scratches have been repaired.	Everything is kept clean at all times.
Windows, window frames, and walls	Windowpanes are missing or cracked, with haphazard repairs.	The panes are dirty, and dust and debris have accumulated on the frames.	The panes are dirty, but the frames are occasionally cleaned.	Both the panes and frames are kept clean.	Window shades are used; walls are kept clean and uncluttered, and no extraneous items are attached to the walls, giving the area a pleasant atmosphere.

(continued)

Worksheet 17

Checkpoints	Evaluation Level				
	1	2	3	4	5

Cleanliness, *continued*

Tools, jigs, and molds	Some items are rusted.	No items are rusted, but some are covered with oil or dirt.	Only the parts that users actually touch are clean.	Grinding dust and other debris have been cleaned off, making tools and implements pleasant to handle.	Devices prevent accumulation of debris in the first place, and any debris is quickly cleaned off.

Standardized Cleanup

Restrooms	Facilities are dirty; supplies run short; unpleasant to use.	Fixtures are rinsed, but they are still dirty and supplies run out sometimes.	Restrooms are cleaned once a day, but they're still a bit dirty.	Restrooms are clean and hygienic, and supply shortages do not occur.	Restrooms are pleasant and well lit, with music piped in.
Cafeteria and employee lounges	These areas are so dirty one doesn't want to sit there.	The areas have been cleaned somewhat, but they're still dirty.	One wouldn't mind sitting there in work clothes, but not when dressed up.	Chairs have been cleaned so clothes don't get dirty.	The areas are extremely clean, sanitary, and attractively decorated; guests can be taken there.
Implements and jigs (low-cost improvements)	Items have been patched together with tape and the like.	Signs and signal light stands are made of flimsy materials such as cardboard.	Makeshift implements look weak and fragile.	Some equipment is crude-looking and handmade.	Well-made mechanisms use simple automation.
Overall layout	The room is dark, making detailed work difficult.	Light and illumination have been provided, reducing the possibility of on-the-job injuries.	The room is bright and safe, with lighting, illumination, and shades.	Ventilation is sufficient, giving the work area a refreshingly airy feel.	The plant is obviously a healthy working environment.

Discipline

Workplace attitude	Employees avoid eye contact and don't say anything, even when they bump into them.	People say "Excuse me" when they bump into someone, but otherwise ignore others.	Employees make eye contact with and greet only about 10 percent of the other people.	Employees acknowledge about half of the others they encounter.	Everyone is polite and pleasant and at least smiles and nods to others.
Smoking	Employees smoke openly, even in front of first-time visitors. More than 50 percent smoke on the job.	Employees smoke even while being addressed by a supervisor. Less than 50 percent smoke on the job.	Employees light up immediately after meals, without asking permission of visitors. Less than 40 percent smoke on the job.	Employees don't smoke if their visitor doesn't smoke. Less than 30 percent smoke on the job.	Employees don't smoke. Less than 20 percent smoke on the job.
Manner of speech	People use unnecessary jargon. They generally ignore others' ideas and are not good listeners.	People have a know-it-all attitude toward others and do not actively pay attention when others are speaking.	People occasionally speak politely and are receptive to others' ideas about half of the time, paying attention when others are speaking.	People always speak politely, with respect for their leaders. They are receptive to others' ideas about 60 percent of the time and nod in assent while actively listening.	People generally speak politely and respectfully to everyone. They are receptive to others' ideas about 70 percent of the time and are positively supportive of others who are speaking.

(continued)

Worksheet 17

Checkpoints	Evaluation Level				
	1	2	3	4	5

Discipline, *continued*

	1	2	3	4	5
Clothing	Many employees wear soiled clothes.	Many people have buttons missing or wrinkled clothes.	Most people are cleanly but carelessly dressed and don't wear their name tags.	The employees look stylish and well put together.	Most employees wear even their work uniforms with pride and flair.
Punctuality	Employees are often more than 30 minutes late. No one is concerned about time.	Employees are often up to 20 minutes late. Some people are often lax about time.	Employees phone ahead when they'll be 10 minutes late.	Employees often hurry to keep appointments or to arrive within 5 minutes of the scheduled time.	Employees arrive 5 minutes ahead of the scheduled time and are never late.
Instilling the 5S spirit	Employees are unaware of 5S conditions and ignore colleagues' mistakes or infractions.	They notice when conditions aren't maintained, but mention it casually if at all.	They don't talk to the person responsible, but try to fix the situation themselves or tell the other person's boss about it.	They talk to the other person on the spot, speaking softly. The other person does not accept the advice positively.	They talk to the other person on the spot, and he or she accepts the advice positively.
Subtotals	points	points	points	points	points
Total	*Goal:* Try to raise the total by 20 points within the next evaluation period.				
Comments					

Pareto Chart

Analysis of Axial Minor Stoppages

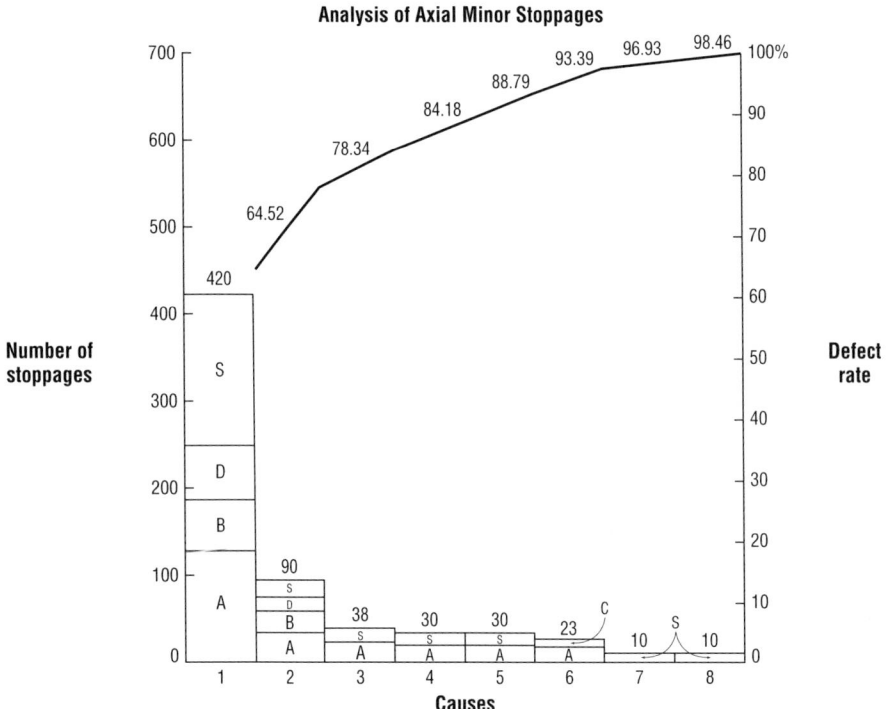

Causes of stoppages

1. Parts shortages
2. Head molding level
3. Insertion guide
4. Rubber pusher
5. Play in the main unit
6. Jaw sharpness, attachment
7. Absorption rubber
8. Other

Worksheet 18

Pareto Chart

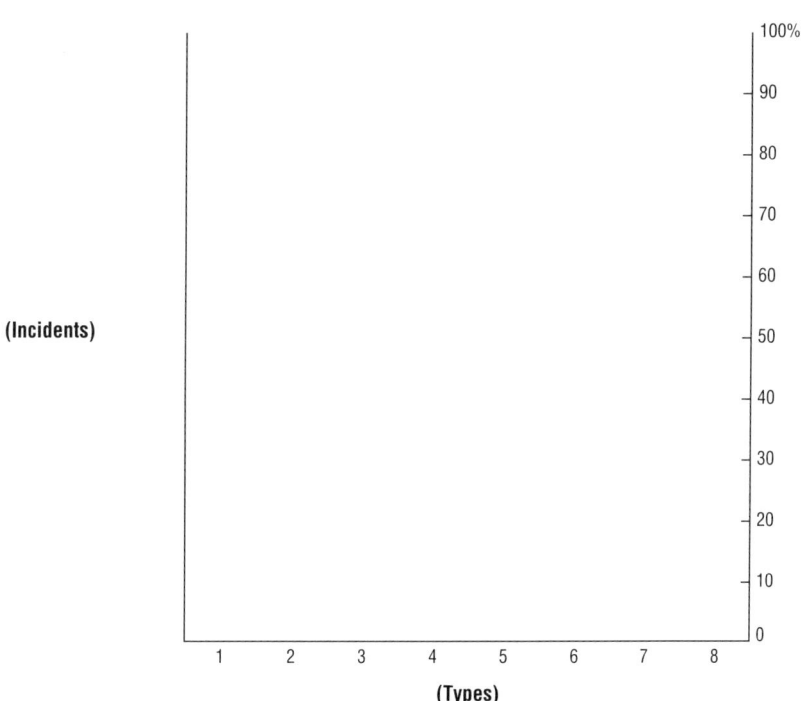

Worksheet 18a

Monthly Line Efficiency Report

Line	A Employee time on the job (min.)	B Actual employee time worked (min.)	A − B Time lost at employee level (min.)	G Test setup total (min.)	C Number of cases	Overall efficiency (C ÷ A)			Operating efficiency (C ÷ B)			Operating rate (B ÷ A)		
						This month	Last month	Highest value	This month	Last month	Highest value	This month	Last month	Highest value
1	140,880	133,870	7,010		169,943	▼120.6	127.5	127.5	▼126.9	135.5	135.5	△95.0	94.1	95.0
2	176,910	169,939	6,971		166,661	▶94.2	103.7	107.1	▶98.1	108.0	112.6	△96.1	96.0	96.1
3	161,220	155,820	5,400		130,234	▶80.8	93.7	108.0	▶83.6	100.2	119.3	△96.7	93.5	96.7
4	119,340	114,513	4,827		156,762	▼131.4	146.4	146.4	▼136.9	151.7	151.7	▼96.0	96.5	96.5
5	279,500	263,225	8,825	7,450	250,154	▶89.5	91.3	91.3	▶95.0	96.9	96.9	▶94.2	94.2	94.6
6	162,435	145,559	16,875		190,987	▼117.6	119.8	119.8	▼131.2	132.8	132.8	▶89.6	90.2	90.8
7	45,840	44,076	1,764		39,111	△85.3	84.8	99.9	▶88.7	89.0	104.2	△96.2	95.3	96.3
8	21,105	19,730	1,375		22,007	▼104.3	116.7	116.7	▼111.5	125.4	125.4	△93.5	93.1	93.9
9	173,164	158,264	6,940	7,960	149,557	▶86.4	93.0	93.0	▶94.5	100.6	100.6	▼91.4	92.5	92.5
10	161,390	156,104	5,286		150,489	▶93.2	97.0	106.0	▶96.4	100.2	110.3	▼96.7	96.8	97.1
11	189,515	183,790	5,725		211,331	△111.5	107.7	111.5	△115.5	110.6	115.0	▶97.0	97.3	97.4
12	75,480	72,687	2,793		75,002	△99.4	93.7	99.4	△103.2	96.4	103.2	▶96.3	97.2	97.5
13	102,905	99,377	3,528		117,737	△114.4	111.5	114.4	△118.5	115.8	118.5	△96.6	96.2	96.6
14	141,330	135,776	5,554		164,933	△116.7	116.3	116.7	△121.5	121.2	121.5	△96.1	96.0	96.6
15	124,045	117,297	6,748		140,625	△113.4	108.8	113.4	△119.9	115.2	119.9	△94.6	94.5	95.8
16	127,910	118,365	7,395	2,150	123,247	▶96.4	97.5	100.2	▶104.1	104.6	108.3	▶92.5	93.2	93.8
17	100,504	95,914	4,590		145,258	△144.5	128.0	144.5	△151.4	133.5	151.4	▶95.4	95.8	97.0
18	115,100	111,384	3,716		131,350	▼114.1	115.1	118.7	▼117.9	119.0	123.4	△96.8	96.7	96.8
19	84,444	81,249	3,195		96,552	▼114.3	119.2	128.3	▼118.8	123.8	133.8	▶96.2	96.3	96.7
20	83,220	76,791	6,429		96,970	▼116.5	117.8	117.8	▼126.3	131.8	131.8	△92.3	89.4	92.3
21	75,788	70,962	4,826		89,738	△118.4	113.7	118.9	△126.5	121.2	127.0	▼93.6	93.8	95.4
22	112,275	106,670	2,025	3,580	150,364	△133.9	132.2	133.9	△141.0	140.2	141.0	△95.0	94.3	95.0

Worksheet 19

264

Monthly Line Efficiency Report

Line	A Employee time on the job (min.)	B Actual employee time worked (min.)	A – B Time lost at employee level (min.)	G Test setup total (min.)	C Number of cases	Overall efficiency (C ÷ A)			Operating efficiency (C ÷ B)			Operating rate (B ÷ A)		
						This month	Last month	Highest value	This month	Last month	Highest value	This month	Last month	Highest value

Equipment Operation Analysis Chart

Process: _____ Plant: _____ Team: _____ Operator: _____ Created by: _____ Date created: _____

No.	Measure	Machine #2	Machine #3	Machine #4	Machine #5	Machine #6	Total
1	Working days	21	24	24	24	24	117
2	Possible operating time (minutes)	9,480	33,670	33,615	33,625	33,615	144,005
3	Standard strokes per minute	160.0	230.0	210.0	230.0	210.0	216.1
4	Actual operating time (minutes)	8,432	29,486	28,577	29,114	27,759	123,368
5	Unit output at 100% capacity (2) × (3)	1,516,800	7,744,100	7,059,150	7,733,750	7,059,150	31,112,950
6	Actual good units produced (P number)	1,376,792	6,425,936	5,553,048	5,897,052	5,185,716	24,438,544
7	Machine counter	1,425,144	6,433,756	5,763,670	6,125,548	5,393,327	25,141,445
8	Intermediate counter	1,391,353	6,031,785	5,589,185	5,916,928	5,208,858	24,138,109
9	Packaging machine counter	0	0	0	0	0	0
10	Main unit weight loss (kg)	0.0	0.0	0.0	0.0	0.0	0.0
11	Packaging machine weight loss (kg)	0.0	0.0	0.0	0.0	0.0	0.0
12	Availability (operating rate) (4) ÷ (2)	88.9%	87.6%	85.0%	86.6%	82.6%	85.7%
13	Efficiency rate (6) ÷ (7)	96.6%	99.9%	96.3%	96.3%	96.2%	97.2%
14	Quality rate (6) ÷ (5)	90.8%	83.0%	78.7%	76.3%	73.5%	78.5%
15	Main unit performance loss rate	2.4%	6.2%	3.0%	3.4%	3.4%	4.0%
16	Packaging machine performance loss rate	1.0%	−6.1%	0.6%	0.3%	0.4%	−1.2%
17	Total loss rate	3.4%	0.1%	3.7%	3.7%	3.8%	2.8%

Worksheet 20

Equipment Operation Analysis Chart

Process: _____ Plant: _____ Team: _____ Operator: _____ Created by: _____ Date created: _____

No.	Measure	Machine #2	Machine #3	Machine #4	Machine #5	Machine #6	Total
1	Working days						
2	Possible operating time (minutes)						
3	Standard strokes per minute						
4	Actual operating time (minutes)						
5	Unit output at 100% capacity (2) × (3)						
6	Actual good units produced (P number)						
7	Machine counter						
8	Intermediate counter						
9	Packaging machine counter						
10	Main unit weight loss (kg)						
11	Packaging machine weight loss (kg)						
12	Availability (operating rate) (4) ÷ (2)						
13	Efficiency rate (6) ÷ (7)						
14	Quality rate (6) ÷ (5)						
15	Main unit performance loss rate						
16	Packaging machine performance loss rate						
17	Total loss rate						

Worksheet 20a

267

Minor Stoppage Cause Sheet

Process: _____ Plant: _____ Team: _____ Operator: _____ Created by: _____ Date created: _____

Month: _____

Machine number	Machine #1		Machine #2		Machine #3		Machine #4		Machine #5		Total		Average per machine per day	
Working days per month	24		24		24		24		24		120			
	Freq.	Time	Freq.	Time	Freq.	Time	Freq.	Time	Freq.	Time	Freq.	Time	Freq.	Time
Mechanical problems (by machine area)														
Dust collector	1	10	1	10	5	45	5	45	0	0	12	110	0.1	0.9
Processor 1 and 2	52	712	30	255	39	228	43	251	39	261	203	1,707	1.7	14.2
P part	5	78	3	27	1	20	2	2	3	38	14	165	0.1	1.4
Liner	9	25	17	71	12	44	11	19	8	16	57	175	0.5	1.5
P-1 part	11	265	5	98	2	7	2	7	8	21	28	398	0.2	3.3
DE	0	0	2	5	28	136	2	18	28	170	60	329	0.5	2.7
Bending part	19	247	28	143	36	149	16	46	30	128	129	713	1.1	5.9
P-8 part	0	0	0	0	3	5	2	5	6	27	11	37	0.1	0.3
NL part	4	11	37	131	13	20	13	38	13	36	80	236	0.7	2.0
Cutter	4	36	6	104	4	62	2	9	3	20	19	231	0.2	1.9
FS part	5	83	25	298	2	53	4	192	2	18	38	644	0.3	5.4
Preheater	4	62	39	480	3	56	1	5	2	25	49	628	0.4	5.2
F part		601	32	247	31	275	14		2				1.0	11.2
Defects in materials														
Cut off	118	398	109	354	49	125	87	230	90	298	453	1,405	3.8	11.7
Defective	0	0	2	10	5	13	23	57	26	76	56	156	0.5	1.3
LT defect	6	17	15	103	2	20	17	43	8	30	48	213	0.4	1.8
N defect	15	53	23	216	19	59	50	143	8	18	115	489	1.0	4.1
S defect	0	0	6	109	4	22	2	12	36	259	48	402	0.4	3.4
B sheet defect	0	0	2	125	1	2	0	0	7	47	10	174	0.1	1.5
P defect	0	0	0	0	0	0	0	0	1	8	1	8	0.0	0.1
F defect	2	4	0	0	4	7	1	2	0	0	7	13	0.1	0.1
FP polyethylene defect	1	40	1	5	0	0	0	0	0	0	2	45	0.0	0.4
Human error														
Forgetfulness	12	73	31	142	10	23	69	192	24	88	146	518	1.2	4.3
Error	30	65	82	312	83	284	114	233	92	211	401	1,105	3.3	9.2
Other	33	460	12	311	14	121	7	30	10	163	76	1,085	0.6	9.0
Downtime														
Morning, noon, end of shift	1	30	1	30	1	30	1	30	1	30	5	150	0.0	1.3
Sampling	0	0	0	0	1	10	0	0	1	10	2	20	0.0	0.2
Power outage	0	0	0	0	0	0	0	0	0	0	0	0	0.0	0.0
Total	549	5,366	721	5,435	584	2,791	645	2,527	608	3,693	3,107	19,812	26	165
Average per day	22.9	223.6	30.0	226.5	24.3	116.3	26.9	105.3	25.3	153.9	25.9	165.1		

Worksheet 21

Minor Stoppage Cause Sheet

Process: _____ Plant: _____ Team: _____ Operator: _____ Created by: _____ Date created: _____

Month: _____

Machine number	Machine #1		Machine #2		Machine #3		Machine #4		Machine #5		Total		Average per machine per day	
Working days per month	24		24		24		24		24		120			
	Freq.	Time	Freq.	Time	Freq.	Time	Freq.	Time	Freq.	Time	Freq.	Time	Freq.	Time
Mechanical problems (by machine area)														
Defects in materials														
Human error														
Downtime														
Total														
Average per day														

Worksheet 21a

269

Summary of Defects by Line

Date created: 2/10 Created by: SS

Line	1	2	3	4	5	6	7	Subtotal	8	9	10
Number of inspections	1,200	866	531	797	500	180	624	4,698	356	961	524
Number of defects	12	7	1	11	2	1	3	37	8	13	9
Rate (%)	1.00	0.80	0.18	1.38	0.40	0.55	0.48	0.78	2.24	1.35	1.71
Ball sound				1				1	1		1
Getting stuck	1	1	1	2		1		6			
Nicks in the gears				1			1	2	1	2	
Housing		1						1	1	3	
Differences in current											
Coating peeling off											
Generator defect											
Absent-minded errors	2	1		1				4	1	1	2
Errors in operation	2	3		1	1			7	1		
Parts defects	2							2	1	2	2
Shaft vibration									2	1	2
Voltage	1			3	1			5			
Broken wires	1	1		1				3		1	1
Defect in lead wire											
Gears meshing	2			1				3			
Thrust	1						2	3		1	1
Unclear										2	

Line	28	29	30	31	Subtotal	32	33	Subtotal	Total
Number of inspections	605	485	515	475	2,528	568	614	1,182	21,921
Number of defects	4	4	4	2	20	8	8	16	247
Rate (%)	0.66	0.82	0.77	0.42	0.79	1.40	1.30	1.35	1.12
Ball sound							1	1	10
Getting stuck	2				4	1	1	2	25
Nicks in the gears							1	1	21
Housing									13
Differences in current									24
Coating peeling off									1
Generator defect									
Absent-minded errors					2		2	2	27
Errors in operation				1	1				12
Parts defects					1	2	2	2	17
Shaft vibration						1		1	20
Voltage						1		1	7
Broken wires									6
Defect in lead wire	1			1	2	2		2	6
Gears meshing									3
Thrust	1				2	1		1	14
Unclear	1	4	3		8	4	1	5	39

Worksheet 22

Summary of Defects by Line

Date created: _____ Created by: _____

Line																					Total
Number of inspections																					
Number of defects																					
Rate (%)																					
Defects																					

Worksheet 22a

Minor Stoppage Analysis Sheet (Hand-Fed Machines)

Machine or equipment: _____

Date: _____ / _____ / _____

No.	Date	Shift	Number produced units	Overall points	Insertion error	Out of parts	Insertion rate	Operating rate	A	B	C	D	E	F	G	H	I	J	K	L	Total min.	Times switched
1	12/9	1	440	29,480	6	4	59.0	48.4	17	12	4						270		30		333	1
		2	580	20,900	8	4	41.8	50.6		2	1	1		3			272		38		317	2
	12/10	1	343	42,440	10	3	84.9	65.3	47		2		1	2			125		47		224	2
		2	530	30,740	10	3	61.5	67.0		3	1	1		3			205				213	0
	12/11	1	590	33,020	6	4	66.0	35.7		1	1			2			230		18		252	1
		2	660	34,980	9	3	70.0	82.6		2	4			6			70		30		112	2
	12/12	1	500	42,000	15	3	84.0	79.7		9		10		2					110		131	2
		2	450	42,600	9	2	85.2	77.1	3					5			20	20	100		148	3
	12/13	1	550	28,838	7	2	57.7	24.0			4			4			155		70		233	3
		2	475	20,125	4	4	40.3	48.1			10			5			120		200		335	6
	12/14	1	288	17,722	2	4	35.7	44.0						1			230		130		361	3
	Subtotal		5,406	342,845	86	34	62.3	56.7	67	29	27	12	1	33			1,697	20	773		2,659	25
8	12/9	1	455	80,080	6	8	117.5	96.3						5				22			27	0
		2	615	75,060	6	18	110.1	91.8	20					8	12	8			27		59	2
	12/10	1	326	80,153	7	6	118.1	95.4			3		1	3				5	24		33	2
		2	547	76,033	7	6	115.5	95.1		1	2		3	8				20			35	0
	12/11	1	597	79,299	7	4	116.3	86.7			1			5				2	12		21	1
		2	618	79,036	5	4	115.9	96.3		1	2			5					20		27	1
	12/12	1	525	78,775	4	7	115.5	95.8	2		1			4					24		30	1
		2	412	74,868	4	5	109.8	90.8	1	3			2	5							66	3
	12/13	1	568	66,122	5	3	97.0	86.1	1		2		2	15				22	55		100	4
		2	308	58,020	4	5	85.1	76.4	1					9					60		170	6
	12/14	2	288	57,272	4	0	84.0	74.2	1		2			12					160		186	4
	Subtotal		5,259	805,078	59	66	107.7	89.5	27	5	12	0	9	79	12	8	0	49	553	0	754	24
	Total		10,665	1,147,923	145	102	85.0	73.1	94	34	39	12	10	112	12	8	1,697	69	1,326	0	3,413	49

Element involved:

A. position
B. cutter
C. head
D. anvil
E. siphon magazine
F. parts, bases
G. programming
H. transfer
I. production adjustments
J. machine adjustments
K. switching
L. other

Problem:

B1. cutter can't cut
B2. feed defect
C1. lever
C2. insertion
C3. chuck
D1. clincher
D2. guide pin
E1. siphon
E2. magazine
E3. loader
L1. power outage
L2. air pressure

Worksheet 23

272

Minor Stoppage Analysis Sheet (Hand-Fed Machines)

Machine or equipment: _____   Date: ___ / ___ / ___

No.	Date	Shift	Number produced units	Overall points	Insertion error	Out of parts	Insertion rate	Operating rate	A	B	C	D	E	F	G	H	I	J	K	L	Total min.	Times switched

Element involved:

A.
B.
C.
D.
E.
F.
G.
H.
I.
J.
K.
L.

Problem:

B1.
B2.
C1.
C2.
C3.
D1.
D2.
E1.
E2.
E3.
L1.
L2.

Worksheet 23a

273

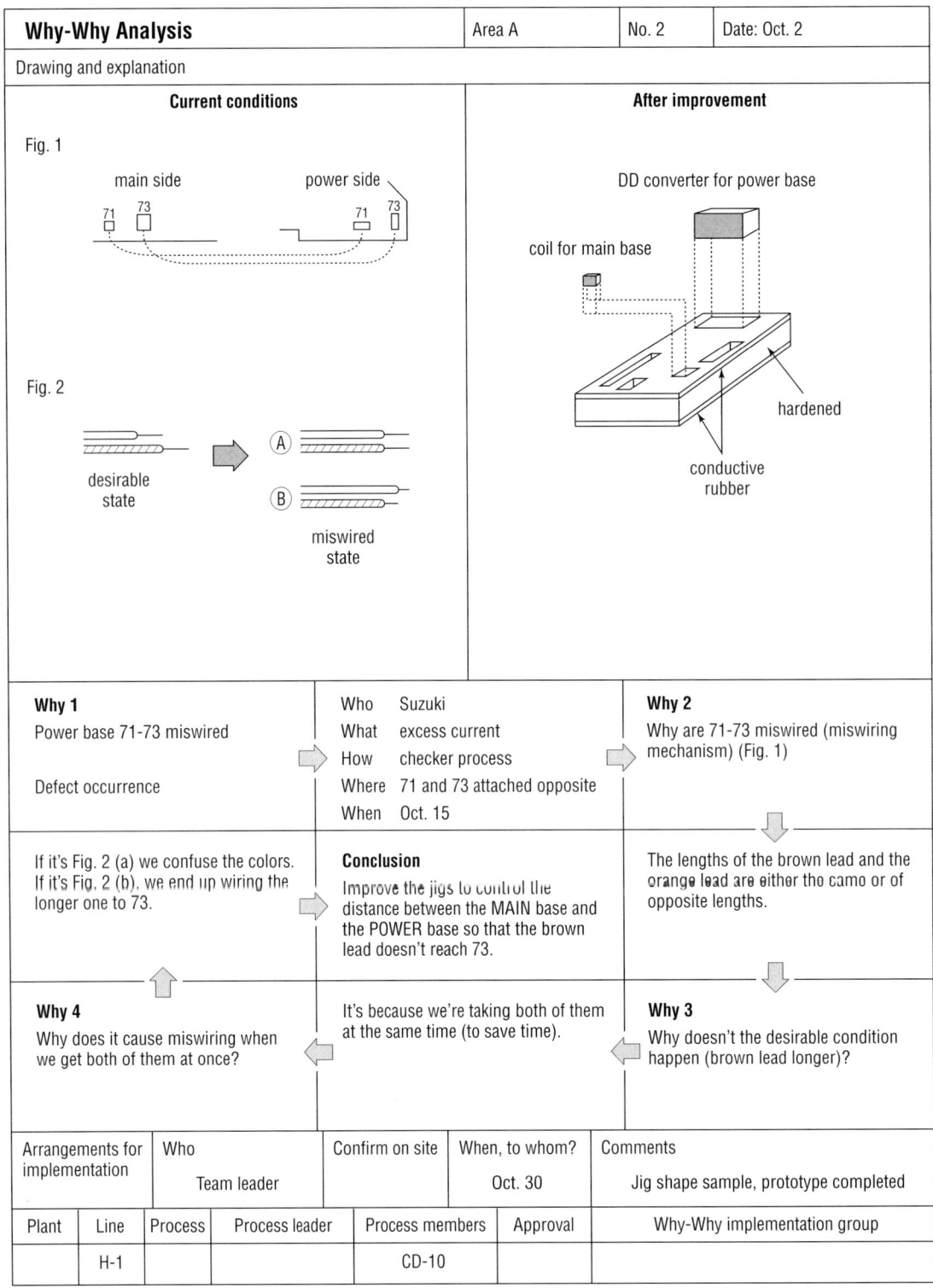

Why-Why Analysis		Area A	No. 2	Date: Oct. 2

Drawing and explanation

Current conditions

Fig. 1

main side   power side

71   73   71   73

Fig. 2

desirable
state

(A) miswired
(B) state

After improvement

DD converter for power base

coil for main base

hardened

conductive
rubber

Why 1

Power base 71-73 miswired

Defect occurrence

Who	Suzuki
What	excess current
How	checker process
Where	71 and 73 attached opposite
When	Oct. 15

Why 2

Why are 71-73 miswired (miswiring mechanism) (Fig. 1)

If it's Fig. 2 (a) we confuse the colors. If it's Fig. 2 (b), we end up wiring the longer one to 73.

Conclusion

Improve the jigs to control the distance between the MAIN base and the POWER base so that the brown lead doesn't reach 73.

The lengths of the brown lead and the orange lead are either the same or of opposite lengths.

Why 4

Why does it cause miswiring when we get both of them at once?

It's because we're taking both of them at the same time (to save time).

Why 3

Why doesn't the desirable condition happen (brown lead longer)?

Arrangements for implementation	Who		Confirm on site	When, to whom?	Comments	
	Team leader			Oct. 30	Jig shape sample, prototype completed	

Plant	Line	Process	Process leader	Process members	Approval	Why-Why implementation group
	H-1			CD-10		

Worksheet 24

Why-Why Analysis		Area	No.	Date: _____ / _____ / _____

Drawing and explanation

Current conditions	**After improvement**

Why 1	Who What How Where When	**Why 2**

	Conclusion	

Why 4		**Why 3**

Arrangements for implementation	Who		Confirm on site	When, to whom?	Comments

Plant	Line	Process	Process leader	Process members	Approval	Why-Why implementation group

Worksheet 24a

Equipment Problem-Solving Plan

Phenomenon	Plan / Cause	Minor improvements	Who	When	Effect	Medium and major improvements	Who	When	Effect
1. One of the screws doesn't tighten.	1. Different type of screw mixed in 2. Change in hose coupler	Confer with supplier and arrange to exchange current item.	HR			Change shape of coupler since nothing can be done with mechanism.	KA		
2. Screws fall off.	1. Blocked filter opening 2. Friction on O-ring of bit 3. Clogged air filter on secondary compresso	Regular cleaning Instant maintenance (3 minutes) Regular cleaning	TK		○	Get rid of metal shavings. I'd like to discontinue this.	KA		
3. Stops within the tube	1. Damage to the hose 2. Hose curvature	Periodically replace hose. Increase curvature of hose.	TK TK		○ ○				
4. Screws caught inside feeder don't flow smoothly.	Vibration, guide position	Adjust the amount of vibration.	TK						
5. Screws don't hold.	Debris on the pressure plate	Periodically clean.	TK						
6. Screws loose	Torque measurement	Adjust to correct value with torque meter.	TK		○				

Worksheet 25

Equipment Problem-Solving Plan

Phenomenon	Plan / Cause	Minor improvements	Who	When	Effect	Medium and major improvements	Who	When	Effect

Worksheet 25a

Small Tool Improvement Sheet

Autonomous Study Group (_____ team)

Date implemented: 8/26

Person in charge: PK–NPS

No.	Operation	Problem point (waste)	Improvement measures (including sketch)
1	Pre-board line	We had gotten the parts boxes in place for the next operation in the pre-setup phase, but replacing the empty boxes with them took time.	We installed a chute under the work bench. (We stopped piling up parts boxes in preparation for the next operation.) holds boxes for next operation / for empty box disposal

Worksheet 26

Small Tool Improvement Sheet

Autonomous Study Group (_____ team)

Date implemented: _____ / _____ / _____

Person in charge:

No.	Operation	Problem point (waste)	Improvement measures (including sketch)

Worksheet 26a

Daily Maintenance Sheet

Date generated: _____ **Name:** _____

Items checked by operators

Label no.	Part inspected	Standard			
△1	Filter inspection	No debris adhering			
△2	Pressure plate	No debris on rubber plate			
△3	Fallen screws	No screws around machine			

Items to be checked by chief engineer

Ⓖ1	Replace O-ring in bit	Degree of screw adsorption
Ⓖ2	Replace hose	No nicks or dirt inside hose
Ⓖ3	Torque adjustment	Check with torque meter

Devices for chief engineer to inspect and measure

Indicator lamp circuit

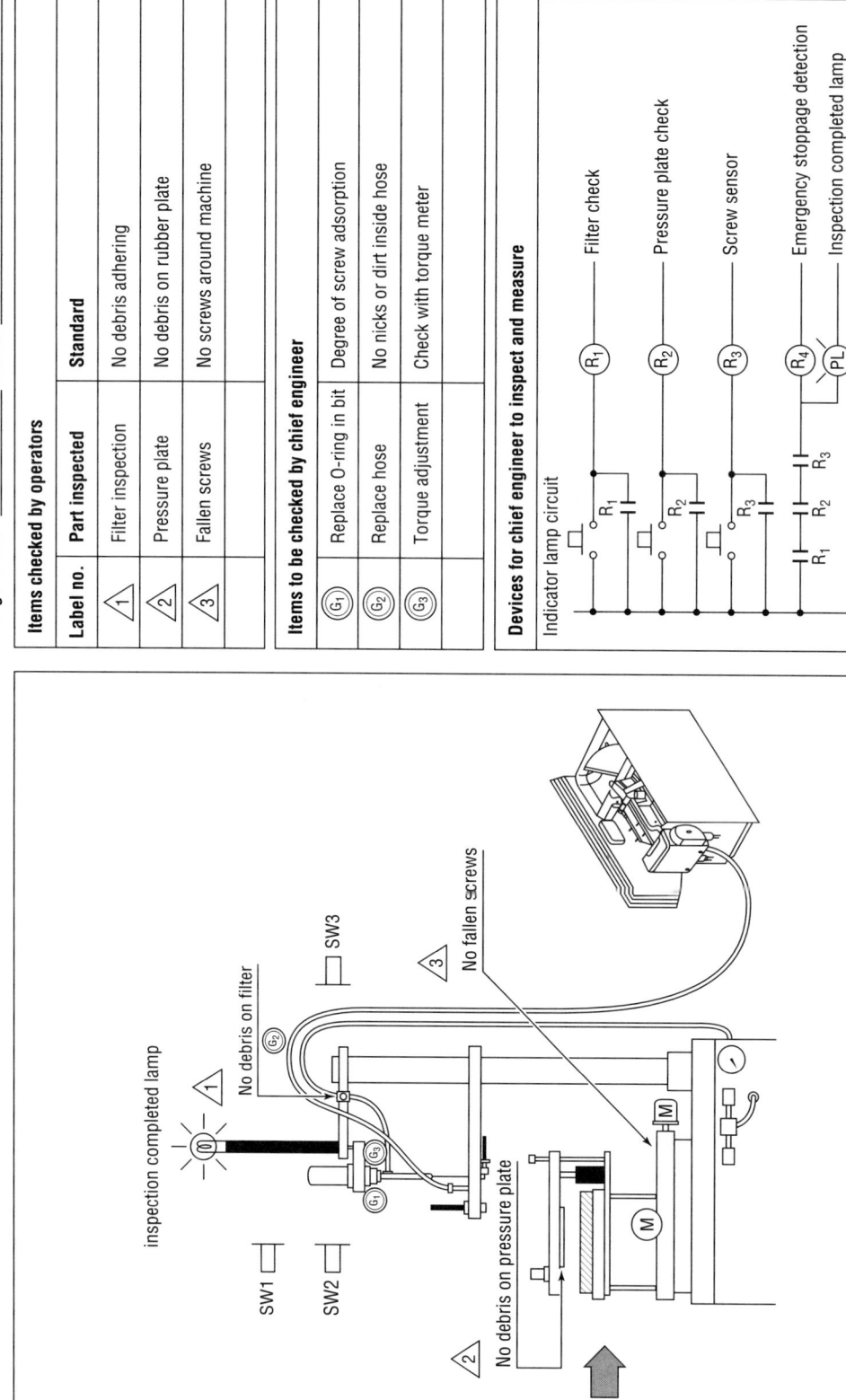

R_1 — Filter check

R_2 — Pressure plate check

R_3 — Screw sensor

R_4 — Emergency stoppage detection

PL — Inspection completed lamp

SW1

SW2

SW3

inspection completed lamp △1

No debris on filter

Ⓖ2

Ⓖ3

Ⓖ1

△2 No debris on pressure plate

△3 No fallen screws

M

M

Daily Maintenance Sheet

Date generated: _____ Name: _____

Items checked by operators

Label no.	Part inspected	Standard				

Items to be checked by chief engineer

Devices for chief engineer to inspect and measure

Instant Maintenance Labels

(front—red)	(back—blue)

Pre-shift Inspection

Line: A-1

Machine: screw tightener

Check items:

1. Is the original air pressure 6 kg/cm^2?

2. Are there screws in the feeder?

3. Is the screw sensor lamp on?

Inspection Completed

Turn back to the front side
at the end of the shift.

(front—red)

Repair Record (Maintenance Note)

Equipment:	Dryer
Linc:	A4
Situation:	Work cannot be confirmed; mechanism operates continuously due to sensor stuck in "on" position.
Countermeasures:	Replace sensor
Repair time:	15 minutes
Person in charge:	C

(back—blue)

Points Noticed

Worksheet 28

Instant Maintenance Labels

(front—red) **Pre-shift Inspection** Line: _____ Machine: _____ ┌─────────────────────────┐ │ Check items: │ │ 1. _____ │ │ 2. _____ │ │ 3. _____ │ │ └─────────────────────────┘	**(back—blue)** **Inspection Completed** Turn back to the front side at the end of the shift.

⟺

(front—red) **Repair Record (Maintenance Note)** Equipment: _____ Line: _____ Situation: _____ _____ _____ _____ Countermeasures: _____ Repair time: _____ Person in charge: _____	**(back—blue)** **Points Noticed**

⟺

Worksheet 28a

Pre-Shift Check Sheet

| Safety department | → | Safety office (maintenance) | → | Supervisor |

Pre-shift Check Sheet

		Section manager	Center manager	Team leader	Supervisor
Month					
Machine number					

Daily inspection items

Inspection item

Summary	Inspection date																															
	1	2	3	4	5	6	7	8	9	10	11	12	13	14	15	16	17	18	19	20	21	22	23	24	25	26	27	28	29	30	31
1. Filter inspection	✓	✓	✓	✓	✓	✓	✓																								
2. Pressure plate	✓	✓	✓	✓	✓	✓	✓																								
3. Fallen screws	✓	✓	✓	✓	✓	✓	✓																								
4. Original air pressure, 6 kg/cm^2	✓	✓	✓	✓	✓	✓	✓																								
5. Presence of screws in feeder	✓	✓	✓	✓	✓	✓	✓																								
6. Screw sensor lamp lit	✓	✓	✓	✓	✓	✓	✓																								

Defective areas	Special notes about defect locations
	Processing conditions
1	Screws blocked

Summary of notations

X = Needs repair
⊗ = Repair adjustment completed
✓ = No abnormality
(Report any abnormalities discovered to the team leader.)

How to use this check sheet

1. The supervisor will record the daily inspections.
2. Team leaders will approve the check sheets once a week.
3. The safety office will keep this record for one year.

Worksheet 29

Pre-Shift Check Sheet

Pre-shift Check Sheet

Safety department	→	Safety office (maintenance)	← →	Supervisor

Section manager	Center manager	Team leader	Supervisor

Month _____ Machine number _____

Daily inspection items

Inspection date

Inspection item	Summary	1	2	3	4	5	6	7	8	9	10	11	12	13	14	15	16	17	18	19	20	21	22	23	24	25	26	27	28	29	30	31

Defective areas	Special notes about defect locations

Summary of notations

X = Needs repair
⊗ = Repair adjustment completed
✓ = No abnormality
(Report any abnormalities discovered to the team leader.)

How to use this check sheet

1. The supervisor will record the daily inspections.
2. Team leaders will approve the check sheets once a week.
3. The safety office will keep this record for one year.

Worksheet 29a

285

Record of Equipment Problems

Date	Line	Equipment	Phenomenon	Cause	Countermeasure	Time stopped	Super-visor
10/3	A4	Burr removal	Brush coming off	Screw loose, insufficient pre-shift inspection	Change from M3 to M5 screws	10 min.	A
10/3	A4	Screw tightening	Indication of abnormal screw height	Wrong screw	Clean selection	2 min.	A
10/3	A4	Screw tightening	Sudden stoppage	Galling	Replace guide	8 min.	A
10/3	A4	Screw tightening	Screw collapse	Unclear	Instant maintenance	10 min.	B
10/3	A4	Drying	Starting up even with no work in place	Fiber sensor broken	1. For now, attach an ON/OFF switch. 2. Purchase fiber sensor.	15 min.	C
10/3	A4	Grinding	Roller doesn't stop in original position	Sequencer defect	1. Ask manufacturer to replace. 2. For now, change the programming.	Repairs 3 hrs.	D

Worksheet 30

Record of Equipment Problems

Date	Line	Equipment	Phenomenon	Cause	Countermeasure	Time stopped	Super-visor

Worksheet 30a

Equipment: Screw tightener

Instant Maintenance Sheet (Type 1)

Problem	Phenomenon	Factors	Disposition	Instant maintenance procedures
Doesn't start	• Doesn't start when the "start" button is pressed	• The safety stop signal is stuck in the "on" position • Not in original position	• Readjustment • Lead wire processing • Check each sensor	Doesn't start when the switch is turned on → Check the starting point sensor → [Off] Adjust sensor / [On] Is there a screw on the end of the bit? → [No] Screw supply / [Yes] Check screw sensor → [Off] Adjust sensor / [On] OK
Sudden stoppage	• Stops even though the work lamp is lit	• Screws loose in the rising tip sensor • The workpiece is defective and doesn't get pressed in • Low pressing pressure	• Tighten and attach after position adjustment • Eliminate defect • Air pressure adjustment	Stops suddenly after screws drop → Check original air pressure → [OK] Adjust pressing pressure / [OK] Check bit tip → [Damaged] Replace bit / [OK] Check for blockage in air nozzle → [Blocked] Clean nozzle / [Clear] OK
Defects	• Screw head crushed (difference in screw position)	• Table lock cylinder loose	• Tighten lock screw; modify programming	Defect (screw fails to hold) → Check bit tip → [Damaged] Replace bit / [OK] Check driver tightening force scale → [Strong] Adjust driver tightening force / [OK] Check driver tightening force scale → [Fast] Adjust / Driver speed → [OK] OK
		• Work feed cylinder lock screw is loose	• Tighten lock screw; modify programming	
		• Bit bent	• Replace bit	
		• Cap ring damaged	• Replace	
	• Screw broken	• Bit dropping speed too fast	• Adjust speed	

Worksheet 31

Instant Maintenance Sheet (Type 1)

Equipment:

Problem	Phenomenon	Factors	Disposition	Instant maintenance procedures

Worksheet 31a

289

Instant Maintenance Sheet (Type 2)

Causes of minor stoppage	**Instant Maintenance Sheet for Tape Lead Cutters**	**Phenomenon:** Axial part insertion error

Phenomenon:

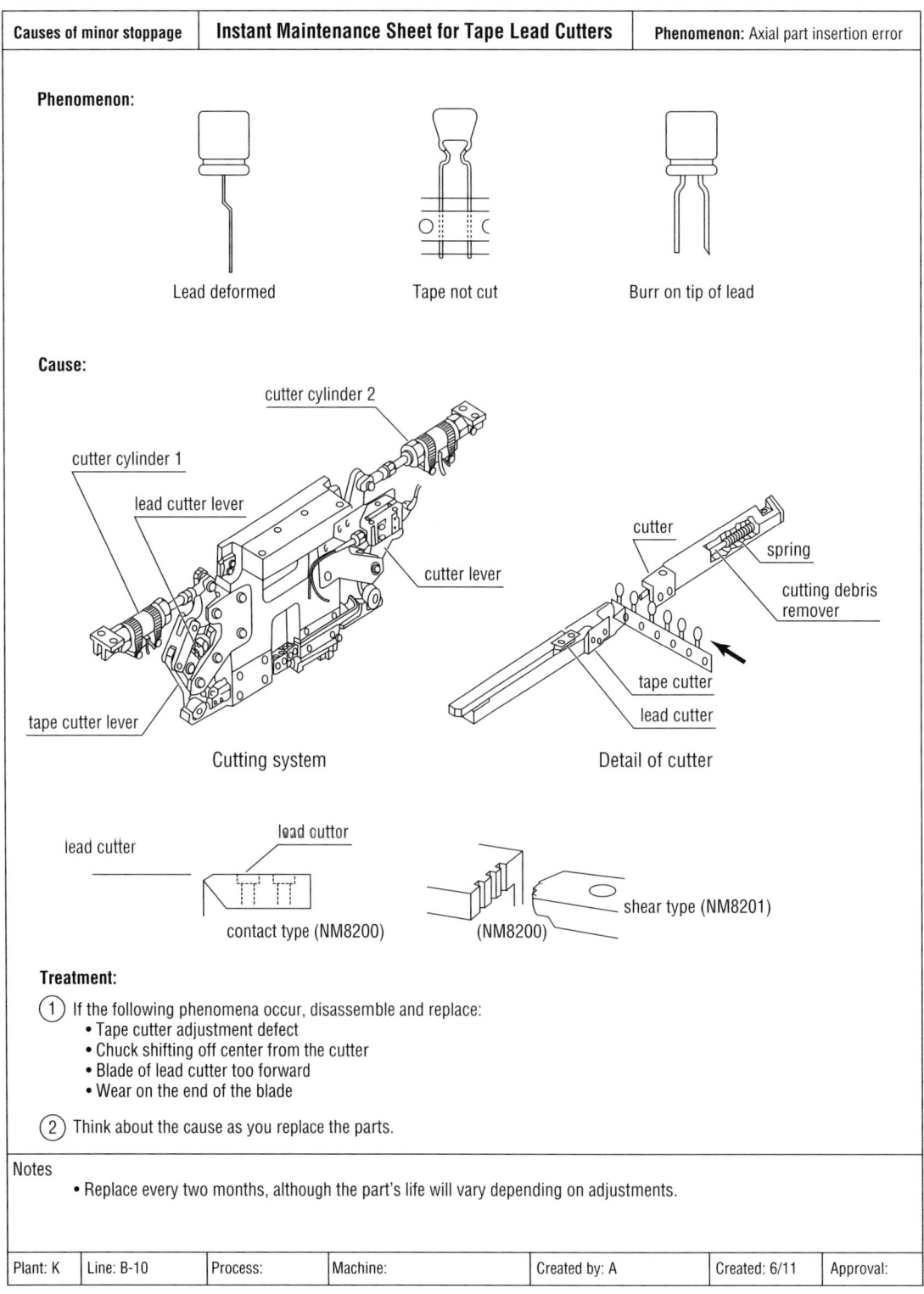

Lead deformed   Tape not cut   Burr on tip of lead

Cause:

cutter cylinder 2
cutter cylinder 1
lead cutter lever
cutter lever
tape cutter lever

Cutting system

cutter
spring
cutting debris remover
tape cutter
lead cutter

Detail of cutter

lead cutter   lead cutter

contact type (NM8200)

(NM8200)   shear type (NM8201)

Treatment:

① If the following phenomena occur, disassemble and replace:
 • Tape cutter adjustment defect
 • Chuck shifting off center from the cutter
 • Blade of lead cutter too forward
 • Wear on the end of the blade

② Think about the cause as you replace the parts.

Notes
 • Replace every two months, although the part's life will vary depending on adjustments.

Plant: K	Line: B-10	Process:	Machine:	Created by: A	Created: 6/11	Approval:

Worksheet 32

Instant Maintenance Sheet (Type 2)

Causes of minor stoppage	Instant Maintenance Sheet		Phenomenon:

Phenomenon:

Cause:

Treatment:

Notes

Plant:	Line:	Process:	Machine:	Created by:	Created:	Approval:

Worksheet 32a

Machining Center Stoppage Cause Sheet

Date: ___ / ___ / ___

Item	1 Freq.	1 Time	2 Freq.	2 Time	3 Freq.	3 Time	4 Freq.	4 Time	5 Freq.	5 Time	6 Freq.	6 Time	7 Freq.	7 Time	8 Freq.	8 Time	9 Freq.	9 Time	10 Freq.	10 Time	11 Freq.	11 Time
1. Insertion error	279		595		396		932		837		678		1164		617		333		173		234	
2. No parts	95		113		82		132		81		153		122		157		62		117		66	
3. Pitch feed	2	30	2	30							1		1	30					1	30		
4. Cutter	2	30	5	60							1		1	10			3	30	2	10		
5. Pallet											1		1	60					1	120		
6. Correction																						
7. Transfer (1)									2	20												
8. Lead cutter																						
9. Transfer (2)																						
10. Guide																						
11. Push rod					1	10																
12. Clinch									9	60	10	150	15	200	2	60	1	30			1	40
13. Air trouble	4	20					8	40														
14. Timing related																						
15. Electrical problem																						
16. System problem																						
17. Alarm went off																						
18. Base																						
19. Taping defect									2	5												
20. Base Defect																						
21. Operating error																						
22. No base supply																						
23. Magazine full																	8	20				
24. No magazine																	3	20				
Total	8	180	7	95	1	13	8	60	13	85	10	150	18	300	2	60	15	100	4	160	1	40

Worksheet 33

Machining Center Stoppage Cause Sheet

Date: ____ / ____ / ____

Item	1		2		3		4		5		6		7		8		9		10		11	
	Freq.	Time	Freq.	Time	Freq.	Time	Freq.	Time	Freq.	Time	Freq.	Time	Freq.	Time	Freq.	Time	Freq.	Time	Freq.	Time	Freq.	Time
Total																					

Worksheet 33a

293

December	Department	PM Notes		Supervisor	Approval	Approval

Date	PM Record	Why-Why Analysis Theme	Why-Why Analysis (Conclusion)	Items Referred to Other Departments
1	Individual differences in anvil detection methods	10 mm missing JW parts	Notation on the record sheet	
2				
3	Due to alternating operations, differences in visual power	Insufficient pins (YK 146 H 1704, 7 incidents)	Insufficient pins, checker manufacture	Ask Manufacturer A
4				
5				
6				
7				
8	Program tape overshoots the block	Program tape defect	Replacement of NC generator puncher	Repair by Company M
9				
10	Taping NG, excessive accumulation	Insertion error in EA 106016	Warning about parts handling	Referred to the materials manager
26	RT machine position error, 2-3 pulse	Insertion error (radial)	Bearing failure in DC servomotor	Get motor repaired
27				
28				
29				
30				
31				

Comments: • Taping failure is especially noticeable.

• Titanium coating process quality test on the 5 mm axial insertion guide.

Worksheet 34

December	Department	PM Notes		Supervisor	Approval	Approval

Date	PM Record	Why-Why Analysis Theme	Why-Why Analysis (Conclusion)	Items Referred to Other Departments
1				
2				
3				
4				
5				
6				
7				
8				
9				
10				
11				
12				
13				
14				
15				
16				
17				
18				
19				
20				
21				
22				
23				
24				
25				
26				
27				
28				
29				
30				
31				
Comments:				

Worksheet 34a

P–M Analysis Chart

Date created: ___ / ___ / ___

Problem:	Line:	Process:	Equipment:	Waste:

Phenomenon
- When key groove cutter is replaced, the key grooves are larger than the standard
- This happened on 4 or 5 of 25 cutter replacements
- The widths of the cutters in which the defects occurred tend to be within the standard width, so they fall within the standard width for key grooves.

Physical analysis

The cutting point a' • b' deviates in the X direction more than the cutter's cutting width a • b

Department: Cutter   **Section:** 6

Section mgr.	Tech.	Section clerk	Plant mgr.	Proj. mgr.
IY	MR	NO	SJ	SJ

Constituent conditions — Connections among equipment, tools and fixtures, materials, and methods — Plan for countermeasures

Unit-level conditions	Primary correlations (subassembly items)	Measurement method	Measured value / Standard value	Finding	Secondary correlations (structural component level)	Measured value / Standard value	Finding	Nature of countermeasures	Who	Date implemented	Result
During machining, the cutting tool moves in the X direction.	1-1 The cutter vibrates at the blade attachment point	① ②	Unit : μm ① 30/10 ② 9/10	① × ② ○	1-1-1 Cutter shaft vibration	Unit : μm 2/3 (n=5)	○	Measurement			
					1-1-2 Cutter blade end vibration	2/3 (n = 5)	○				
					1-1-3 Cutter attachment vibration	9/10	○	1-1-6			
					1-1-4 Wear on inner diameter of holder	9/10–11	○				
					1-1-5 Holder tip vibration	8/5	×	Holder replacement	T7	8/1	OK
					1-1-6 Main shaft holder attachment vibration	8/5	×	Bearing replacement	S4	8/13	OK

Worksheet 35

296

P–M Analysis Chart

Date created: ___ / ___ / ___

Problem:		Line:	Process:	Equipment:		Waste:	Section:

Phenomenon

Physical analysis

	Department:					
	Section mgr.	Tech.	Section clerk	Plant mgr.	Proj. mgr.	

Constituent conditions	**Connections among equipment, tools and fixtures, materials, and methods**						
Unit-level conditions	Primary correlations (subassembly items)	Measurement method	Measured value / Standard value	Finding	Secondary correlations (structural component level)	Measured value / Standard value	Finding

Plan for countermeasures

Nature of countermeasures	Who	Date implemented	Result

Cause and Effect Diagram

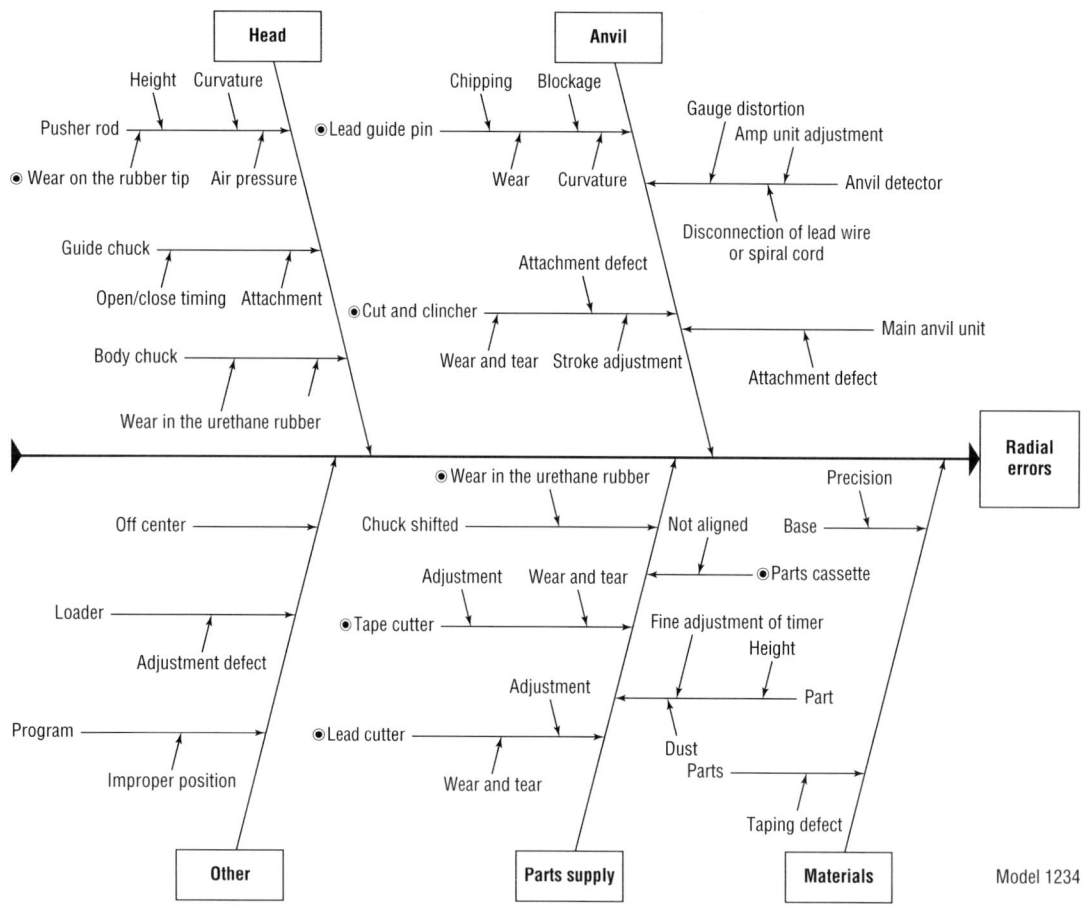

Model 1234

Worksheet 36

Cause and Effect Diagram

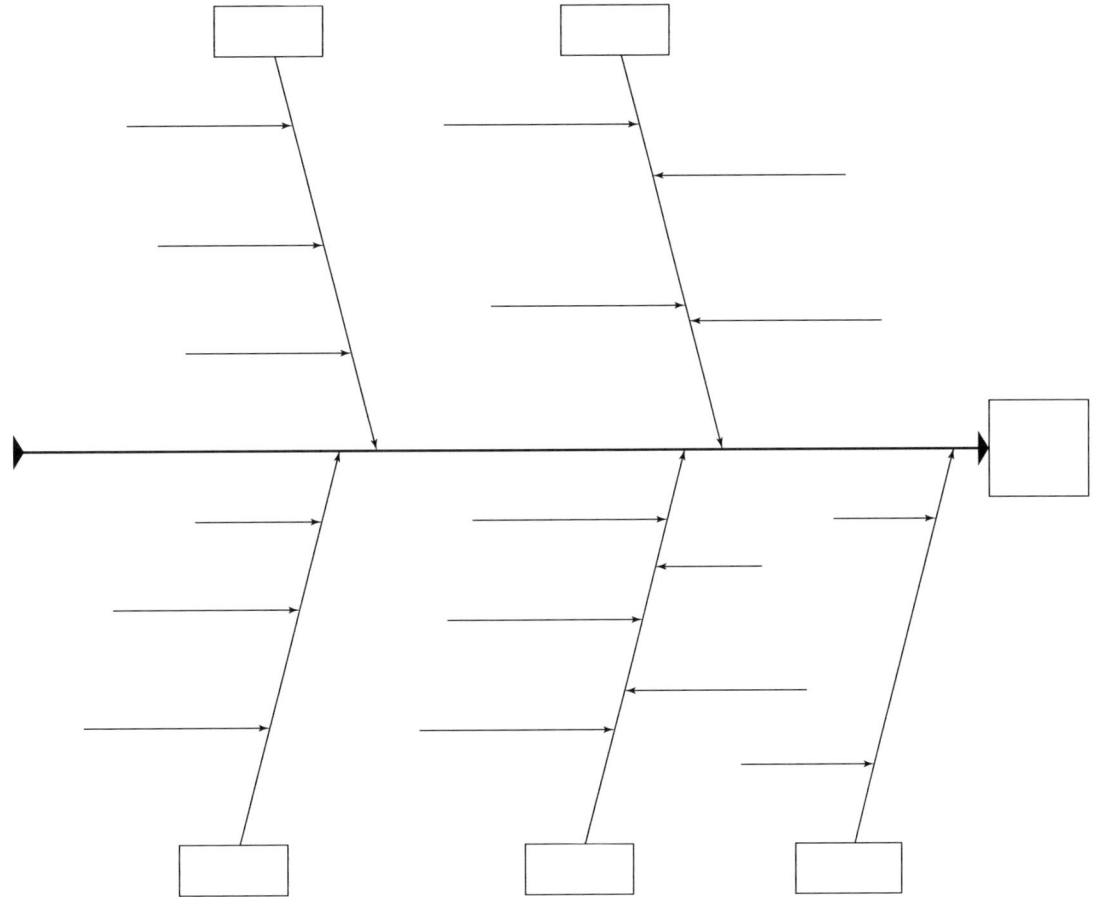

Equipment Maintenance Card

Note: Items marked ※ should be filled in at the issuing office

Log no.

	0→10	11→20	21→30	31→40	41→50	51→60
	0.2H	0.3H	0.5H	0.7H	0.8H	1H

※ Section mgr.	Center head	Team leader	Person in charge	※ Machine no.
IT	FR	MA	SZ	B211

Card arrangement no.
3 | 4 | 8
S 3 2

Date or request and defect occurrence
18 | 19 | 23 24
9 1 8 1 0 1 0 2 0

※ Asset no.

Type of request
※ Sporadic, preventive, other

Time operations wanted
※ Month | Day | Time

Case no.

Name of equipment (manufacturer)
Chilling Equipment

Original price code (incl. group)

	Staff	Center head	Team leader
Main responsibility			
TD			SZ

Operation indicated

Replacing motor

	Estimated employee hours	Actual employee hours
Operator		
SZ		
NS		
MY		

Parts used
Name of part	Model	Qty.	Unit price	Part commonly used
Cord	Motor on hand			

Breakdown phenomena
Due to overheating the equipment is overloaded and burns out

Breakdown cause
Overload

Repair method (nature of operation)
Since we had a replacement on hand, all the operators performed instant maintenance

Future countermeasures
If it doesn't go any farther than this, we'll get by with just keeping a replacement on hand. We sent the seized-up motor to the manufacturer for repairs.

※ Confirmation by issuing source
	Completed	Provisional	Incomplete
	41 42	43 44	45 46	47 48 49
	0 0	0 1	3 5 6

Breakdown cause | Breakdown component | Part used | Broken down parts

Maintenance operation times (time stamp)
Item	Date/Time
Received in maintenance dept.	Month/Day/Year/Time
Start	Month/Day/Year/Time 8 10 98 10:22
Parts arrangement	Month/Day/Year/Time
Parts received	Month/Day/Year/Time
Start re-repairing	Month/Day/Year/Time
Provisional repairs completed	Month/Day/Year/Time
Repairs completed	Month/Day/Year/Time 8 10 98 10:25
Total repair hours	3

Item (employee hours)
Item	(H)
Machine maintenance	68 70
Electrical maintenance	71 73
Departments used	74 76
From outside	77 0 6 80

Item (production time disrupted)
Item	
Reception waiting	50 52
Waiting for start	53 55
Waiting for parts	56 59
Hours of repair	60 63
Non-disrupted hours	64 67

30

※31 PM rank

※32 Working conditions
1. Production possible
2. Production impossible
3. Production decrease
4. Setup
5. Planned stoppage
9. Other

※34 Working format
1. Day
2. Day/night
3. Two-day shifts
4.
34. Repetitiveness
1. First breakdown
2. Recurrence (in 1 mo.)
3. Recurrence (in 3 mos.)
4. Recurrence (in 6 mos.)
5. Recurrence (in more than 6 mos.)

35 Class of operations
1. Machine maintenance
2. Electrical maintenance
3. Departments used
4.
5. From outside
6. Joint operation
36. Maintenance class
1. Sporadic
2. Preventive

37 Nature of operations
1. Inspection
2. Emergency repairs
3. Restorative repairs
4. Improvement
5. Rebuilding
6. Preparation
7. Making parts
9. Other

38 Division of responsibility
1. Manufacturer
2. Production tech
3. Department using
4. Machine maintenance
5. Electrical maintenance
6.
7. From outside
8. Unclear
9. Other

39·40 Breakdown phenomenon
1. Wear
2. Corrosion
3. Sagging—falling off
4. Deformation
5. Cracks breakage
6. Leakage
7. Running out of oil
8.
9. Other
10. Noise
11. Vibration
12. Disconnection
13. Wire connection defect
14. Contact defect
15. Blockage
16. Short circuit
17. Overheating
18. Burn-out

41·42 Breakdown causes
1. Design defect
2. Parts defect
3. Turning defect
4. Operation defect
5. Repair defect
6. Inspection defect
7. Lubrication defect
8. Cleaning defect
9. Adjustment defect
10. Environmental defect
11. Specifications defect
12. Useful life
13. Work defect
14. Unclear

41·42 Breakdown component
1. Drive revolution
2. Slide
3. Blade base
4. Braking
5. Oil pressure, air pressure
6. Lubricant, cutting water
7. Positioning
8. Transport
9. Heat control
10. General electric
11. N/C equipment
12. Chute

45 Parts used
1. None needed
2. Purchased (have)
3. Purchased (don't have)
4.
5. Made (have)
6. Made (don't have)
7. Add-on (in-house)
8. Add-on (outside)
9. Other

46-49 Parts broken
11. Bearing	20. Hose
12. Cylinder	21. Gear rack
13. Pipes	22. Bolts, nuts
14. Pump	23. Knock pin
15. Chain belt	24. Pulley
16. Slide surface	25. Brake
17. Valve	26. Brick
18. O-ring seal	27. Coupling
19. Shaft	28. Roller wheel
51. Limit switch	60. Open-shut
52. Photo-electric tube	61. Push button
53. Proximity switch	62. Solenoid
54. Relay	63. Connector
55. Timer	64. Capacitor
56. Motor	65. Print base
57. Wiring	66. Lamp
58. Thermal	67. Heater
59. Fuse	68. Brush
69. Charger	
70. Breaker	
71. Pressure switch	
72. Connector	
73. IC transistor	
74. Clutch	
75. Measuring instruments	
77. Other	

Worksheet 37

300

Note: Items marked ※ should be filled in at the issuing office

Equipment Maintenance Card

Log no. ____

	0→10	11→20	21→30	31→40	41→50	51→60	1H
	0.2H	0.3H	0.5H	0.7H	0.8H		

1 / 3 Card arrangement no. [4] [8]

Name of equipment (manufacturer)

Date of request and defect occurrence — Original price code (incl. group) [18][19] [23][24] — Asset no.

9

Type of request
※ Sporadic, preventive, other

Time operations wanted
※ Month — Day — Time

Case no.

	Section mgr.	Center head	Team leader	Person in charge
	※	※	※	※

Main responsibility — Staff — Center head — Team leader

Operation indicated

Breakdown phenomena

Breakdown cause

	Estimated employee hours	Actual employee hours
Operator		
Date		

Parts used

Name of part	Model	Qty.	Unit price

Confirmation by issuing source

	Completed	Provisional	Incomplete
※			

Machine no. ※

Repair method (nature of operation)

Future countermeasures

Maintenance operation times (time stamp)

Item	Date/Time
Received in maintenance dept.	Month/Day/Year/Time
Start	Month/Day/Year/Time
Parts arrangement	Month/Day/Year/Time
Parts received	Month/Day/Year/Time
Start re-repairing	Month/Day/Year/Time
Provisional repairs completed	Month/Day/Year/Time
Repairs completed	Month/Day/Year/Time
Total repair hours	Month/Day/Year/Time

Central coding grid (columns 30–80):

PM rank (31) · Working conditions disrupted (32) · Working form (33) · Tendency to repeat (34) · Operation classification (35) · Maintenance classification (36) · Nature of operation (37) · By responsibility (38) · Breakdown phenomenon (39·40) · Breakdown cause (41·42) · Breakdown component (43·44) · Part used (45) · Broken down parts (46–49)

Item		(H)
Reception waiting	50	52
Waiting for start	53	55
Waiting for parts	56	59
Hours of repair	60	63
Non-disrupted hours	64	67
Production time disrupted		
Employee hours	68 / 71 / 74 / 77	70 / 73 / 76 / 80

※32 Working conditions
1. Production possible
2. Production impossible
3. Production decrease
4. Setup
5. Planned stoppage
9. Other

※34 Working format
1. Day
2. Day/night
3. Two-day shifts

34 Repetitiveness
1. First breakdown
2. Recurrence (in 1 mo.)
3. Recurrence (in 3 mos.)
4. Recurrence (in 6 mos.)
5. Recurrence (in more than 6 mos.)

35 Class of operations
1. Machine maintenance
2. Electrical maintenance
3. Departments used
4.
5. From outside
6. Joint operation

36 Maintenance class
1. Sporadic
2. Preventive

37 Nature of operations
1. Inspection
2. Emergency repairs
3. Restorative repairs
4. Improvement
5. Rebuilding
6. Preparation
7. Making parts
9. Other

38 Division of responsibility
1. Manufacturer
2. Production tech
3. Department using
4. Machine maintenance
5. Electrical maintenance
6.
7. From outside
8. Unclear
9. Other

39·40 Breakdown phenomenon
1. Wear
2. Corrosion
3. Sagging—falling off
4. Deformation
5. Cracks breakage
6. Leakage
7. Running out of oil
8. Overheating
9. Burn-out
10. Noise
11. Vibration
12. Dis-connection
13. Wire connection defect
14. Contact defect
15. Blockage
16. Short circuit
18. Other

41·42 Breakdown causes
1. Design defect
2. Parts defect
3. Turning defect
4. Operation defect
5. Repair defect
6. Inspection defect
7. Lubrication defect
8. Cleaning defect
9. Adjustment defect
10. Environmental defect
11. Specifications defect
12. Useful life
13. Work defect
14. Unclear

41·42 Breakdown component
1. Drive revolution
2. Slide
3. Blade base
4. Braking
5. Oil pressure, air pressure
6. Lubricant, cutting water
7. Positioning
8. Transport
9. Heat control
10. General electric
11. N/C equipment
12. Chute

45 Parts used
1. None needed
2. Purchased (have)
3. Purchased (don't have)
4. Made (have)
5. Made (don't have)
6. Add-on (in-house)
7. Add-on (outside)
9. Other

46–49 Parts used
11. Bearing
12. Cylinder
13. Pipes
14. Pump
15. Chain belt
16. Slide surface
17. Valve
18. O-ring seal
19. Shaft

Parts broken
20. Hose
21. Gear rack
22. Bolts, nuts
23. Knock pin
24. Pulley
25. Brake
26. Brick
27. Coupling
28. Roller wheel
51. Limit switch
52. Photo-electric tube
53. Proximity switch
54. Relay
55. Timer
56. Motor
57. Wiring
58. Thermal
59. Fuse
60. Open-shut
61. Push button
62. Solenoid
63. Resistance
64. Capacitor
65. Print base
66. Lamp
67. Heater
68. Brush
69. Charger
70. Breaker
71. Pressure switch
72. Connector
73. IC transistor
74. Clutch
75. Measuring instruments
77. Other

Worksheet 37a

Sporadic/Planned Equipment Repair Report

Team: 219	Line: Rotor			Machine number: GR 5170	

Circumstances of malfunction:	Date: 8/19	
Due to a defect in the amount of corrective incision in the grindstone, remnants in the slit width	Requesting department	
	Section clerk	Team leader
		KT

Cause	Handling (replacement adjustment parts, etc.)	Requests to team from Maintenance Dept. (arrangements, parts, etc.)
Rattling/play in the worm and foil	Tightening cylinder screw, Radial matt jaw repair and adjustment	

Implementing departments				Operation of machine begun: month/day/time	Managers		
Section mgr.	Section clerk	Team leader	Super-visors	Net stoppage time: _____ hours	Assistant mgr. Section mgr.	Section clerk	Clerk
		TS	TK MR	Employee hours for repairs: 1 person × 3 hrs 1 person × 7 hrs = 10 hrs	SN	SS	TH

Requesting dept. → Implementing dept. → Parts mfg. section (Section mgr., section clerk) → Parts technology, Section 2

Worksheet 38

Sporadic/Planned Equipment Repair Report

Team:	Line:		Machine number:

Circumstances of malfunction:

	Date: / /
	Requesting department

	Section clerk	Team leader

Cause	Handling (replacement adjustment parts, etc.)	Requests to team from Maintenance Dept. (arrangements, parts, etc.)

Implementing departments				Operation of machine begun: ____ / ____ / ____	Managers		
Section mgr.	Section clerk	Team leader	Super-visors	Net stoppage time: ____ hours	Assistant mgr. Section mgr.	Section clerk	Clerk
				Employee hours for repairs:			

Requesting dept. → Implementing dept. → Parts mfg. section (Section mgr., section clerk) → Parts technology, Section 2

Worksheet 38a

Setup Operation Analysis

Plant: _____   Date: _____ / _____ / _____

Process: I	Person responsible:	Product: 3956
Machine: 150-ton press setup and changeover	Skilled operator: B	Setup time:
Machine no:	Frequency (setups and changovers/month):	Observer:

No.	Setup/changeover step	Time		Setup			Improvement noted
		Reading	Elapsed	Internal	External	Waste	
1	Undo upper mold bolt	44.00 45.00	60 sec.	✓			
2	Undo lower mold bolt	46.30	90 sec.	✓	✓		
3	Look for lift	47.35	65 sec.			✓	Special changeover dolly
4	Transport on lift	49.20	105 sec.			✓	
5	Take out metal mold	50.00	40 sec.		✓		
6	Mold	50.25	25 sec.	✓			
7	To mold position	51.05	40 sec.			✓	
8	Set mold position	52.32	87 sec.		✓		
9	New mold	53.25	53 sec.		✓		
10	Transport on lift	55.00	95 sec.			✓	
11	To bolster	56.00	60 sec.	✓			
12	Upper metal	57.00	60 sec.	✓			
13	Tighten on to upper metal	58.22	82 sec.	✓			
14	Clean mold	58.48	26 sec.			✓	Cover with plastic bag
15	Mount lift	59.27	39 sec.			✓	
16	Correct position on lift tip	59.59	32 sec.		✓		
17	Return the lift	1:01.00	61 sec.			✓	
18	Attachment place	1.28	28 sec.			✓	
19	Stroke bottom dead center	1.44	16 sec.	✓			
20	Zero point adjustment	2.37	53 sec.	✓			
21	Position adjustment	3.05	28 sec.			✓	
22	Attach upper mold bolt	3.35	30 sec.	✓			
23	Attach lower mold bolt	3.47	12 sec.	✓			
24	Firmly tighten	5.10	83 sec.	✓			

Suggestions:
1. You may analyze in as much detail as you wish, but put observations in units that will let you see the waste.
2. It's best to observe in pairs. One person watches the time, while the other looks for waste.
3. At the beginning it's best to videotape as you observe. If you happen to overlook something, the tape can supplement your observations.

Worksheet 39

Setup Operation Analysis

Plant: _____   Date: _____ / _____ / _____

Process:	Person responsible:	Product:
Machine:	Skilled operator:	Setup time:
Machine no:	Frequency (setups and changovers/month):	Observer:

No.	Setup/changeover step	Time		Setup			Improvement noted
		Reading	Elapsed	Internal	External	Waste	

Suggestions:
1. You may analyze in as much detail as you wish, but put observations in units that will let you see the waste.
2. It's best to observe in pairs. One person watches the time, while the other looks for waste.
3. At the beginning it's best to videotape as you observe. If you happen to overlook something, the tape can supplement your observations.

Worksheet 39a

Improvement Deployment Chart

Topic: Automatic Equipment Changeover

Dept.: 3rd Mfg. Date: 4/14

Line graphs (min. axis: 20, 15, 10, 5; months: March, April, May, June, July, Aug): Small improvements (6.5 → 10); Medium improvements (6.5 → 6); Major improvements (5 → 5); overall trend graph (17 → 9 → 9).

Implementation items

Trend graphs (min. axis; month axis: 4, 6, 8):
- *A. Waste in preparation: 3.5 → 2 → 2*
- *B. Waste in replacement: 7.7 → 5 → 5*
- *C. Waste in adjustment: 5.8 → 2 → 2*

A. Waste in preparation
1. Transporting parts
2. Not enough parts cassettes
3. Human actions

B. Waste in replacement
1. Loss during program input
2. Parts replacement
3. Parts cassette receiving

C. Waste in adjustment
1. Adjustment of standard pin
2. Adjustment of backup pin

Small improvements

No.	Item	Period	In charge	Results
1	Make sure the same amount is on the top and bottom of the transport shelf	4/30	Op.	0.6M
2	Increase 8mm parts cassettes (15 pieces)	5/15	MY	On order
3	Create an action chart	4/14	Op.	2.1M
4	Standard operations combination chart	4/14	SZ	2.1M
5	Cord disconnection	4/15	ST	1.2M
1	Change input order	4/4	Op.	1.4M
2	Improve setup carts (vertical type)	5/30	MY	Under study
1	Mark according to machine model	4/15	Op.	1.2M
2	Unify offset values	4/30	MY	Being implemented

Medium improvements

No.	Item	Period	In charge	Results
1	Add setup carts	5/25	NK	
2	Wind-up reel arrangements	4/30	MY	On order
3	Mechanize top tape remover	5/20	KS	0.6M
4	Change keyboard height	4/30	Op.	Re-quested
5	Add stairs	5/10	ST	
1	Use hypothetical origin	4/20	MY	5.9M
2	Common use of CD-10MK2•CD-15 — 1. Fix parts cassette	5/30	MY	
	2. Make screen replacement a one-touch operation	5/20	MY	
3	Special NC tape cart	5/30	SZ	on order
4	Make input cable switching a one-touch operation	4/20	SZ	Under study
1	Make the standard pin easy to install and remove	6/20	MY	Under study
2	Make screen setting a one-touch operation	5/30	SZ	Under study

Major improvements

No.	Item	Period	In charge	Results
1	Cassettes should be in blocks	6/30	ST	
2	Parts numbers should be bar-coded	6/20	DG	
1	Backup pins should be in a cassette	7/20	MY	

Worksheet 40

Improvement Deployment Chart

Topic: _____   Dept.: _____   Date: ____ / ____ / ____

No.	Item	Small improvements				No.	Item	Medium improvements				No.	Item	Major improvements			
		Period	In charge	Results				Period	In charge	Results				Period	In charge	Results

Implementation items

Standard Work Combination Sheet

Process: Setup for polishing

	Before improvement
	After improvement

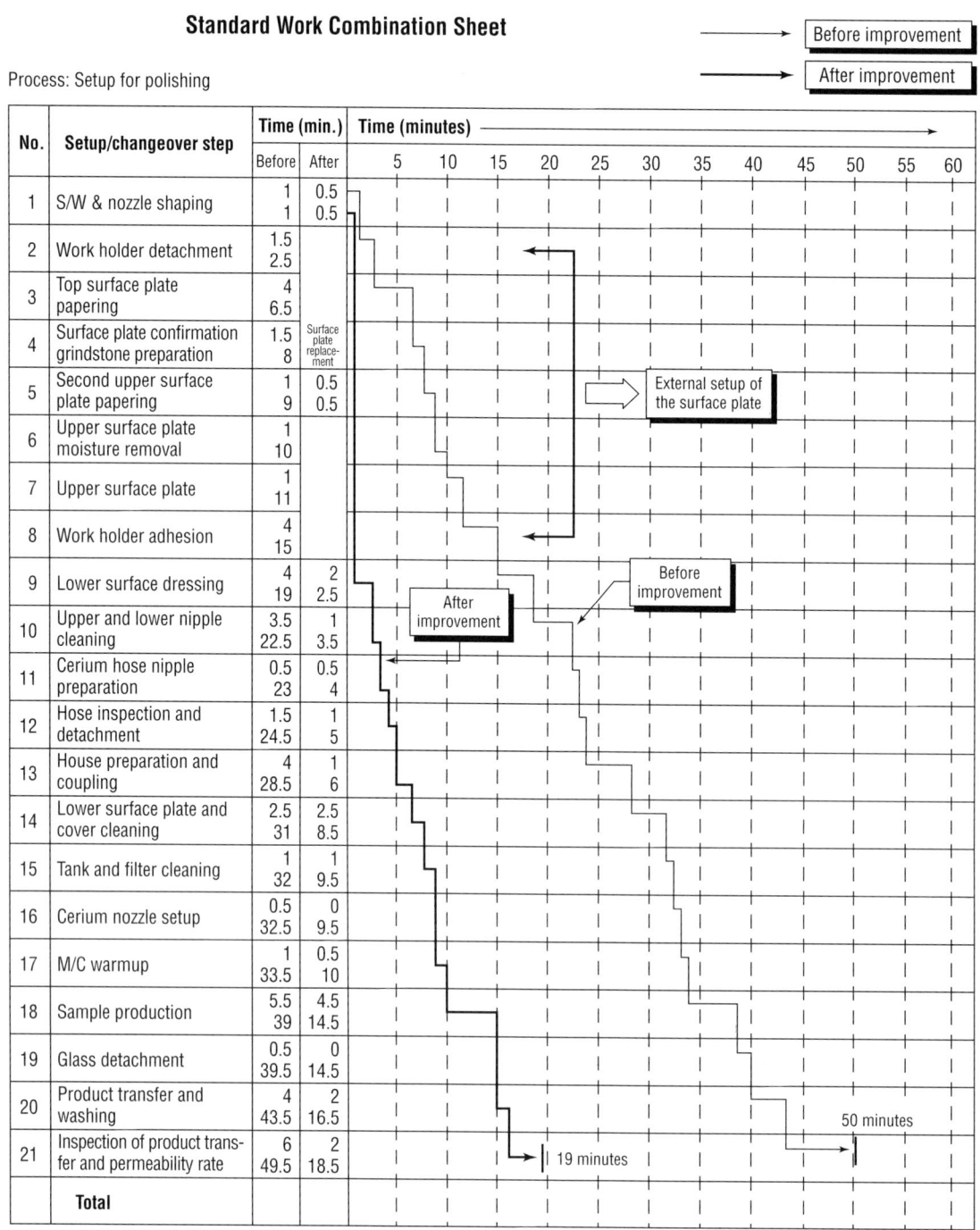

No.	Setup/changeover step	Time (min.) Before	Time (min.) After	Time (minutes)
1	S/W & nozzle shaping	1 / 1	0.5 / 0.5	
2	Work holder detachment	1.5 / 2.5		
3	Top surface plate papering	4 / 6.5		
4	Surface plate confirmation grindstone preparation	1.5 / 8	Surface plate replacement	
5	Second upper surface plate papering	1 / 9	0.5 / 0.5	External setup of the surface plate
6	Upper surface plate moisture removal	1 / 10		
7	Upper surface plate	1 / 11		
8	Work holder adhesion	4 / 15		
9	Lower surface dressing	4 / 19	2 / 2.5	Before improvement
10	Upper and lower nipple cleaning	3.5 / 22.5	1 / 3.5	After improvement
11	Cerium hose nipple preparation	0.5 / 23	0.5 / 4	
12	Hose inspection and detachment	1.5 / 24.5	1 / 5	
13	House preparation and coupling	4 / 28.5	1 / 6	
14	Lower surface plate and cover cleaning	2.5 / 31	2.5 / 8.5	
15	Tank and filter cleaning	1 / 32	1 / 9.5	
16	Cerium nozzle setup	0.5 / 32.5	0 / 9.5	
17	M/C warmup	1 / 33.5	0.5 / 10	
18	Sample production	5.5 / 39	4.5 / 14.5	
19	Glass detachment	0.5 / 39.5	0 / 14.5	
20	Product transfer and washing	4 / 43.5	2 / 16.5	50 minutes
21	Inspection of product transfer and permeability rate	6 / 49.5	2 / 18.5	19 minutes
	Total			

Worksheet 41

308

Standard Work Combination Sheet

Process: _____

	Before improvement
	After improvement

No.	Setup/changeover step	Time (min.) Before	Time (min.) After	Time (minutes) 5	10	15	20	25	30	35	40	45	50	55	60
1															
2															
3															
4															
5															
6															
7															
8															
9															
10															
11															
12															
13															
14															
15															
16															
17															
18															
19															
20															
21															
	Total														

Worksheet 41a

Revised: ____ / ____ / ____   Created ____ / ____   pages of ____

Process Capacity Table

Section manager	Project manager

Product number	17111–24060		Model	KE	Affiliation	Name
Name	Intake manifold		Number of items	1	532	SZ
					542	AT

① Order	② Process I	③ Machine number	④ Basic time Manual operation min. | sec.	Automatic feed min. | sec.	Completion min. | sec.	⑤ Cutting tool Items replaced	Total replacement time	⑥ Production capacity (16 hrs.)	⑦ Time graph
1	Booster attachment surface cutting	Mi1764	3	25	28	100	1'00"	2,013	3" 25"
2	Booster hole polishing	DR2424	3	21	24	1,000	30"	2,397	3" 21"
3	Booster hole screw insertion	TP1101	3	11	14	1,000	30"	4,105	3" 11"
4	Quality check (measure screw diameter)		5	—	5	—	—	11,520	
	Total		14						

⑦ Time graph
——— = manual
– – – = automatic

Worksheet 42

Process Capacity Table

Revised: ___ / ___ / ___   Created ___ pages of ___

Section manager	Project manager

	Product number		Model	Affiliation	Name
	Name		Number of items		

① Order ② Process I ③ Machine number ④ Basic time ⑤ Cutting tool ⑥ Production capacity (___ hrs.) ⑦ Time graph

④ Basic time

Manual operation		Automatic feed		Completion	
min.	sec.	min.	sec.	min.	sec.

⑤ Cutting tool

Items replaced	Total replacement time

⑦ Time graph
——— = manual
– – – = automatic

Total

Worksheet 42a

311

Part number	17111-24060		Standard Work Combination Sheet		Affiliation department	1/1		Daily quota	1920	Units/day	Manual operations
Part name	Intake manifold				Machining department	ET		(5) Takt time	30″		Automatic feed
											Walking

(4) **Operation time** (in seconds)

(1) Order of operation	(2) Name of operation	(3) Time		
		Manual operation	Automatic feed	
1	Get raw materials	2	—	
2	Mi 1764 Detach and attach work. Turn on feed.	3	25	
3	DR2424 Detach and attach work. Turn on feed.	3	21	
4	TP1101 Detach and attach work. Turn on feed.	3	11	
5	Measure screw diameter	5	—	
6	Place completed item	2	—	

Operation time (in seconds) — axis markings: 5″ 10″ 15″ 20″ 25″ 30″ 35″ 40″ 45″ 50″ 55″ 60″

Worksheet 43

312

Standard Work Combination Sheet

Part number		Affiliation department		Daily quota		Units/day	
Part name		Machining department		⑤ Takt time			

① Order of operation	② Name of operation	③ Time		④ Operation time (in seconds)
		Manual operation	Automatic feed	5″ 10″ 15″ 20″ 25″ 30″ 35″ 40″ 45″ 50″ 55″ 60″
Total				

Legend:
— Manual operations
······· Automatic feed
〰〰 Walking

Operation time (in seconds): 5″ 10″ 15″ 20″ 25″ 30″ 35″ 40″ 45″ 50″ 55″ 60″

Worksheet 43a

313

Defect Reduction Activity Chart

Month: April Promoter: KR Date: 4/30

Assembly process performance Goal: 0.5 %

Inspector: MD, Auto parts dept. 2, parts manuf.

Characteristic	Phenomenon	Cause	Team	Line	1	2	3	4	5	6	7	8	9	10	11	12	13	14	15	16	17	18	19	20	21	22	23	24	25	26	27	28	29	30	31	Total
Pump	Cam assembly surface leakage	Plate surface nicks	224	Lock-in press																																0
		Front surface nicks	224	Lock-in press					‖										‖			‖		‖‖		‖											6
		Pin hitter burr	215	Lock-in press												‖					‖																1
		Foreign matter	224	Cleaning	‖								‖	‖				‖‖‖		‖‖‖	‖																8
		Broken—hole taper small	224	Honing																																	0
	Broken lock	Broken—nicks	212	Broken lock-in	‖	‖						‖							‖							‖	‖	‖									7
		Broken—inserted wrong	215	Lock-in press															‖				‖														3
		Foreign matter	224	Cleaning																																	0
Pressure doesn't rise	Rotor side clearance large	Cam measurement error	219	Rank measurement																																	0
		Rotor measurement error	219	Rank measurement																																	0
		Joining defect	219	Rank measurement																																	0
Amount of flow small	Rotor defect	Grooves wide	219	Slit polishing	‖	‖	‖‖	‖‖		‖‖	‖‖‖	‖‖	‖‖	‖	‖‖‖			‖‖‖	‖‖‖				‖			‖											28
		Groove width defect	219	Rank measurement																																	0
Varying pressure		Groove polishing defect	219	Slit polishing																	‖																1
Large amount of torque		Rotor surface degree defect	219	Surface polishing			‖				‖	‖																									4
	Plate fit defect	Front pin defect	215	Lock-in press																																	0
		Plate fitting defect	215	Lock-in press																							‖										1
		Vane missing	215	Lock-in press																						‖											1
	Vane defect	Vane thickness defect	Vend.	Lock-in press																																	0
		Vane length error	Vend.	Lock-in press																																	0
		Wrong vane	215	Lock-in press																																	0
	Cam ring defect	Surface roughness defect	219	Lap		‖‖‖	‖‖‖						‖																								8
		Stepping defect	219	Inner diameter polishing				‖					‖						‖		‖‖‖			‖													9
		Surface degree defect	219	Surface polishing lap																																	0
	Plate leakage	Plate o-ring missing	215	Lock-in press																		‖															1
		Plate ring out	215	Lock-in press	‖																																1
Other				**Total**																																79

Worksheet 44

314

Defect Reduction Activity Chart

Month: _____

Assembly

Promoter:

Goal:

Inspector:

Date: ____ / ____ / ____

Characteristic	Phenomenon	Cause	Team	Line	1	2	3	4	5	6	7	8	9	10	11	12	13	14	15	16	17	18	19	20	21	22	23	24	25	26	27	28	29	30	31	Total

Other

Total

Daily Inspection Chart

Mold   13651

June   Molding machine   3

	Rank			Inspection time		Key to results		Revision	Nature of and reason for revisions	Date	Approval	Approval	Approval
A	Every 3 hours	○		= At startup	○	= Passed		1					
B	Weekly	△		= During repairs	×	= Failed		2					
C	Quarterly				⊕	= Passed after treatment		3					
					|	= Operations halted		4					

Responsible	Line	Inspection items		Areas (drawing)	Time	6 ①	②	3	4	5	6	7	⑧	⑨	10	11	12	13	14	⑮	16	17	18	19	20	21	㉒	㉓	24	25	26	27	28	㉙	㉚
Op.	A	Stripper	Dirty PL surface	1		○	○	○																													
	A		Scoring on PL surface	1		○	○	○																													
Super.	B		PL surface blow check	1		○	○	○																													
Op.	A		Dirty taper lock	1		○	○	○																													
	A	Cavity	Dirty PL surface	2		○	○	○																													
	A		Scoring on PL surface	2		○	○	○																													
Super.	B		PL surface blow check	2		○	○	○																													
Op.	A		Dirty taper lock	2		○	○	○																													
Super.	C	Guide pin	Scoring check	3		△				○																											
	C		Bending check	3		△				○																											
	C	Cooling water pipes	Blockage check	4		△				×																											
	C	Ejector lot	Scoring check	5		△				○																											
	C		Bending check	5		△				○																											
	C		Check for wear on tip	5		△				○																											
	C	Spoor bush	Scoring check	6		△				○																											
	C		Sagging check	6		△				○																											
	C		Deformation check	6		△				○																											
	C		Resin leakage check	6		△				○																											
	C	Junction box	Teminal loosening check	7		△				○																											
	C		Lead check	7		△				○																											
	C		Wire deterioration check	7		△				○																											
Op.	B	Wear ring	Dirty taper lock	8		○	○	○																													
Super.	B		Taper lock sagging	8		○	○	○																													
		Inspection	Operator, supervisor	1/mo.		S		S		S																											
		Confirmation	Supervisor, person in charge of molding	2/mo.		U		U		U																											

No. 7
No. 8
No. 5
No. 6
No. 4
No. 1   No. 2   No. 3

Inspection Standards

No.	Areas to be inspected	Inspection items	Inspection methods	Measurement tools	Inspection standards	Treatment	Cycle	Inspector
1	Stripper	Dirty PL surface	Visual	Sight	No plastic gas	Wipe off	1/3 hr.	Op.
		PL surface scoring	Visual	Sight	Not out of grease, no scoring	Increase grease	1/3 hr.	Op.
		PL surface blow check	Look at striking surface during blow check	Blowing	Not striking air vent	Shim panel adjustment	1/wk.	Super.
		Dirty taper lock	Visual	Sight	No plastic gas	Increase grease	1/3 hr.	Op.
2	Cavity	Dirty PL surface	Visual	Sight	No plastic gas	Shim panel adjustment	1/3 hr.	Op.
		PL surface scoring	Visual	Sight	Not out of grease, no scoring	Increase grease	1/3 hr.	Op.
		PL surface blow check	Look at striking surface during blow check	Blowing	Not striking air vent	Shim panel adjustment	Once a week	Super.
		Dirty taper lock	Visual	Sight	No plastic gas	Wipe off	1/3 hr.	Oper.
3	Guide pin	Scoring check	Visual	Sight	Not out of grease, no scoring	Treat surface with sandpaper, increase grease	1/3 mo.	Super.
		Bending check	Measure the gap gauge	Gap gauge	Bent less than 0.5 mm	Replacement	1/3 mo.	Super.
4	Cooling water pipe	Blockage check	Run water through	Flow meter	Over 3.2 kg/cm^3	Disassemble and check	1/3 mo.	Super.
5	Ejector lot	Scoring check	Visual	Sight	No scoring	Check nuts and bolts	1/3 mo.	Super.
		Bending check	Visual	Sight	No bending	Replacement	1/3 mo.	Super.
		Check for tip wear	Visual	Sight	No wear	Replacement	1/3 mo.	Super.

Worksheet 45

Daily Inspection Chart

Machine: _____

Date: _____

	Rank			Inspection time		Key to results		Revision	Nature of and reason for revisions	Date	Approval	Approval	Approval
A	Every 3 hours	◯		= At startup	◯	= Passed		1					
B	Weekly	△		= During repairs	✕	= Failed		2					
C	Quarterly				⊕	= Passed after treatment		3					
						= Operations halted		4					

Responsible	Line	Inspection items		Areas (drawing)	Time	6 1	2	3	4	5	6	7	8	9	10	11	12	13	14	15	16	17	18	19	20	21	22	23	24	25	26	27	28	29	30	
		Inspection	Operator, supervisor																																	
		Confirmation	Supervisor, person in charge																																	

Inspection Standards

No.	Areas to be inpected	Inspection items	Inspection methods	Measurement tools	Inspection standards	Treatment	Cycle	Inspector

Worksheet 45a

Minor Stoppage Investigation Sheet (Automated Equipment)

Name of equipment: _____ Plant l: _____

Stoppage frequency times/shift-unit (minutes): _____

Process l: _____

Date: ___/___/___
Time: 8:30–16:30
Recorded:
Total machine units: 5
Machines/operation units:

Machine no.	1	2	3	4	5	6	7	8	Total	Mean Events time	Share	Comments
Cycle time sec/each	2.2	2.2	2.4	2.1	2.2							
Specification formula	DR4 × 3 E0606	DR4 × 3 EL0606	DR4 × 3 EL0606	DR4 × 3 EL0606	DR4 × 3 EL0606							
Mode of stoppage (Part, phenomenon)												
1 Wire runs out	Twice (5")	6 times (28")	Twice (5")						10 times 38 min.	Twice 7.6 min.	19.8	
2 Silicon replacement due to wire running out		2 times (70")							Twice 70"	0.4 times 14"	36.5	
3 Cut piece doesn't fit in chuck		5"							Once 5"	0.2 times 1"	2.6	
4 Foreign matter adhering to rail				5"					Once 5"	0.2 times 1"	2.6	
5 Debris adhering to chute photoelectric tube			2"						Once 2"	0.2 times 0.4"	1.0	
6 Core doesn't fit in chuck		15"							Once 15"	0.2 times 3"	7.8	
7 Pasteboard sagging		2"							Once 2"	0.2 times 0.4"	1.0	
8 Lot replacement			5"						Once 5"	0.2 times 1"	2.6	
9 Iron tip replacement	2"		2"	2"	2"				5 times 10"	Once 2"	5.2	
10 Bucket cleaning	8"	8"	8"	8"	8"				5 times 40"	Once 8"	20.8	
Total	15"	130"	22"	15"	10"				192 min.	38.4 min.		
Units/day	10,972	8,361	10,005	11,609	11,070							
Comments												

Worksheet 46

Minor Stoppage Investigation Sheet (Automated Equipment)

Name of equipment: _____

Date: ___/___/___
Time: _____
Recorded _____
Total machine units: _____
Machines/operation units: _____

Plant l: _____

Stoppage frequency
times/shift-unit (minutes): _____

Process l: _____

Mode of stoppage (Part, phenomenon)																	
Machine no.	1	2	3	4	5	6	7	8	Total	Mean		Comments
Cycle time *sec/each*										Events time	Share	
Specification formula	2.2	2.2	2.4	2.1	2.2							
1												
2												
3												
4												
5												
6												
7												
8												
9												
10												
Total												
Units/day												
Comments												

Worksheet 46a

319

TPM Top Management Diagnosis—Self-Diagnosis Sheet

Cycle: 12

Inspection date: 11/13

Dept./office	Bull parts section		Inspector:	Keisuke Arai

Evaluation items							Rating (out of 5)	Points to be proud of	Points for improvement
Key indicators		5	4	3	2	1		Reduction in amount of money lost to defects in the process and a reduction in sporadic breakdowns.	Clarify ways of dealing with the worst line (production).
Productivity (personnel reduction performance rate setup)		1	5	7	2	1	3.6	Planned maintenance has been thoroughly inculcated.	Further promote measures against minor stoppages.
QA (Defects-adjustments, yield)			2	8		3	2.7	Changes in line layout due to reductions in employee hours and 1-unit flow have continued, and all lines have shown increased productivity.	Promote revision and strengthening of plans that authorize autonomous maintenance.
Instant maintenance			3	10			3.3	Production, quality, etc. and other forms of daily management are ingrained.	Ensure an amount of work that will have the machining center running at 100%.
New 5Ss				7		4	2.7		
Continuous improvement challenges									
Productivity, QA, instant maintenance, new 5Ss							4.0		
Concreteness of plans to implement revised TPM							3.5		
Visual management of problems (Minor stoppages, medium stoppages, major stoppages, daily inspections, planned stoppages, over production, production startup problems)							5.0		
Design TPM	Design						4.0		
	Concurrent engineering						4.0		
Easy comprehensibility of explanatory materials and demonstrations							4.0		
Total							36.8		

Worksheet 47

TPM Top Management Diagnosis—Self-Diagnosis Sheet

Cycle:

Inspection date: _____ / _____ / _____

Dept./office		Inspector:

Evaluation items							Rating (out of 5)	Points to be proud of	Points for improvement
Key indicators		5	4	3	2	1			
Productivity (personnel reduction performance rate setup)									
QA (Defects-adjustments, yield)									
Instant maintenance									
New 5Ss									
Continuous improvement challenges									
Productivity, QA, instant maintenance, new 5Ss									
Concreteness of plans to implement revised TPM									
Visual management of problems (Minor stoppages, medium stoppages, major stoppages, daily inspections, planned stoppages, over production, production startup problems)									
Design TPM	Design								
	Concurrent engineering								
Easy comprehensibility of explanatory materials and demonstrations									
Total									

Worksheet 47a

Line Improvement Follow-up Sheet

Plant: D

Date: 9/17

No.	Items for improvement	Deadline	Who	Summary
1	Side expansion 1 • 2	9/19	A	
	Side expansion 3 • 4	9/20	B	
	Side expansion 6 • 7	9/21	C	
2	Install new drive belts	9/18	B	
3	Try reducing rpm on machine	9/21	D	Confirm the presence or lack of a quality problem by changing the sewing pitch from 5mm to 8mm
4	Try to detect running out of thread by using a limit switch	9/21	A	
5	Have manufacturer change formula for spinning thread	9/17	E	Machine linkage → main linkage

Worksheet 48

Line Improvement Follow-up Sheet

Plant: _____   Date: ____ / ____ / ____

No.	Items for improvement	Deadline	Who	Summary

Worksheet 48a

TPM Team Presentation Evaluation

Evaluator: K

Presentation: ____ / ____ / ____

	Evaluation items	Team Scale	A	B	C						
			Points	Points	Points	Points	Points	Points	Points	Points	Points
1	Have they analyzed the current situation well and chosen their themes accordingly? (Recent themes, reasons behind themes, links to supervisors and policies, management traits, goal values)	15 12 9 6 3	9	12	10						
2	Did they describe team activities and give accounts of experiences? (Concerning the meetings: frequency, history of concerns, efforts to develop original ideas, autonomous activities, cooperation within team)	15 12 9 6 3	6	12	10						
3	Do they understand QC techniques and put them into action correctly? (Causal factors chart, Pareto diagram, histogram, scatter stratification diagram, graph, management graphs, check sheets)	15 12 9 6 3	12	10	8						
4	Are the factors and countermeasures logically connected and well implemented? (Confirmed effectiveness of countermeasures)	10 8 6 4 2	2	8	10						
5	Did they achieve their goals? (How results were confirmed, tangible results vs. intangible results)	10 8 6 4 2	2	8	8						
6	Have the effective measures been standardized? (Standardization, revision of standards, maintaining effects)	10 8 6 4 2	2	10	10						
7	Have they differentiated their effective and ineffective activities and reconsidered them? (Problems, future plans)	15 12 9 6 3	6	12	8						
8	Was the presentation organized, logical, and easy to understand? (Materials, time, attitude, QC stories)	10 8 6 4 2	6	8	6						
	Total	100 possible	45	80	70						

Worksheet 49

TPM Team Presentation Evaluation

Evaluator: _____   Presentation: ____ / ____ / ____

	Evaluation items	Team Scale	Points	Points	Points	Points	Points	Points	Points	Points	Points
1	Have they analyzed the current situation well and chosen their themes accordingly? (Recent themes, reasons behind themes, links to supervisors and policies, management traits, goal values)	15 12 9 6 3									
2	Did they describe team activities and give accounts of experiences? (Concerning the meetings: frequency, history of concerns, efforts to develop original ideas, autonomous activities, cooperation within team)	15 12 9 6 3									
3	Do they understand QC techniques and put them into action correctly? (Causal factors chart, Pareto diagram, histogram, scatter stratification diagram, graph, management graphs, check sheets)	15 12 9 6 3									
4	Are the factors and countermeasures logically connected and well implemented? (Confirmed effectiveness of countermeasures)	10 8 6 4 2									
5	Did they achieve their goals? (How results were confirmed, tangible results vs. intangible results)	10 8 6 4 2									
6	Have the effective measures been standardized? (Standardization, revision of standards, maintaining effects)	10 8 6 4 2									
7	Have they differentiated their effective and ineffective activities and reconsidered them? (Problems, future plans)	15 12 9 6 3									
8	Was the presentation organized, logical, and easy to understand? (Materials, time, attitude, QC stories)	10 8 6 4 2									
	Total	100 possible									

Worksheet 49a

TPM Team Line Contest Evaluation

Concern (specific)
Reducing the time for wrapping operations and secondary testing of the LB10

Group name: LB10 team Division: Special production section

	Time of inspection		Time of registration	
	Approval	Created	Approval	Created
	A	B	C	D

Key to rate of increase (decrease) and ranks

Item no.	Item	Ranks 5	4	3	2	1	0	Units
1	Setup	≥50	40–	30–	20–	10–	<10	%
2	WIP	≥50	40–	30–	20–	10–	<10	%
3	Idleness	≥50	40–	30–	20–	10–	<10	%
4	Movement	≥25	20–	15–	10–	5–	<5	%
5	Breakdowns	≥80	65–	50–	40–	20–	<20	%
6	Results	≥50	40–	30–	20–	10–	<10	%
7	Effective amount of money	≥56	40–	24–	16–	8–	<8	$1,000
8	On-site inspection	≥30	25–	20–	15–	10–	<10	

Inspection seal	Inspection	Registration
G	E	F

Item no.	Item	Definition 1	Definition 2 (Each team fill in specifically)	At registration	At inspection	Rate of increase (decrease) (B–A)/A	Rank 5 4 3 2 1 0	Weight	Score
1	Setup/ changeover time	Time of switching between standard model (A) to standard model (B)	During setup and changeover between other models and the LB10	97 sec./ event	97 sec./ event	0%	○(0)	2	0
2	WIP	Unfinished goods in the process (amt. of money or number of units)	—	—	—	—		1	—
3	Idleness	Accumulated time per unit time	Idleness of 2nd degree ester per sheet (5 units)	2 sec./ sheet	0 sec./ sheet	100%	○	1	5
4	Movement	Time and distance for moving people and things	Motion distance for preparing the materials to wrap 100 units	30m/ 1000 units	10m/ 1000 units	67%	○	1	5
5	Breakdowns	Rate of defects, on-the-spot adjustments (time or money)	Jig defects, rate of product loss due to misoperation: units lost / units put on the line × 100	0.05%	0	100%	○	3	15
6	Results per unit	Examples of definitions (Units produced per hour, processing tickets per day, evaluation items per day)	Number of employee hours for secondary tests and wrapping per sheet (5 units)	50 sec./ sheet	18.5 sec./ sheet	63%	○	3	15
7	Effective monetary amount	Predicted effective monetary amount per year	The six most improved amounts of money, 10,000 units/month production (computed as average of last six months)	—	Operation time = $5,800 Other improvements = about $8,200/year	$14,000		3	3
8	On-site inspection	Based on inspection and evaluation charts that deal with concepts of wasted motion, 5S, and U-shaped lines.	—	—				6	
								Total	

Write clearly when registering

Worksheet 50

TPM Team Line Contest Evaluation

Time of inspection		Time of registration	
Approval	Created	Approval	Created

Concern (specific)

Group name: _____ Division: _____

Item no.	Item	Definition 1	Definition 2 (Each team fill in specifically)	At registration	At inspection	Rate of increase (decrease) (B − A)/A	Rank 5 4 3 2 1 0	Weight	Score
1	Setup/ changeover time								
2	WIP								
3	Idleness								
4	Movement								
5	Breakdowns								
6	Results per unit								
7	Effective monetary amount								
8	On-site inspection								
								Total	

Key to rate of increase (decrease) and ranks

Item no.	Item	Ranks 5	4	3	2	1	0	Units
1	Setup	≥50	40–	30–	20–	10–	<10	%
2	WIP	≥50	40–	30–	20–	10–	<10	%
3	Idleness	≥50	40–	30–	20–	10–	<10	%
4	Movement	≥25	24–	20–	15–	10–	5–	<5	%
5	Breakdowns	≥80	65–	50–	40–	20–	<20	%
6	Results	≥50	40–	30–	20–	10–	<10	%
7	Effective amount of money	≥56	40–	24–	16–	8–	<8	$1,000
8	On-site inspection	≥30	25–	20–	15–	10–	<10	

Inspection seal	Inspection	Registration

Write clearly when registering

Worksheet 50a

References

Books by the Authors Available in English

Sekine, Ken'ichi, *One-Piece Flow* (Productivity Press, 1992).

Sekine, Ken'ichi, and Keisuke Arai, *Design Team Revolution* (Productivity Press, 1994).

Sekine, Ken'ichi, and Keisuke Arai, *Kaizen for Quick Changeover* (Productivity Press, 1992).

Additional Sources

"Creating Written Standards for Cleaning and Lubrication," *TPM Age* 2 (4).

"From a Collection of Ideas for Rationalizing and Improving Press Operations," *Puresu Gijitsu* (Press Technology) 28 (3).

Esaki and Kanemori, "High-Efficiency Machining and Setup Improvement in an NC Lathe," *Tool Engineer* 28 (8).

Fukunichi Seiichi, "Revising Inspection Charts," *TPM Age* 2 (3).

Fukunichi Seiichi, "Standards for Autonomous Maintenance of a Die Cast Machine," *TPM Age* 2 (1).

Gotoh Fumio, *Development and Design of Equipment for TPM* (Japan Institute of Plant Maintenance).

Hanawa Seijiro, "Manage Machining Oil on Site," *TPM Age* 3 (6).

Hibi Shigeru, "A Good Diagram from a Measurer's Point of View," *Tool Engineer* 28 (8).

Imoto Akira, "Points to Remember in the Strip Layout of Ordered Molds for Medium and Large Parts," *Gata Gijutsu* 4 (12).

Kawasaki Steel Corporation, *On-Site Preparation for Maintenance Personnel* (Japan Institute of Plant Maintenance).

Kimura Yoshifumi, "Our Role," *TPM Age* 1 (1).

Nawada Kunihiko, "Daily Inspections of High-Pressure Gas Production Installations," *TPM Age* 1 (6).

Ohtsu Wataru, Quality Management of Design (Nikkagiren [Association of Chemical Engineers]).

Osada Takashi, "What Is TPM Aiming for? What Does It Accomplish?" *Kojo Kanri* (Plant Management) 37 (3).

Sekiguchi Keizo, "How to Conduct Daily Inspections," *TPM Age* 1 (8).

Sekine and Arai, "What is the Biggest Waste in Design? Part II," *Kojo Kanri* (Plant Management) 36 (2).

Sekine and Arai, *Zero Setup Improvement Techniques* (Nikkan Kogyo Shinbunsha).

Sekine and Iwasaki, *Zero Setup Know-how* (Nikkan Kogyo Shinbunsha, Audiovisual Affairs Department).

Sekine and Yamazaki, *Sekine-Type Workplace Improvement: Questions and Answers* (Shin Gijutsu Kaihatsu Sentaa [New Technology Development Center]).

Sekine et al., *The New 5Ss at Mynac* (video) (Nikkan Kogyo Shinbunsha, Audiovisual Affairs Department).

Sekine, Arai, and *Yamazaki, Factory Reorganization Know-how* (Nikkan Kogyo Shinbunsha, Audiovisual Affairs Department).

Sekine, Arai, and Yamazaki, *Process Razing Know-how* (Nikkan Kogyo Shinbunsha, Audiovisual Affairs Department).

Sekine, Arai, and Yamazaki, *Toyota Implementation Formulas: The Tiger Volum*e (Nikkan Kogyo Service Center).

Sekine, Ken'ichi, ed., *Process Razing Manual* (Shin Gijutsu Kaihatsu Sentaa [New Technology Development Center]).

Sekine, Ken'ichi, *Manual for Cutting the Design Period and the Number of Personnel in Half* (Shin Gijutsu Kaihatsu Sentaa [New Technology Development Center]).

So Takashi, *Know-how for Machine Design Based on Experience* (Nikkan Kogyo Shinbunsha).

Totoki Keigo, "Shortening Delivery Periods by Taking Advantage of Standardized Parts," *Gata Gijutsu* (Mold Technology) 3 (5).

Watanabe Hidenori, *More Notes for Understanding Machine Design* (Nikkan Kogyo Shinbunsha).

Watanabe Hidenori, *Notes for Understanding Machine Design* (Nikkan Kogyo Shinbunsha).

About the Authors

Ken'ichi Sekine is director of the Institute of Value-Added Management Research. He began working in quality control in 1949, serving Bridgestone Tire as a QC promoter when it won the Deming Prize. After becoming an independent consultant, he established the Institute of Value-Added Management Research, where he continues his consulting work focused on shopfloor improvements. In addition to this book, he is the author of *One-Piece Flow* (Productivity Press, 1992), and the coauthor (with Keisuke Arai) of *Kaizen for Quick Changeover* (Productivity Press, 1992), and *Design Team Revolution* (Productivity Press, 1994), in addition to many other books published in Japan related to the Toyota Production System, value-added management, and other manufacturing improvement topics.

Keisuke Arai is chief consultant of the Institute of Value-Added Management Research. Following his graduation from the machine engineering department of Osaka Furitsu University in 1957, he worked for Mitsui Engineering, where he served as project manager for design, assembly, construction, and production control of plant and nuclear power engineering. After leaving Mitsui in 1982, he consulted and delivered training for plastic molding industries. Currently he is a chief consultant, management engineer, and diagnostician of small to mid-sized companies at the Institute of Value-Added Management Research. He is the coauthor (with Ken'ichi Sekine) of *Kaizen for Quick Changeover* and *Design Team Revolution*.

Index

BOOKS FROM PRODUCTIVITY PRESS

Productivity Press publishes books that empower individuals and companies to achieve excellence in quality, productivity, and the creative involvement of all employees. Through steadfast efforts to support the vision and strategy of continuous improvement, Productivity Press delivers today's leading-edge tools and techniques gathered directly from industry leaders around the world. Call toll-free (800) 394-6868 for our free catalog.

40 TOP TOOLS FOR MANUFACTURERS
A GUIDE FOR IMPLEMENTING POWERFUL IMPROVEMENT ACTIVITIES
Walter Michalski

We know how important it is for you to have the right tool when you need it. And if you're a team leader or facilitator in a manufacturing environment, you've probably been searching a long time for a collection of implementation tools tailored specifically to the needs of manufacturers. Well, look no further. Based on the same principles and user-friendly design of the Tool Navigator's *The Master Guide for Teams*, here is a group of 40 dynamic tools to help you and your teams implement powerful manufacturing process improvement. Use this essential resource to select, sequence, and apply major TQM tools, methods, and processes.

ISBN 1-56327-197-4/ 160 pages / $25.00 / Order NAV2-B8006

BECOMING LEAN
INSIDE STORIES OF U.S. MANUFACTURERS
Jeffrey Liker

Most other books on lean management focus on technical methods and offer a picture of what a lean system should look like. Some provide snapshots of before and after. This is the first book to provide technical descriptions of successful solutions and performance improvements. The first book to include powerful first-hand accounts of the complete process of change, its impact on the entire organization, and the rewards and benefits of becoming lean. At the heart of this book you will find the stories of American manufacturers who have successfully implemented lean methods. Authors offer personalized accounts of their organization's lean transformation, including struggles and successes, frustrations and surprises. Now you have a unique opportunity to go inside their implementation process to see what worked, what didn't, and why. Many of these executives and managers who led the charge to becoming lean in their organizations tell their stories here for the first time!

ISBN 1-56327-173-7/ 350 pages / $35.00 / Order LEAN-B8006

EQUIPMENT PLANNING FOR TPM
MAINTENANCE PREVENTION DESIGN
Fumio Gotoh

This practical book for design engineers, maintenance technicians, and manufacturing managers details a systematic approach to the improvement of equipment development and design and product manufacturing. The author analyzes five basic conditions for factory equipment of the future: development, reliability, economics, availability, and maintainability. The book's revolutionary concepts of equipment design and development enables managers to reduce equipment development time, balance maintenance and equipment planning and improvement, and improve quality production equipment.

ISBN 0-915299-77-1 / 337 pages / $65.00 / Order ETPM-B8006

TOTAL PRODUCTIVE MAINTENANCE
MAXIMIZING PRODUCTIVITY AND QUALITY (VIDEO)
Japan Management Association (ed.)

Introduce TPM to your work force with this accessible two-part video program, which explains the rationale and basic principles of TPM to supervisors, group leaders, and workers. It explains five major developmental activities of TPM, includes a section on equipment improvement that focuses on eliminating chronic losses, and describes an analytical approach called PM Analysis to help solve problems that have complex and continuously changing causes. (Approximately 45 minutes long.)

ISBN 0-915299-49-6 / 2 videos / $799.00 / Order VTPM2-B8006

TPM CASE STUDIES
FACTORY MANAGEMENT SERIES
Nikkan Kogyo Shimbun (ed.)
Total Productive Maintenance (TPM) combines the best features of productive and predictive maintenance with innovative management strategies and total employee involvement. This collection of foundational articles and classic implementation case studies culled from *NKS Factory Management Journal* details how TPM has helped prize-winning companies in Japan achieve remarkable results. It includes in-depth explorations of the approach to loss reduction and plantwide implementation; a classic essay on the relationship between JIT and TPM by Seiichi Nakajima, the "father" of TPM; and numerous detailed examples of equipment modifications addressing specific types of losses.
ISBN 1-56327-066-8 / 200 pages / $30.00 / Order TPMCS-B8006

TPM FOR AMERICA
WHAT IT IS AND WHY YOU NEED IT
Herbert R. Steinbacher and Norma L. Steinbacher
As much as 15 to 40 percent of manufacturing costs are attributable to maintenance. With a fully implemented TPM program, your company can eradicate all but a fraction of these costs. Co-written by an American TPM practitioner and an experienced educator, this book gives a convincing account of why American companies must adopt TPM if they are to successfully compete in world markets. Includes examples from leading American companies showing how TPM has changed them into more efficient and productive organizations.
ISBN 1-56327-044-7 / 169 pages / $25.00 / Order TPMAM-B8006

TPM IN PROCESS INDUSTRIES
Tokutaro Suzuki (ed.)
Process industries have a particularly urgent need for collaborative equipment management systems like TPM that can absolutely guarantee safe, stable operation. In *TPM in Process Industries*, top consultants from JIPM (Japan Institute of Plant Maintenance) document approaches to implementing TPM in process industries. They focus on the process environment and equipment issues such as process loss structure and calculation, autonomous maintenance, equipment and process improvement, and quality maintenance. Must reading for any manager in the process industry.
ISBN 1-56327-036-6 / 400 pages / $85.00 / Order TPMPI-B8006

TPM TEAM GUIDE
Kunio Shirose (ed.)
This book makes TPM team activities understandable to everyone in the company. *TPM Team Guide* gives simple explanations of basic TPM concepts like the 6 big losses, and emphasizes the integration of TPM activities with production management. Chapters describe the team-based improvement process step by step, from goal setting to standardization of the improved operations. Team leaders will learn how to hold effective meetings and work with the human issues that are a big part of success. The tools for team problem solving and the steps for preparing a good presentation of results are detailed here as well. Written in straightforward, easy to digest language, with abundant illustrations and cartoon examples. Front-line supervisors, operators, facilitators, and trainers in manufacturing companies will want to use this practical guide to improve company performance and build a satisfying workplace for employees.
ISBN 1-56327-079-X/175 pages / $25.00 / Order TGUIDE-B8006

TRAINING FOR TPM
A MANUFACTURING SUCCESS STORY
Nachi-Fujikoshi (ed.)
A detailed case study of TPM implementation at a world-class manufacturer of bearings, precision machine tools, dies, industrial equipment, and robots. The book details how the company trained managers and workers and shows the improvements they achieved in reducing breakdowns and defects while revitalizing the workforce. In just 2-1/2 years the company was awarded Japan's prestigious PM Prize for its program. Here is a detailed account of their improvement activities—and an impressive model for yours.
ISBN 0-915299-34-8 / 274 pages / $50.00 / Order CTPM-B8006

UPTIME
STRATEGIES FOR EXCELLENCE IN MAINTENANCE MANAGEMENT

John Dixon Campbell

Campbell outlines a blueprint for a world class maintenance program by examining, piece by piece, its essential elements – leadership (strategy and management), control (data management, measures, tactics, planning and scheduling), continuous improvement (RCM and TPM), and quantum leaps (process reengineering). He explains each element in detail, using simple language and practical examples from a wide range of industries. This book is for every manager who needs to see the "big picture" of maintenance management. In addition to maintenance, engineering, and manufacturing managers, all business managers will benefit from this comprehensive and realistic approach to improving asset performance.

ISBN 1-56327-053-6 / 204 pages / $35.00 / Order UP-B8006

TO ORDER: Write, phone, or fax Productivity Press, Dept. BK, P.O. Box 13390, Portland, OR 97213-0390, phone 1-800-394-6868, fax 1-800-394-6286.
Outside the U.S. phone (503) 235-0600; fax (503) 235-0909
Send check or charge to your credit card (American Express, Visa, MasterCard accepted).

U.S. ORDERS: Add $5 shipping for first book, $2 each additional for UPS surface delivery. Add $5 for each AV program containing 1 or 2 tapes; add $12 for each AV program containing 3 or more tapes. We offer attractive quantity discounts for bulk purchases of individual titles; call for more information.

ORDER BY E-MAIL: Order 24 hours a day from anywhere in the world. Use either address:

To order: **service@ppress.com**

To view the online catalog and/or order: **http://www.ppress.com/**

QUANTITY DISCOUNTS: For information on quantity discounts, please contact our sales department.

INTERNATIONAL ORDERS: Write, phone, or fax for quote and indicate shipping method desired. For international callers, telephone number is 503-235-0600 and fax number is 503-235-0909. Prepayment in U.S. dollars must accompany your order (checks must be drawn on U.S. banks). When quote is returned with payment, your order will be shipped promptly by the method requested.

NOTE: Prices are in U.S. dollars and are subject to change without notice.

About the Shopfloor Series

Put powerful and proven improvement tools in the hands of your entire workforce!

Progressive shopfloor improvement techniques are imperative for manufacturers who want to stay competitive and to achieve world class excellence. And it's the comprehensive education of all shopfloor workers that ensures full participation and success when implementing new programs. The Shopfloor Series books make practical information accessible to everyone by presenting major concepts and tools in simple, clear language and at a reading level that has been adjusted for operators by skilled instructional designers. One main idea is presented every two to four pages so that the book can be picked up and put down easily. Each chapter begins with an overview and ends with a summary section. Helpful illustrations are used throughout.

Books currently in the Shopfloor Series include:

5S for Operators
The Productivity Press Development Team
ISBN 1-56327-123-0 /
incl. applic. questions / 133 pages
Order 5SOP-B289 / $25.00

Quick Changeover for Operators
The Productivity Press Development Team
ISBN 1-56327-125-7 /
incl. applic. questions / 93 pages
Order QCOOP-B289 / $25.00

Mistake-Proofing for Operators
The Productivity Press Development Team
ISBN 1-56327-127-3 / 93 pages
Order ZQCOP-B289 / $25.00

Just-in-Time for Operators
The Productivity Press Development Team
ISBN 1-56327-133-8
incl. applic. questions / 96 pages
Order JITOP-B289 / $25.00

TPM for Every Operator
Japan Institute of Plant Maintenance
ISBN 1-56327-080-3 / 136 pages
Order TPMEO-B289 / $25.00

TPM Team Guide
Kunio Shirose
ISBN 1-56327-079-X / 175 pages
Order TGUIDE-B289 / $25.00

TPM for Supervisors
The Productivity Press Development Team
ISBN 1-56327-161-3 / 96 pages
Order TPMSUP-B289 / $25.00

Autonomous Maintenance for Operators
Japan Institute of Plant Maintenance
ISBN 1-56327-082-X / 138 pages
Order AUTMOP-B289 / $25.00

Focused Equipment Improvement for TPM Teams
Japan Institute of Plant Maintenance
ISBN 1-56327-081-1 / 138 pages
Order FEIOP-B289 / $25.00

Continue Your Learning with In-House Training and Consulting from Productivity, Inc.

Productivity, Inc. offers a diverse menu of consulting services and training products based on the exciting ideas contained in the books of Productivity Press. Whether you need assistance with long term planning or focused, results-driven training, Productivity's experienced professional staff can enhance your pursuit of competitive advantage.

Productivity, Inc. integrates a cutting edge management system with today's leading process improvement tools for rapid, measurable, lasting results. In concert with your management team, we will focus on implementing the principles of Value Adding Management, Total Quality Management, Just-In-Time, and Total Productive Maintenance. Each approach is supported by Productivity's wide array of team-based tools: Standardization, One-Piece Flow, Hoshin Planning, Quick Changeover, Mistake-Proofing, Kanban, Problem Solving with CEDAC, Visual Workplace, Visual Office, Autonomous Maintenance, Equipment Effectiveness, Design of Experiments, Quality Function Deployment, Ergonomics, and more.

Productivity is known for significant improvement on the shopfloor and the bottom line. Through years of repeat business, an expanding and loyal client base continues to recommend Productivity to their colleagues. Contact us to learn how we can tailor our services to fit your needs.

Telephone: 1-800-966-5423 (U.S. only) or 1-203-846-3777
Fax: 1-203-846-6883